EMBRYOLOGY
OF
THE PIG

By BRADLEY M. PATTEN
Professor of Anatomy in the University of Michigan Medical School

THIRD EDITION

WITH COLORED FRONTISPIECE
AND 186 ILLUSTRATIONS IN THE TEXT (CONTAINING 412 FIGURES)
OF WHICH 6 ARE IN COLOR

1948

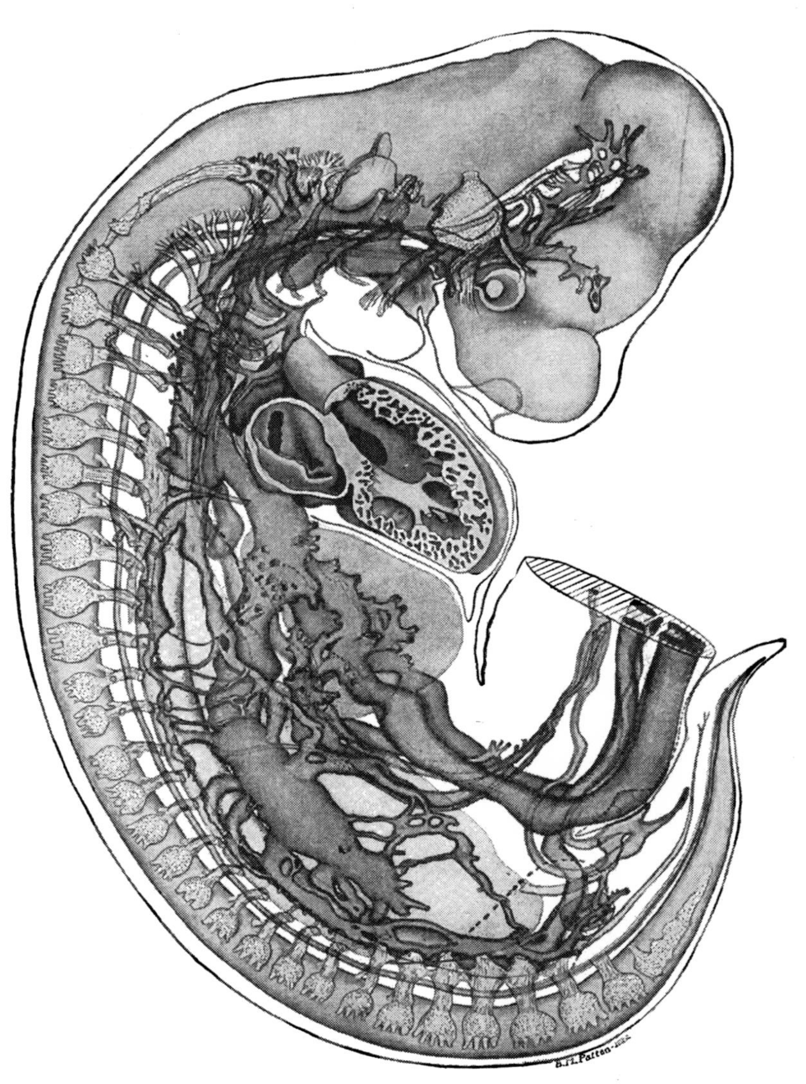

FRONTISPIECE

Reconstruction (× 17.5) showing the organ systems of a 9.4 mm. pig embryo. For explanation see figures 60 and 66.

Contents

CHAPTER 1

Foreword to the Student

You and I are to start out together to explore some of the regions of embryology. Still vivid memories of the erratic progress of my own first expeditions of that kind have led me to offer my services to you. Perhaps I can help you to avoid some of the difficulties I encountered and lead you to points of interest by routes less devious than you might otherwise find. But it is your own expedition. I am merely a guide. I can show you the passes to understanding but you must climb them. I can lead you to worth-while things but you yourself must unearth them and carry them away.

Objectives in the Study of Embryology

Most courses in embryology are planned with one of two ends in view. They aim either to teach undergraduate students the fundamental facts and principles of development, or to show medical students the processes by which the human body reaches its adult form.

The student of zoology rarely finds embryology a "duty course." To one at all interested in the how and why of living things, the processes by which animals grow from a single fertilized egg cell to their fully elaborated adult structure seldom fail to be fascinating.

The medical student is likely to be more utilitarian. He sees ahead a definite objective. Between him and his objective stand many courses to be surmounted. He expects to work hard and is quite willing to do so, but he quite properly expects each course to contribute definitely to his progress toward his goal. In the harassed transitional period of his first year in medical school he rarely takes time to think out just why he is required to study embryology. This is not surprising, for the applications of embryology are, for the most part, indirect. It has no obvious bedside value. Its service lies rather in the rational interpretation of other subjects that it makes possible.

The entering student readily sees the necessity for a thorough knowledge of anatomy. Not until he becomes enmeshed in a maze of anatomical details does he realize his need of some knowledge of how adult conditions became as they are to lead him beyond anatomical memorizing to a comprehension of anatomy. Having no familiarity with the structural and functional abnormalities encountered in clinical practice, he can scarcely be expected to realize how many of these conditions can be interpreted only in the light of embryological phenomena. Knowing pathology, neurology, and obstetrics only by name or by a catalogue prospectus, one cannot expect him to see in advance how much his embryological background will help him in these subjects. It is only fair to the student confronted by a heavy medical curriculum to point out to him that embryology is more than just an interesting field of knowledge—that it is in reality a subject which will be of constantly increasing value to him as he goes on with his training and practice.

The Importance of Laboratory Study

Whatever the purpose and scope of a course in embryology may be, it must necessarily be based on laboratory study of actual material. No amount of didactic instruction, however forcefully it may be presented, can take the place of work done by the student himself. In the laboratory, with critical and encouraging guidance, students can be led to become themselves active acquisitors, instead of passive recipients of information.

It is because of its value as laboratory material that we have woven the thread of our story about the development of the pig. If it is desired to illustrate the fundamental processes of mammalian embryology as the basis for a general course, there is no better form than the pig. If a knowledge of human development is the end in view, the pig is equally serviceable. Human embryos in the vitally important stages during which the various organ systems are being established, cannot be procured in sufficient numbers for class use. During these early stages, the development of all the mammals is fundamentally the same. The specific characteristics of any form emerge but slowly, and relatively late. It is, therefore, quite possible, by using young pig embryos, to give students the opportunity of studying at first hand the same processes which go on in the early development of the human body.

There are several other reasons why pig embryos are so frequently utilized as laboratory material. They can readily be obtained in all but the very earliest stages of development in sufficient numbers for class use. The older embryos are large enough to permit gross dissection—a method of study not possible with the embryos of smaller animals and one which is of great value in linking embryology with adult anatomy. Furthermore the pig has been the subject of many important embryological investigations. The information thus accumulated opens an uncommonly wide field to both teacher and student.

The Plan of Presentation Used in This Book

In planning the subject matter of this book, one of the first problems to present itself involved the question of the preliminary training the student of embryology might be expected to possess. It is of the utmost importance that new work should start on firm ground. On the other hand nothing but wasted effort results from unnecessary repetition of work already covered and it seems reasonable to assume that one taking up the study of mammalian embryology will not be without a considerable background of biological information. The exact extent of this background necessarily will be variable and is exceedingly difficult to evaluate. I have assumed that it was unnecessary to give any preliminary review of such essentials as protoplasmic activity, cell structure, cell division, or the aggregation and specialization of cells to form the fundamental tissues. Certainly most students will be well grounded in these things, and be familiar as well with some aspects of evolution and heredity, sexual reproduction, the recapitulation theory, and in a general way at least with the adult structure of vertebrates.

Many students will have had some preliminary work in embryology before taking up the development of mammals. Even in the case of such readily available forms as the pig, very young embryos are difficult to obtain in large numbers. For that reason, some other embryos such as frogs or chicks are commonly utilized to illustrate early embryological processes. In fact, the mammalian type rarely is studied by students until it is relatively advanced in development. Under these circumstances it is important that the stage of development at which work on the preliminary form is discontinued, be equivalent to that of the mammalian embryos to which attention is transferred. A common procedure is to make the transition from chicks of about three or four days' incubation to pig embryos of approximately 5 to 10 mm. in length. This works well because the degree of development in the various organ systems is readily comparable in both embryos, and so the story of development can be carried forward with little break in continuity.

Because preliminary work is usually done on the early stages of some other form, and the study of the pig begun on fairly well-advanced embryos, it has not seemed necessary to go into great detail as to its early stages. Nevertheless it would be unwise to start the consideration of any developmental process elsewhere than at its beginning. Embryology is a progressively constructive series of events. A knowledge of preceding stages and an appreciation of the trend of the developmental processes by which conditions at one stage become transmuted into different conditions in the next, are direct and necessary factors in acquiring a real comprehension of the subject. Just as historical events are led up to by preparatory occurrences and are followed by results which in turn affect later events, so in embryology events in development are presaged by preliminary changes and, when consummated, affect in turn later steps in the process.

The stage of mammalian development which commonly receives most attention in the laboratory is that in which the body form is well defined and the various organ systems are just appearing. This, for the beginner, is the most critical part of the subject. If the student successfully follows the emergence of the various organ systems from undifferentiated primordial tissues to a point where he can recognize the beginnings of familiar adult structures, his troubles are largely passed. The chief difficulty in embryology lies in getting a start among strange names and unfamiliar structural conditions. It is in recognition of this fact that so large a proportion of the laboratory work in most courses is spent on embryos of this stage. Nothing but actual contact with these unfamiliar structures can fix them in the student's mind and give him a clear understanding of their origin and their relation to one another. In this text, for the same reason, the critical stages in the early differentiation of the organ systems have been especially emphasized.

In dealing with the later phases of development it has seemed best to pass lightly over, or omit altogether, certain things. In doing this I have been guided by two considerations, first, the degree of general interest of the subject, and second, the ease with which it can be presented in the laboratory with actual material as a basis for study. Thus a discussion of the later development of the organs of special sense has been omitted because the preparation of adequate laboratory material of these organs involves a greater expenditure of time than is practical in most laboratories. The later changes in the muscular system have been omitted because it was believed that the space necessary for their consideration could advantageously be devoted to things of more general interest. Similarly the detailed configuration of the intestinal tract and mesenteries has been dismissed with but a word of comment.

Selection of the histogenesis of bone and the organogenesis of teeth for special consideration was made because these processes as they occur in the pig are essentially typical for all mammals. Moreover laboratory material covering the more important phases of these phenomena can be prepared without undue expenditure of time. I offer no apology for presenting these examples of histogenesis in some detail nor for failure to present additional examples of this type. Either of these processes carefully studied from actual material will give the alert student a first-hand acquaintance with fundamental cytological activities, and with the development of specialized cell products. If such acquaintance be an appreciative one, even though it be limited to a single process, it will be worth far more than superficial contact with many facts categorically presented.

The figures in this book have been planned primarily to tell a graphic story running parallel to that of the text. In many cases, however, it has been found possible to include in the figures details not discussed in the text. My purpose in so doing was to provide a readily available basis for supplementary instruction, and to encourage further study by the student on his own initiative.

CHAPTER 2

The Reproductive Organs; Gametogenesis

I. The Reproductive Organs

Any logical account of prenatal development in mammals must start with a consideration of the phenomena which initiate that development. It is necessary to know more than the mere structure of the conjugating sex cells. We must know something of how they are produced and of the extraordinary provisions which ensure their union in such a place and at such a time that each is capable of discharging its function. Of vital importance, also, are the changes in the body of the mother that provide for the nutrition of the embryo during its intra-uterine existence, and for its feeding during the relatively long period after birth when it cannot subsist on food such as that eaten by its parents. Before it is possible to deal with any of these things intelligibly, or to appreciate the underlying factors in the periodic recurrence of the breeding impulse in animals, it is necessary to become familiar with the main structural features of the reproductive organs.

The Female Reproductive Organs. The location of the reproductive organs of the sow and their relations to other structures in the body are shown in figure 1. The paired gonads of the female, the *ovaries*, lie in the pelvic portion of the abdominal cavity. Each ovary is nearly completely enwrapped by a funnel-like dilation of the end of the corresponding *uterine tube*. This relationship tends to ensure that the ova, when they become mature and are discharged from the ovary, will find their way into the uterine tubes and thence into the *uterus*. There, if they have been fertilized, they become attached and nourished during prenatal development (Fig. 52). The uterus of the sow is of the bicornate type, that is, it has two limbs or horns (Fig. 2). These right and left limbs, which are enlarged continuations of the oviducts, unite with each other mesially to form an unpaired portion known as the body (corpus) of the uterus. The body of the uterus is continuous caudally with the neck or cervix, a region characterized

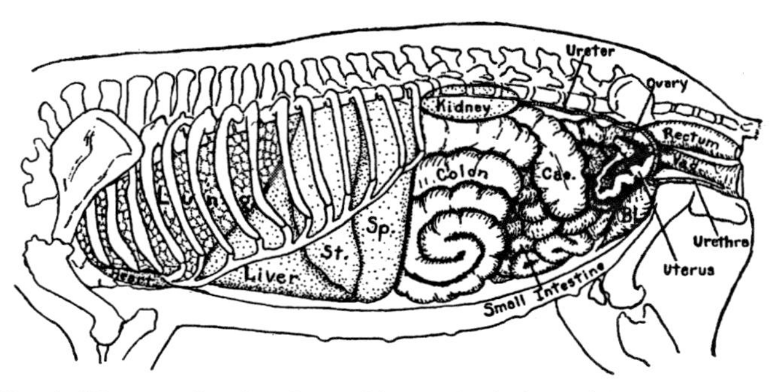

FIG. 1. Diagram showing the position and relations of the more important viscera of the pig. (Modified from Sisson.) Adult female viewed from the left.

Abbreviations: Bl., urinary bladder; Cae., cecum; Sp., spleen; St., stomach; Vag., vagina.

by an attenuated lumen and thick walls (Fig. 126, C). The cervix of the uterus projects into the anterior part of the *vagina*, which serves the double function of an organ of copulation and a birth canal.

The Male Reproductive Organs. The general arrangement and relationships of the male reproductive system are shown in figure 3. The *testes* do not lie in the abdominal cavity as do the ovaries, but are suspended in a pouch-like sac called the *scrotum*. The sex cells produced in the testes must pass over an exceedingly long and elaborate series of ducts before reaching the outside (Fig. 3). From the convoluted or

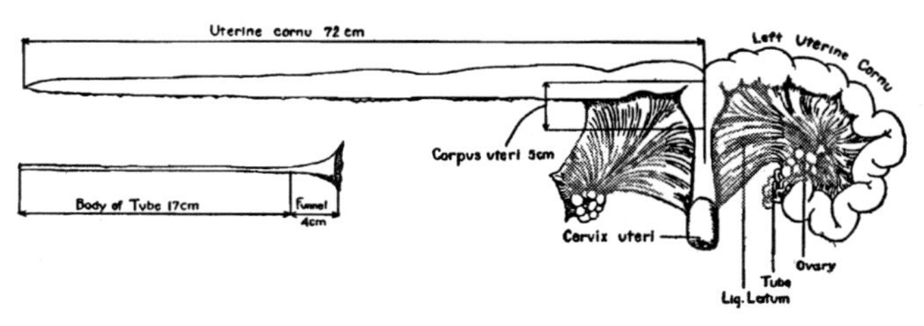

FIG. 2. Diagram of uterus and adnexa of sow. (After Corner.) The right horn of the uterus and the right oviduct (uterine tube) have been cut away from the broad ligament (lig. latum) and pulled out straight to show their dimensions.

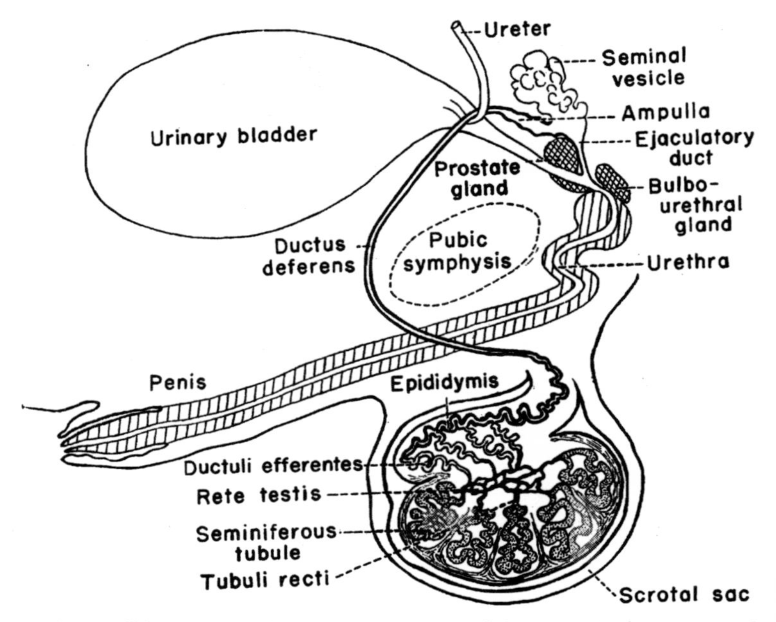

FIG. 3. Diagram showing the arrangement of the reproductive organs of the adult boar. The testes and their ducts up to the entrance of the ejaculatory ducts into the urethra are paired structures, but for simplicity, only those on the left side have been represented.

seminiferous tubules where the spermatozoa (spermia)[1] are formed, they find their way through short straight ducts (the tubuli recti) into an irregular network of slender anastomosing ducts known as the *rete testis*. From the rete testis the spermia are collected by the *ductuli efferentes* which in turn pass them on by way of the much coiled duct of the epididymis, into the *ductus deferens*. At the distal end of the ductus deferens is a glandular dilation known as the *seminal vesicle*. As their name implies, it has been believed that the seminal vesicles served as sort of reservoirs in which the spermatozoa were stored pending their ejaculation. Recently it has become known that the

[1] There are many instances in embryology where two or more synonymous words are in common usage. To facilitate collateral reading some of the most frequently encountered synonyms have been inserted parenthetically.

spermatozoa collect in the somewhat dilated distal ends (ampullae) of the ducti deferentes, and that the seminal vesicles are primarily glandular organs which produce a secretion serving as a vehicle for the spermatozoa.

When, during coitus, the spermatozoa are discharged they enter the urethra by way of the *ejaculatory ducts* (Fig. 3). Coincidently the contents of the seminal vesicles, the prostate gland, and the bulbo-urethral glands (Cowper's glands) are evacuated into the urethra providing a fluid medium in which the spermatozoa are actively motile. This mixture of secretions with spermatozoa suspended in it (*semen*)[2] is swept out along the urethra by rhythmic muscular contractions.

II. Gametogenesis

The manner in which the sex cells or gametes are produced in the gonads demands attention less cursory than that accorded the sex organs as a whole. Nevertheless it is unnecessary to emphasize the peculiarities of gametogenesis in the pig. The processes can more profitably be traced in a manner sufficiently broad to be applicable to mammals generally.

Continuity of Germ Plasm. The cells which give rise to the gametes in any individual, collectively, are said to constitute its "germ plasm." The cells which take no direct part in the production of gametes and which cease to exist with the death of the individual are called somatic cells (collectively, somatoplasm in antithesis to germ plasm).

The germ plasm is of paramount interest not only to the biologist but to all thinking persons, because some of its cells are all that are preserved from one generation to the next. A single cell from the germ plasm of the male parent unites with a single cell from the germ plasm of the female parent. From the cell formed by their fusion a new individual develops. Some of the germ plasm of the parents thus lives on in succeeding generations. All the other cells in their bodies die. The two conjugating gametes alone pass on the entire hereditary dowry of the species, not only from the immediate parents, but from all their ancestors.

[2] In addition to being used as a convenient means of giving synonyms, parentheses have been used to introduce technical terms which may be unfamiliar. In such cases the term, as has been done here, is placed in parentheses following a characterizing phrase.

It is difficult to realize fully the implication carried in this simple statement of the continuity of the germ plasm. The entire future of any species depends on the germ plasm held in trust within the bodies of the individuals now living. Whatever changes for good or ill the germ plasm undergoes will inevitably be written into the history of the species. Fortunately, very early in the life of an individual, the germ plasm is segregated in the gonads and not subject to most of the diseases from which the somatic cells suffer. But the germ plasm, even though not directly affected, may nevertheless suffer indirectly through a poor environment forced on it by an unhealthy body.

Of greater importance still is the nature of the combination of germ plasm which occurs in each generation when the two gametes fuse. As surely as either gamete brings into the new combination defective germ plasm, so surely will both the body and the germ plasm of the new individual suffer therefrom.

Early History of the Primordial Sex Cells. It becomes a matter of fundamental interest, therefore, to go back of the production of gametes in the sexually mature individual. One naturally wants to know the whole story of how the germ plasm is handed on from one generation to the next, not just its last chapter. Obviously the germ plasm of the individual under consideration must have come to it from its parents by way of the ovum and spermium which united to initiate its development. But how and where is it cared for by the individual? When did it first become possible to recognize the germ plasm? When and how did it become segregated in the gonad, more or less protected from the accidents of injury or disease from which the body so commonly suffers? What is it doing during the long time before sexual maturity? All these questions naturally occur to us as leading up to the final maturation and liberation of the gametes in the adult.

The earliest part of the history of the germ plasm in an individual is as yet imperfectly known. The cells which are destined to give rise to the gametes are, however, definitely recognizable at a surprisingly early stage in development. Long before it is possible to tell whether an embryo is to become a male or a female, certain large cells become differentiated from their neighbors. By working backward from older stages where conditions are more clear cut, it has been possible to identify these cells as the progenitors of the gametes. In other words they constitute the germ plasm as it exists in an embryo of that age. These large cells are called the primordial sex cells. Since their early

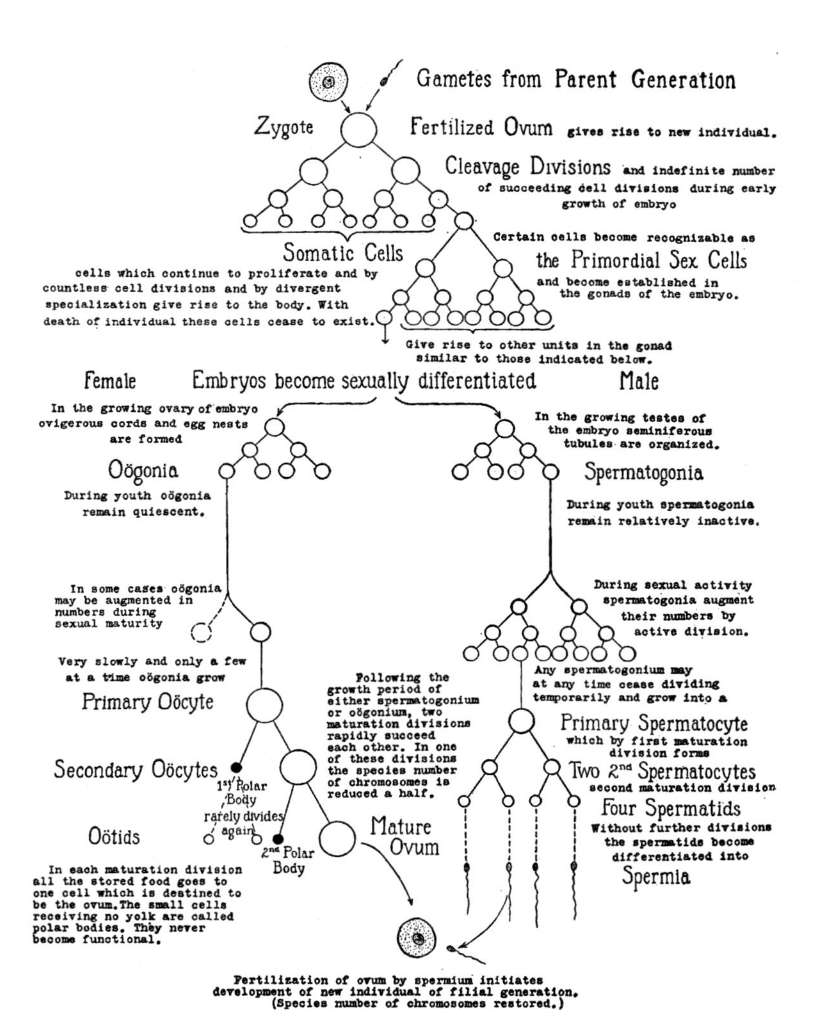

FIG. 4. Chart outlining, for one generation, the history of the gametes and the germ plasm from which they are derived.

history is the same whichever sexual type of gamete they will ultimately produce, no account needs to be taken at first of sex differences in the individual (Fig. 4).

The primordial sex cells are first easily identifiable in the mammalian embryo when they lie in the epithelial covering of the gonad. Until comparatively recently it was believed that they could not be recognized at all as germ cells any earlier in their history. Of late, however, much more detailed studies have been made and there is cogent evidence being brought forth indicating that they can be identified prior to their appearance in the gonad. This recent work seems to show that the primordial germ cells become recognizably differentiated in most vertebrate embryos in the yolk-sac entoderm, and that they migrate thence to establish themselves in the gonads. How much farther back toward the fertilized ovum the lineage of the primordial sex cells may, in the future, be traced with definiteness it would be rash to predict. Even now, in some of the invertebrates, it is believed that the individual cell which gives rise to all the sex cells can be identified as far back as the early cleavage divisions.

Early Differentiation of the Testes. Shortly after the primordial sex cells become established in the gonads, sexual differentiation begins. It is then necessary to trace the course of events separately for the two sexes (Fig. 4). During the embryonic life of the young male individual the primordial sex cells grow from the epithelial covering of the testis into its substance and there become organized to form the seminiferous tubules. Many cell generations are of course occupied with the fabrication of a seminiferous tubule, but the cells which eventually constitute its wall can, nevertheless, be traced back to the primordial sex cells (Fig. 4). When they become established in the walls of a seminiferous tubule the cells are known as "sperm mother cells" or *spermatogonia.* At this stage of development the spermatogonia constitute the individual's "germ plasm." During early postnatal life and the growth period, these future gamete producers remain relatively quiescent and undeveloped. Their inactivity stands in sharp contrast to the rapid proliferation and differentiation of the remainder of the cells which go to make up the body of the growing individual. It is as if the spermatogonia, thus set apart, were hoarding their energy for the next generation. Only when the individual becomes sexually mature do they begin intensive activity.

Spermatogenesis. The mature testes contain a large number of much convoluted seminiferous tubules. Their position and relations

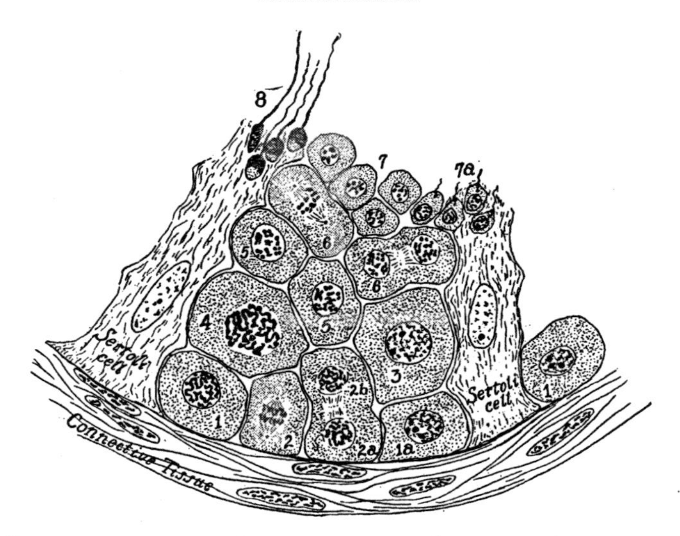

FIG. 5. Semischematic figure showing small segment of the wall of an active seminiferous tubule. The sequence of events in the production of spermia is indicated by the numbers. A spermatogonium (1) goes into mitosis (2) producing two daughter cells (2a and 2b). One daughter cell (2a) may remain peripherally located as a new spermatogonium eventually coming to occupy such a position as 1a. The other daughter cell (2b) may grow into a primary spermatocyte (3), being crowded meanwhile nearer the lumen of the tubule. When fully grown the primary spermatocyte will divide again (4) and produce two secondary spermatocytes (5, 5). Each secondary spermatocyte at once divides again (6, 6), producing spermatids (7). The spermatids become embedded in the tip of a Sertoli cell (7a), there undergoing their metamorphosis and becoming spermia (8), which when mature are detached into the lumen of the seminiferous tubule.

are schematically indicated in figure 3. This figure, however, gives no conception of the astonishing total length of the gamete-producing tubules crowded within the testes. It has been estimated by Osterud and Bascom from a study of serial sections that the total length of the seminiferous tubules from one testis of a mature boar, pulled out straight and placed end to end, would reach 3200 meters. When one realizes the total length of these tubules it is not difficult to understand how each ejaculate of semen contains millions upon millions of fully formed, active spermatozoa.

If we examine the spermatogonia lying at the periphery of an active adult seminiferous tubule we see many mitotic figures (Fig 5, 2). A cell arising from such a spermatogonial division may do on of two things. It may cease dividing for a time and, by growing to a markedly larger size than its parent, become differentiated as a *primary spermatocyte* (Fig. 5, 3). It may remain like its parent and con tinue to produce other spermatogonia. The new cells thus formed take the place of the spermatogonia which have grown into spermatocyte and moved out of the spermatogonial layer toward the lumen of the tubule. Thus some of the cells always remain in the peripheral par of the tubule as spermatogonia and furnish a constant source of new cells ready for conversion into spermatocytes.

Once a cell has undergone the growth phase which differentiate it so that we call it a primary spermatocyte, its future history is very definitely determined. It first undergoes a division resulting in the formation of two smaller daughter cells called *secondary spermatocyte* (Fig. 5, 5). Each of these secondary spermatocytes, without any resting period which might allow the cells to grow to the size attained by their parents, promptly divides again and forms two *spermatids* (Fig. 5, 6, 7). Cell division then ceases and each spermatid is gradually transformed into a fully formed, potentially functional male gamete, the spermium. In the metamorphosis of a spermatid: the nuclear material becomes exceedingly compact to form the bulk of the head of the spermium; the centrosomal apparatus of the spermatid undergoes an elaborate modification to give rise to the motile axial filament of the tail of the spermium; and the cytoplasm is reduced greatly in bulk giving rise to an envelope with a tiny thickened cap (acrosome) about the head of the spermium, and a delicate investment of the axial filament of its middle piece and tail. During their transformation, the spermatids are embedded in the cytoplasm of nurse cells (*supporting cells of Sertoli*) which lie at intervals in the wall of the seminiferous tubule (Fig. 5, 7a, 8). It is believed that the Sertoli cells in some way transfer food material to the metamorphosing spermatids from the small blood vessels in the connective tissue investing the seminiferous tubule. When they are fully mature the spermia free themselves from the Sertoli cells and are carried out along the lumen of the tubule toward the epididymis (Fig. 3).

Early Differentiation of the Ovaries. The origin, migration, and early segregation of the primordial sex cells in the gonads take place, as we have seen, before there is any sexual differentiation observable in the embryo. Consequently in tracing these phenomena we established a common starting point for following later developments in the female as well as in the male (Fig. 4). Even when the indifferent stage is passed and it is possible to say definitely that a given embryo is developing into a female, conditions in the ovary are at first similar in a general way to conditions in the testis at a corresponding stage of development. The sex cells, which have appeared in the epithelium investing the growing ovary, push centripetally into the ovarian connective tissue in a manner very suggestive of the way seminiferous tubules arise in the testis. The cords of sex cells thus invading the ovarian stroma are known as *egg tubes* (ovigerous cords, Pflüger's tubes) (Fig. 6).

From this point on, the structural resemblances of the growing gonads in the two sexes become less and less apparent. A striking homology, however, exists throughout the entire series of changes in the germ cells themselves, and should not be lost sight of even though its later phases occur in organs so divergently differentiated as the adult ovary and testis come to be. Although the inference will undoubtedly have been drawn from what has already been said of the early history of the primordial germ cells in the two sexes, it may be well to emphasize the fact that the cells of the ovigerous cords are homologous with the spermatogonia of the male. It is, of course, because of this homology that they are called *oögonia* (Fig. 4).

Oögenesis. Although the oögonia exhibit a growth period, and then undergo two rapidly succeeding maturation divisions just as do the spermatogonia, the details of these processes in the two sexes are quite unlike. Their differences are correlated with the antithetical nature of the specializations in the gametes themselves. In the male, small, actively motile gametes with no stored food material are produced in enormous numbers. The energy which in the male goes into quantity production, in the female is expressed by more elaborate preparation of the gametes and the storing of food material in their cytoplasm. The ova thus become very large, non-motile cells and, compared with spermatozoa, relatively few of them are brought to maturity.

Very early in the history of the oögonia the tendency to specialize a few cells rather than many, becomes apparent. In the ovigerous cords, and in the egg nests which are formed by the breaking up of the cords, one or two of the cells will almost always be found to have grown larger than their neighbors (Fig. 6). All the cells of the cords or nests are potentially oögonia. The ones which show enlargement

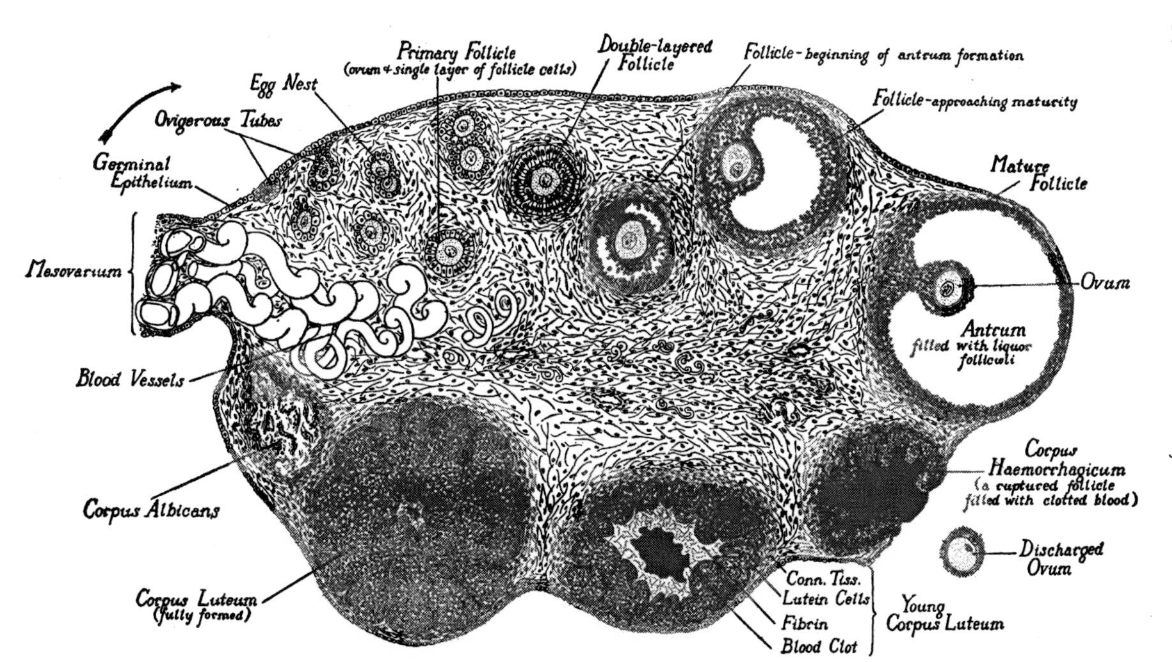

FIG. 6. Schematic diagram of mammalian ovary showing the sequence of events in the origin, growth, and rupture of the ovarian (Graafian) follicle, and the formation and retrogression of the corpus luteum. Follow clockwise around the ovary starting at the arrow.

are already beginning their long slow growth to form primary oöcytes. The cells which lie adjacent to one of these growing oögonia, figuratively speaking, forego their own chances of becoming oöcytes and arrange themselves as a protecting and food-purveying investment about the future ovum. The entire group of cells thus formed is known as a *primary ovarian (Graafian)*[3] *follicle.* (Fig. 6.) The cells surrounding the oöcyte proliferate rapidly and form an increasingly thick covering about it. With continued growth there appears in the layer of follicle cells a cavity which fills with fluid and expands very rapidly. This cavity is called the *antrum*, and the fluid which fills it is known as the *liquor folliculi.*

During these changes the developing follicle has usually pushed its way deep into the connective tissue framework (stroma) of the ovary. When the follicle begins to fill with fluid it starts to work gradually toward the surface of the ovary. As its size is still further increased it comes to protrude from the ovary, appearing in gross, fresh material, much like a water-blister.

Such a follicle is nearly ready to rupture and release the contained egg cell (Fig. 6). In the course of its development through the stages which we have just sketched in brief outline, it has acquired a degree of differentiation which demands closer scrutiny. The egg cell at this stage is commonly called the ovum, but if we used the terminology which emphasizes the homologies of development in male and female gametes we should call it by its more cumbersome name, *primary oöcyte*. It has grown to a size many times that of the follicle cells which surround it, and its abundant cytoplasm is dotted with granules of stored food material (yolk, deutoplasm). The total yolk content in mammalian ova being relatively small and uniformly distributed, the nucleus is not crowded to one side but is centrally located within

---

[3] Formerly it was customary to name structures after the first man describing them. For example the ovarian follicle in all the older literature will be found designated as the Graafian follicle, after the Dutch anatomist Reijnier de Graaf (1641–1673). While this old custom is interesting in that it preserves to us the names of pioneer workers, the present tendency is to make our nomenclature more logical and more easily remembered by replacing proper-name designations with ones descriptive of the structure. Such a change can be accomplished only gradually, however, and in many cases, as is true in the present instance, a proper name has become so firmly established through long usage that it is necessary to know it as a synonymous term in order not to be confused by its constant appearance in reference reading.

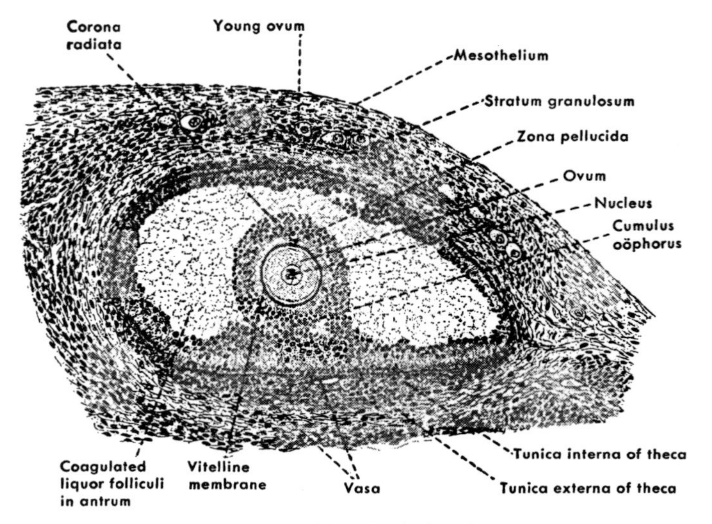

FIG. 7. Drawing of an ovarian (Graafian) follicle approaching maturity, showing details of its structure and relations.

the cytoplasm (Fig. 7). The cell membrane has become considerably thickened. It still keeps its old name of "vitelline membrane," which along with such obsolescent terms as "germinal vesicle" for the nucleus, and "germinal spot" for the nucleolus, was given it before the true significance of "the egg" as a specialized cell was understood.

Surrounding the ovum is a transparent, non-cellular, secreted layer known as the *zona pellucida*.[4] Outside the zona pellucida is an investment of radially elongated follicle cells (Fig. 7) some of which still cling to the ovum when it is discharged (Fig. 6). These cells constitute the corona radiata.

With the increase in size and fluid content which the antrum has by this time undergone, most of the follicle cells are crowded peripherally to constitute the so-called *stratum granulosum* of the follicular wall (Fig. 7). Outside the stratum granulosum, the immediately

---

[4] In some forms this zone exhibits delicate radial striations. When these are conspicuous the term zona radiata instead of zona pellucida is applied to this same layer. The term, although descriptively appropriate, is unfortunate because it is so frequently confused with the totally different cellular investment outside of it called the *corona* radiata.

surrounding ovarian connective tissue has become condensed about the growing follicle. This secondary connective-tissue investment is known as the theca folliculi. It is differentiated into an outer, densely fibrous layer, and an inner layer less conspicuously fibrous and containing many cells and numerous small vessels (Fig. 7).

At the point where the ovum lies among the follicle cells they form a hillock projecting from the stratum granulosum into the antrum. This hillock is known as the *cumulus oöphorus* (Fig. 7). When first formed it is broad and low. As the follicle approaches maturity the cumulus becomes more elevated and somewhat undercut (Fig. 6). Finally the ovum is carried on a slender stalk of cells which readily releases it and allows it to escape in the follicular fluid set free when the follicle ruptures.

Ovulation. The precise mechanism which precipitates the rupture of ovarian follicles is not as yet known with certainty. In all probability there are several factors involved. We know from the way mature follicles protrude at the ovarian surface that the fluid pressure within them is considerable (Fig. 6). As the follicle bulges under this pressure its connective-tissue envelope (*theca folliculi*) is squeezed against the connective-tissue capsule (*tunica albuginea*) of the ovary. It seems not unlikely that this process would result in compressing the small blood vessels where the bulging is most pronounced. This would reduce the nutrition of the region affected and eventually the strength of the tissues. Such a mechanical effect of fluid pressure might well pave the way for the rupture of the follicle. Underlying the development of the follicle and the accumulation of its contained fluid, there is certainly the stimulus of a hormone produced in the anterior lobe of the pituitary. According to the recent work of Joseph Smith, there appears to be a marked increase in the concentration of salts in the liquor folliculi as the time of rupture approaches. The endosmotic effect of such a concentration may well be a precipitating factor in bringing the internal fluid pressure to a point where rupture of the follicle occurs.

However future work may evaluate the importance of the several possible causative factors involved, we know that the rupture of the follicle when it does occur is an abrupt, almost explosive, process. Hill, Allen, and Kramer have succeeded in making a detailed micro-moving-picture record of ovulation in the rabbit, and their film shows with great vividness the rapid terminal bulging of the follicle culminating in sudden rupture with a gush of follicular fluid which brings with it the ovum surrounded by its radiate corona of follicle cells. A slight hemorrhage can be seen to accompany the rupture of the follicle.

Atresia of Follicles. By no means all the follicles that start to enlarge as a given ovulatory period approaches go on to maturity and ovulation. A considerable number of follicles that have started to undergo marked enlargement and give every apparent indication that they are going to continue developing suddenly cease to grow and then begin to degenerate. The process is known as follicular atresia, and a follicle so involved is said to be atretic. The underlying regulatory factors causing the atresia of certain follicles while other neighboring follicles go on to maturity is not known.

Maturation of the Ovum. The maturation of the "ovum" (primary oöcyte) takes place at just about the time of its liberation from the follicle. As in the male, two cell divisions occur in rapid succession, but instead of four functional gametes being formed as an end result, there is only one. At each maturation division two cells are formed. But one of these cells receives practically all the stored food material of the primary oöcyte, while the other receives little or none and soon degenerates. The cell receiving no yolk material was called a "polar body" before its significance was understood. It is of course an oöcyte minus its proper share of cytoplasm.

The gross results of the two maturation divisions in the female are schematically summarized in the chart of figure 4. The primary oöcyte divides to form two *secondary oöcytes*. One of these receives little cytoplasm and is called the first *polar body*. The secondary oöcyte which has preëmpted all the stored food material promptly undergoes another mitosis, the second maturation division. Again the bulk of the cytoplasm goes to one of the two resulting *oötids*. The oötid receiving it is commonly called the "matured ovum" and is now ready to be fertilized by a spermium. The other oötid is the *second polar body*. Occasionally the first polar body undergoes a second division, clearly indicating the homology of the maturation divisions in the two sexes (Fig. 4). Usually, however, it degenerates before such a division occurs. The second polar body likewise degenerates soon after it is formed, leaving, of the four potential oötids, only one which becomes functional.

Significance of Maturation. The events in the maturation of male and female gametes which have just been discussed are but the more evident phases of the process. There have been changes of profound significance going on at the same time in the nuclear material. It would carry us far afield into cytology and genetics, to discuss these changes in detail or to attempt an interpretation of their full meaning. We can, however, indicate briefly wherein their importance lies.

It has already been stated that the inheritance of an individual comes to it by way of the gametes arising from the germ plasm of its parents. We can be more definite. It comes by way of the chromosomes in the nuclei of the gametes. The chromosomal content of the nuclei in the cells which go to make up the body of an individual is definite and constant. The elaborate mechanism of mitosis splits the chromosomes lengthwise into qualitatively and quantitatively equivalent daughter chromosomes. Thus in each of the countless cell divisions involved in the growth of an individual the number of chromosomes remains in the daughter cells what it was in the parent cell. The chromosomal number so maintained is different in different species, but in the various cells of individuals of the same species it is fixed and definite, the *species number of chromosomes* in the parlance of cytologists and geneticists.

While the maintenance of the species number of chromosomes in an individual is dependent on the way each chromosome is split in the process of mitosis, it is preserved from generation to generation by the processes of maturation and fertilization. In the maturation divisions the number of chromosomes in the gametes is reduced to half the number characteristic of the species. When, in fertilization, a male and a female gamete each bearing half the species number of chromosomes unite with each other, the species number of chromosomes is reëstablished in the individual of the new generation.

Cytologists have worked out the mechanism of the maturation divisions with great care in many forms. The process consists essentially of two specialized cell divisions which follow each other in rapid succession without the nucleus returning to a resting stage as happens between ordinary mitoses. To distinguish these maturation divisions they are called *meiotic* (in contrast to mitotic) *divisions*.

Another characteristic feature of maturation is the special behavior of the chromosomes in the prophase preceding the first of the two meiotic divisions. To understand this process, which is known as *synapsis*, it is necessary to realize that the chromosomes making up the species number possessed by each somatic cell, and by all the germ cells before their maturation, are present in pairs. Thus the species number of chromosomes is always an even number, for example, in man it is 48. A skilled cytologist working with favorable material can see that these chromosomes have peculiarities of size and shape which make them individually identifiable. By careful comparison each chromosome can be matched to a similar one so they can all be arranged in pairs. The number of pairs will, of course, be half the total number of chromosomes characteristic for the species. Moreover, the hereditary implications of the fact that the chromosomes can thus be arranged in pairs are of the greatest importance, for we know that one member of each chromosomal pair came from the male parent and the other member from the female parent.

In an ordinary mitosis the members of the chromosomal pairs can be recognized but they appear to lie scattered in haphazard fashion. During the prophase, the chromosomes seem to have aggregated at once, and without any evident scheme of arrangement, at the equator of the spindle. In contrast, during the prolonged prophase of the first maturation division, the members of the chromosomal pairs come to lie close to each other and so remain for some time. It is this pairing off of the chromosomes that is called synapsis.

In one of the meiotic divisions (usually the first) the chromosomes which were brought together in synaptic pairs move apart without being split. Thus each daughter cell receives one member of each chromosomal pair, its chromosomal content thereby being reduced to half the species number. The maturation division which accomplishes this separation of the chromosomal pairs is appropriately enough called a *reduction division*. The daughter cells which contain but half the species number of chromosomes are said to contain the *haploid number* of chromosomes, in contrast to the *diploid or species number*, contained before the reduction division. Following the reduction division, the second maturation division splits the chromosomes in much the manner of an ordinary mitosis. It, therefore, is called an *equational division* and produces daughter cells which still have the haploid number of chromosomes. When these matured male and female gametes, each containing half the species number of chromosomes, unite in fertilization, the diploid or species number is restored.

But there is more in maturation than a mechanism which maintains the species number of chromosomes. In the reduction divisions the chromosomes are not split but distributed, some to one cell, some to the other. The resulting cells contain different hereditary potentialities because they contain different chromosomes, not halves of the same chromosomes as results in an ordinary mitosis. What hereditary possibilities are discarded into the polar bodies thrown off from the female gamete and what retained in the mature ovum is a matter of chance distribution. What potentialities find their way into the particular sperm which alone out of millions of its fellows fertilizes the ovum is likewise fortuitous.

> Thus in the game of life, the maturation processes virtually shuffle the hereditary pack and deal out half a "hand" to each gamete. A full hand is obtained by drawing a partner from the "board,"—by combining with some other gamete of the opposite sex. Hence offspring resemble their parents because they play the game of life with the same kind of cards, but not, however, with the same hands. The minor differences in offspring, or the variations from the standard type that always go with these basic resemblances, are due to variations in the distribution of the genes during maturation, fertilization and cleavage.
>
> Thus sufficient stability and variety is produced to insure continuity and progress. For the offspring will in the main resemble progenitors which have successfully lived in the prevailing conditions of the past, but will exhibit sufficient variability among themselves to insure that some of them shall successfully live in any conditions likely to arise in the future.[5]

Sex Determination. Probably no embryological question has been the subject of so much conjecture as sex determination. From time immemorial theory after theory purporting to explain why this embryo became a male and that one a female has been advanced only to be discredited. Under such circumstances one becomes exceedingly cautious in discussing even a tentative hypothesis. Of late, however, there has been so much discussion of the chromosome theory of sex determination that anyone who would consider himself well informed biologically must know what it is, regardless of its ultimate fate.

In discussing maturation, it was emphasized that the chromosomes present in the cells of a species could be arranged in pairs, the members of which were alike. In a male individual, however, one pair of chromosomes is an exception in that its members are strikingly unlike. The members of this pair are called the "X" and "Y" chromosomes. Although our knowledge is as yet fragmentary and unsatisfactory, there is sufficient evidence indicating that the X-Y pair of chromosomes is associated with the determination of sex, so that they are commonly referred to as *the sex chromosomes*. If the cells of a female individual are examined with reference to this peculiar pair of chromosomes, we find that instead of the large X and the small Y

[5] From William Patten, "Life, Heredity and Evolution."

members characteristically appearing in the male, the female has two large X members.

Careful studies of maturation have yielded the clue as to how this sex difference in chromosomal pattern comes about and is maintained. In the synapsis which occurs in the maturation divisions, the two members of the sex chromosomal pair, as is the case with any other chromosomal pair, will be found associated with each other. In the reduction division which separates the members of the chromosomal pairs in the spermatocytes of the male, the X chromosomes must inevitably go to one cell and the Y chromosomes to another. Since in all the cells of the female there is an X-X combination, in the reduction division which occurs in the maturation of the ova one of the X chromosomes must go to the polar body and one to the maturing ovum, so that all the ova will have an X chromosome.

When an ovum ready for fertilization is surrounded by swarms of spermatozoa, half of which have one chromosomal pattern and half another, it is obvious that there are equal chances as to which of the two types will be the one first to penetrate the ovum. If it is a sperm cell carrying an X chromosome, fertilization will establish in the zygote the X-X combination characteristic of the female. If, on the other hand, the successful sperm cell carried a Y chromosome, the X-Y combination characteristic of the male would result. If, for the sake of simplicity in diagraming, we use germ cells from an animal having a species number of only eight chromosomes, the essential happenings as postulated by the chromosome theory of sex determination may be schematically summarized as indicated in figure 8.

Perhaps the fairest way to assess this particular theory as to the determination of sex is to say that at the present time it accords with more of the known facts than any other theory. We must recognize, however, that as yet we know practically nothing as to the mechanism by which the characteristically different chromosomal pattern present in the two sexes may operate. There is some indication that the chromosomal combination established at the time of fertilization may provide merely the initial impetus toward sexual differentiation in one direction or the other, and that the action of certain internal environmental factors may be important in bringing about full differentiation.

The Corpus Luteum. When the ovarian follicle ruptures and liberates the ovum, its history is by no means closed. There remain in the ovary the great bulk of the follicle cells and the connective-

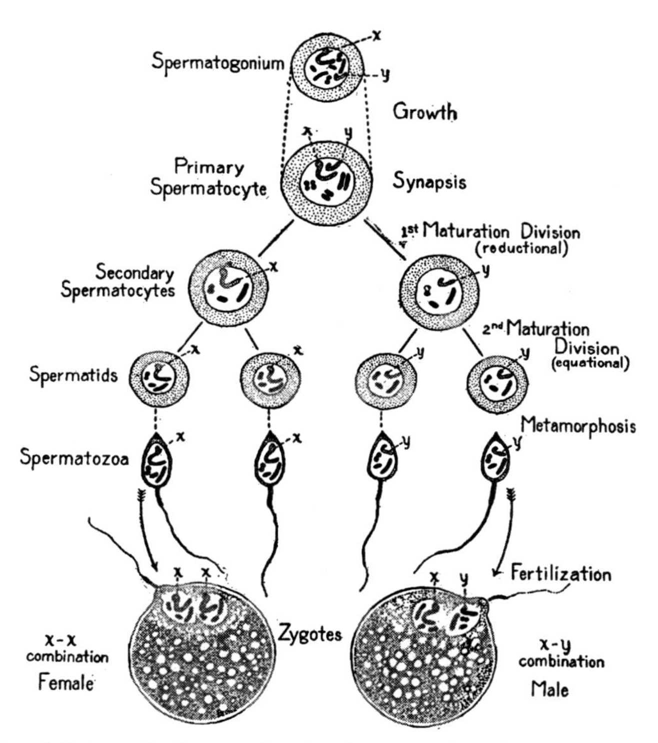

FIG. 8. Schematic diagram showing the separation of the members of the sex chromosome pair in maturation, and their recombination in fertilization. It is assumed that the species number of chromosomes is eight and that it is the male which produces gametes of different potentialities with regard to sex determination. The sex chromosomes are stippled; other chromosomes are drawn in solid black. (Patten: "Early Embryology of the Chick," The Blakiston Company.)

tissue theca which surrounded the follicle before its rupture. These structures do not degenerate at once but become involved in the development of the *corpus luteum*. The corpus luteum, so called because of the yellow color it exhibits in fresh material, grows very rapidly in bulk for a time and becomes an organ of internal secretion (endocrine organ, ductless gland). That is to say it produces a secretion which is not discharged by way of ducts as is the case with ordinary glandular secretions, but which is liberated into the blood stream. The secretion diffused from a ductless gland into blood vessels, and carried by the blood stream to some other place in the body where it exerts a definite physiological effect, is called a hormone. The probable action of the particular hormone produced by the corpus luteum is a subject to which we shall have occasion to return later in connection with the sexual cycle. For the moment we are concerned with the origin and structure of the corpus luteum itself.

The formation and histology of the corpus luteum of the sow have been described with great care by Corner to whose work reference should be made for a detailed account of the subject (see bibliography). Here but a brief sketch can be given.

When the ovarian follicle ruptures, escape of the contained fluid and contraction of the smooth muscle in the stroma of the ovary reduce the size of its lumen. Bleeding of the small vessels injured in the rupture of the follicle fills the partially collapsed antrum with blood which promptly becomes consolidated as a clot. A newly ruptured follicle thus filled with clotted blood is called a *corpus haemorrhagicum* (Fig. 6).

The blood clot is soon attacked peripherally by phagocytic white blood corpuscles and becomes progressively reduced in size. Concomitantly the follicular cells of the stratum granulosum increase greatly in size and are crowded into the area formerly occupied by the blood clot. At the same time small vessels from the connective-tissue theca penetrate the mass of enlarged follicle cells and ramify among them. These vessels bring in with them numerous small cells which become packed in among the more conspicuous cells which originated from the stratum granulosum. Thus both layers of the follicular wall contribute to the corpus luteum although the most conspicuous and characteristic cellular elements are derived from the follicle cells of the stratum granulosum.

Corpora lutea normally develop from all ruptured follicles, but if the liberated ova are not fertilized the corpora lutea soon degenerate. If, however, the ova are fertilized and implanted in the uterus, the corpora lutea undergo a greatly prolonged period of growth and persist much longer before degenerating. This difference in the history of the corpora lutea is recognized by designating the short-lived ones as the *corpora lutea of ovulation* and the ones which persist longer as the *corpora lutea of pregnancy*. Histologically they exhibit the same structural picture, their differences being apparently quantitative rather than qualitative.

When either type of corpus luteum begins to degenerate the retrogressive changes are in the nature of a fibrous involution. That is, the cellular part of the organ disintegrates and fibrous connective tissue takes its place. As this connective tissue grows older and more compact it gradually takes on the characteristic whitish appearance of scar tissue. Finally all that is left in the ovary to mark the site of what was ovarian follicle, and subsequently corpus luteum, is a shrunken patch of scar tissue called a *corpus albicans* (Fig. 6).

CHAPTER 3

The Sexual Cycle; Fertilization

I. The Sexual Cycle

The periodic recurrence of times when the mating impulse becomes the dominating factor in the reactions of an animal has long excited comment and conjecture. Why does the urge to sexual union suddenly manifest itself in an animal but recently indifferent to the opposite sex? Why should it become latent for considerable stretches of time even when pregnancy has not ensued? What changes are going on inside the body correlated with, or causing, the animal's abrupt changes in reactions? To such questions no answers other than surmises have until recently been forthcoming. Many phases of these questions cannot yet be answered with particularity and there are still problems involved on which we have virtually no tested evidence. But the main points of the story have been fitted together and checked experimentally and new details are constantly being added.

As is inevitable in any field where our knowledge is growing rapidly, there is much difference of opinion as to the significance of certain of the observed facts. It is through the discussion and evaluation of just such divergent interpretations that further progress comes. But it is neither possible nor advisable in such a brief account as this to become involved in the points of controversy. We must content ourselves with a mere outline of the facts which seem best established and realize that even some of these facts may be subjected to different interpretations.

Sexual periodicity is much less strongly developed as a rule in the male than in the female. In some species the male is sexually potent without interruption throughout adult life. In many species the sexual instinct in the male seems to be aroused from latency periodically merely by the presence of females manifesting the mating impulse. In either of these cases there are no apparent structural changes in the body or in the reproductive organs of the male at different seasons. In contrast to such conditions there are manifested by the males of many forms, short, well marked periods of intense sexual activity alternating with long periods of sexual impotence. Such a cycle is accompanied by actual structural fluctuations in the state of the reproductive organs themselves. Sometimes the changes in the gonads are accompanied by the development of the very striking secondary sexual characteristics such as the antlers of the buck, which appear only at the breeding season to be shed afterwards and grown again at the next breeding season. A brief period of pronounced sexual activity, when it occurs in males, is known to animal breeders as their "rutting season." It always corresponds in time with the females' period of strong mating impulse which breeders call the "period of heat" and biologists speak of as the *estrus*.

The term estrus originally referred merely to the existence of a period of strong sexual desire made evident through behavior. As more information has been acquired about the concomitant changes going on within the body, it has become evident that this period of desire is but an external indication that all the complicated internal mechanism of reproduction is ready to become functional. If pregnancy does not occur at this time regressive changes follow and another period of preparation must ensue before conditions are again favorable for reproduction. This repeated series of changes is known as the estrous cycle. Its phases in the absence of pregnancy are: (1) a short time of complete preparedness for reproduction accompanied by sexual desire (estrus); (2) a short period of regressive changes in which evidences of a fruitless preparation for pregnancy disappear (post- or metestrum); (3) a relatively long period of rest (diestrum); followed by (4) a period of active preparatory changes (proestrum) leading up to the next estrus when everything is again in readiness for reproduction.

There is wide variation in the length of time occupied by this cycle in different animals. In some it occupies an entire year, the estrus recurring at the same time each year and being so placed seasonally that when in due time the young are born, conditions are favorable for their rearing. The deer family exemplifies this condition. Their mating season comes in the autumn and the fawns are born in the spring. The young then have an entire summer of plentiful food supply before they are subjected to the rigors of winter conditions. Whether or not pregnancy occurs, mating does not take place again until the next autumn. Forms having thus but one breeding season in the year are said to be monestrous.

Many forms, in contrast, exhibit a regularly recurring series of mating periods in a year. Such forms are said to be polyestrous. The pig is an example of this condition. Sows failing to become pregnant will "come in heat" again after an interval of about 21 days, and maintain such a cycle until it is interrupted by pregnancy.

On the basis of our present knowledge it would appear that a polyestrous rhythm such as that occurring in the sow is the underlying condition in mammals generally. Many factors mask or modify it in different cases, but in the forms which have been most fully studied it is unmistakably present. Accepting this proposition tentatively, it is not unreasonable to suppose that an annual estrus such as that exhibited by the deer family has become established through the suppression of other periods, primarily because of the regular recurrence of pregnancies of long duration following what was originally merely the most favorable of several estrous periods. It is well known, furthermore, that the estrous cycle may be interrupted by many things other than pregnancy. Thus starvation, extreme exposure, or severe sickness may cause the suppression of an estrus. A contributing cause in reducing a polyestrous to a monestrous rhythm, operative in the case of females failing to become pregnant, might well be the severity of the conditions under which they live during the winter.

It is known with certainty that many animals, as for example the sheep, which have but one breeding season in the year when living in their wild state, develop a polyestrous rhythm when living in domestication. An underlying polyestrous condition is necessarily obscured when a pregnancy follows each estrus, as occurs normally among wild animals. It becomes apparent, however, when such an animal under conditions of domestication is not permitted to become pregnant. Then there is a brief period of rest and preparation followed shortly by another estrus. Living conditions under domestication being relatively uniform, suppression of an estrus through starvation or exposure does not occur and the estrous periods recur at fairly regular intervals until one of them is consummated by pregnancy.

Before seeking the factors which underlie the recurrent estrous manifestations it will be well to have clearly in mind the changes which are going on in the internal reproductive organs in correlation with the alternate conspicuousness and abeyance of the mating instinct.

In the ovaries a group of Graafian follicles begins rapid enlargement just before the onset of estrus (Fig. 9). These follicles usually rupture during or toward the close of the estrus, in other words at the time when active spermatozoa are most likely to be awaiting them in the oviducts where fertilization usually occurs.

The oviducts in the sow have been shown by Seckinger to exhibit a gradually increasing rate of muscular activity during estrus which reaches its height toward the close of estrus at about the time the discharged ova are in the oviduct (Fig. 9). It is probable that such a condition will be found to exist in other mammals also, when observations are extended to them. It seems clearly correlated with the transportation of the ovum to the uterus. The lining epithelium of the oviduct is also thicker and apparently more active in secretion at this time. It would be interesting to know whether or not there is also an increase in the power of its ciliary action.

Toward the close of estrus, the uterus begins to exhibit a marked congestion which reaches its height during the postestrous period (Fig. 9). Accompanying the excessive congestion there is hypertrophy of the uterine mucous membrane and increased activity of the glands. In the uterus of primates, especially in man, this stage of congestion

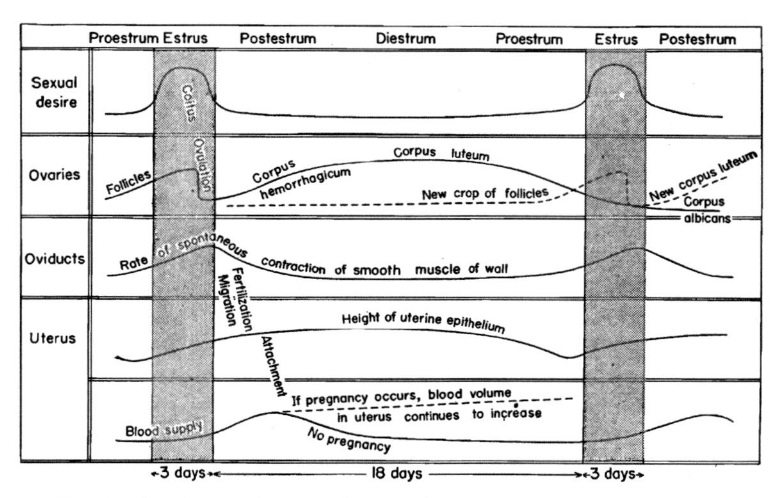

FIG. 9. Graph showing correlation of changes which occur during the estrous cycle in the sow. (Compiled from the work of Corner, Seckinger, and Keye.) Note the coincidence of the important events leading toward pregnancy (coitus, ovulation, fertilization of the ovum and its migration through the oviduct to the uterus, and finally its attachment to the uterine mucosa) with the height of local activity as indicated by the curves.

and hypertrophy is very pronounced and is terminated rather abruptly by hemorrhage and by sloughing of the uterine epithelium (menstruation). In most mammals the analogous changes are more gradual and less extensive, and are accomplished without hemorrhage.

By the time the ovum has traversed the oviduct and reached the uterus, the uterine mucous membrane is at the height of its constructive phase. The experiments of Leo Loeb in transplanting embryonic placental tissue indicate that the uterine mucosa is at this time in its most favorable condition for the attachment of embryos.

One would scarcely expect the mechanism by which such an elaborate series of changes are induced and synchronized to be a simple one. It is indeed exceedingly complex. Recent investigations have, however, thrown much light on it. Although there remains much to be cleared up we are at least finding the nature of the factors involved if not their precise action.

With the reservation that much of the information is recently acquired and therefore not thoroughly tested and subject to revision as new facts come to light, the main thread of the story seems to be unfolding in this manner. The development of the reproductive system as a whole appears to be accelerated at the appropriate phase of bodily maturity through the operation of hormones, the most active of which are derived from the anterior lobe of the pituitary gland.

The factors immediately responsible for the maintenance of the estrous cycle during maturity have long been known to be resident in the gonads. When the ovaries are removed from an immature female the estrous phenomena never develop. When the ovaries are removed from an adult female the estrous cycle ceases. The experimental studies of Edgar Allen and his co-workers gave the first conclusive evidence that the maturing ovarian follicles were the source of a potent estrogenic hormone. They were able by injections of follicular extracts to induce typical estrous reactions in females which had previously ceased to show estrous changes as a result of complete ovariectomy. Such a source for the hormone would account for the height of sexual desire occurring at about the time of the liberation of the ova, when the follicles are turgid with hormone-containing fluid. It would account, also, for the initiation of congestion in the uterus, and at the same time for the subsidence of these phenomena after the escape of the follicular contents (see Fig. 9). The ripening, after a period of rest, of another crop of follicles would account in the same way for the succeeding estrus.

Extensive experimental work by many investigators, following the pioneering studies by P. E. Smith, has shown that the ovaries cannot function effectively unless they are activated by hormones produced in the anterior lobe of the pituitary gland. These hormones are essential for initiating the growth of the gonads which occurs with the onset of puberty, and for maintaining the effective function of the gonads during sexual maturity. They are designated as *gonadotropic hormones*.

From ovarian follicles after their rupture, corpora lutea are formed. Constantly increasing evidence has been brought out by a large number of workers that the corpus luteum also plays an important part in the reproductive cycle. The injection of corpus luteum extract has repeatedly been shown to inhibit ovulation. This effect is entirely consistent with the time relations between the maximal development of the corpora lutea of ovulation and the other events of the estrous cycle (Fig. 9). Corpora lutea attain a marked degree of development and give every histological indication of active secretion a few days after the rupture of the follicles from which they are formed. They begin to exhibit retrogressive changes on microscopical examination, and to decrease visibly in size at about the time a new group of follicles first begins to show rapid growth. Such findings point to the corpora lutea of ovulation as the source of a second hormone which retards the development of the next group of follicles with their activating hormone, until a period of rest has been enforced.

During pregnancy the corpora lutea attain a greater growth and persist longer than they do in the absence of pregnancy (Fig. 10). Their persistence throughout pregnancy and for a variable time thereafter would seem to account for the inhibition of ovulation which occurs during pregnancy and the early part of the period of lactation.

In addition to their inhibitory effect on the development of the follicles, the corpora lutea of ovulation produce a hormone (*progesterone*) which acts on the uterine mucosa in such a way that the implantation of the fertilized ovum is facilitated. Moreover in their extended development during pregnancy they have, in conjunction with the anterior lobe of the pituitary gland, an activating influence on the mammary glands.

Thus the sexual desire manifested at estrus is merely one event in a closely coördinated series of changes involving all the reproductive organs. All these changes are synchronized with one another by the regulatory action of hormones. All of them are preparatory for pregnancy. The recurrence of the estrus when pregnancy does not

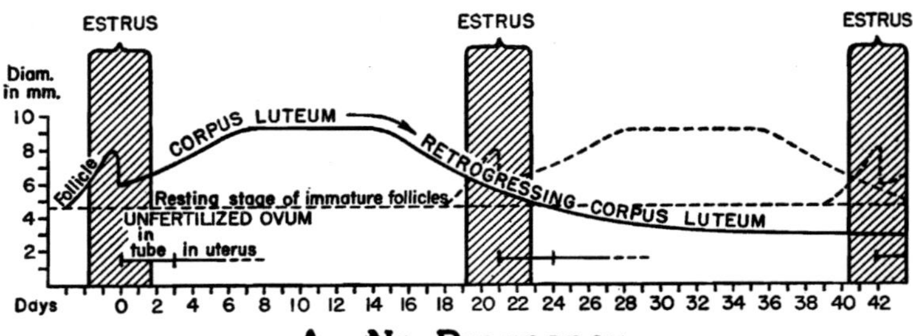

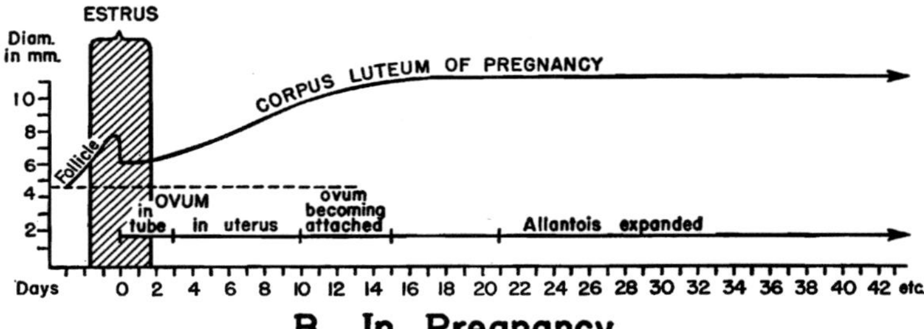

FIG. 10. Graphs showing the difference in history of the corpus luteum of ovulation and the corpus luteum of pregnancy in the sow. (After Corner.)

ensue is but the outward sign that the entire female mechanism is again, after a period of rest, in a state of preparedness to carry through the rearing of offspring. If pregnancy occurs the estrous rhythm is suppressed until the young have been born and suckled.

II. Fertilization

Direct observations of the processes involved in fertilization in mammals are exceedingly fragmentary. Nevertheless, interpreting these observations in the light of the much more detailed information available from the study of water-living forms where fertilization normally occurs outside the body of the mother, it is possible to piece out a fairly circumstantial story of the main events.

Even though coitus in most mammals occurs only at the phase of the estrous cycle when mature follicles are ready to rupture and release ova, actual fertilization of the ova does not inevitably follow. The immediate result of copulation is merely the deposition of semen in the vagina (*insemination*). Thence the spermia must make their way by their own motility through the uterus and into the oviducts where fertilization ordinarily takes place. In comparison with the size of the spermia the distance they must travel is exceedingly great. The enormous numbers of spermia contained in an ejaculate of semen give great probability, but not positive assurance, that some of them will reach the oviduct while they are still capable of penetrating and fertilizing the ovum.

Ordinarily within a few hours of the coitus of healthy individuals great numbers of spermia will have made their way to the oviducts and surround the ova as they are carried through the oviducts toward the uterus. However great may be the numbers present, normally only a single spermium enters the ovum (Fig. 11, A). As soon as it has been entered by a spermium an ovum appears to undergo at once changes which result in the cessation of penetrating activity on the part of other spermia in its neighborhood. This phenomenon may readily be observed in manv marine forms where fertilization can be

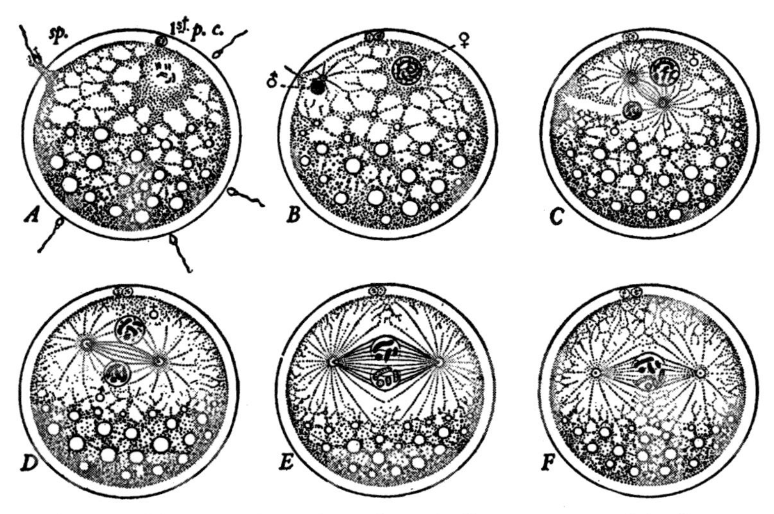

Fig. 11. Diagrams illustrating schematically tne process of fertilization and the formation of the first cleavage spindle. (After William Patten.) As in figure 8 the species number of chromosomes is assumed to be eight. Abbreviations: sp., spermium; p.c., polar cell (polar body). The male and female symbols designate the male and female pronuclei respectively.

carried out in a dish of sea-water under the microscope. When the spermia are first introduced into a dish containing ova, one sees swarms of them surrounding each ovum. Even the relatively enormous bulk of the egg cell may actually be set in rotation by their combined activity. Abruptly, when one spermium has penetrated the ovum, its surface membrane becomes thickened and less readily penetrated; coincidently the remaining spermia appear to lose their directive activity, and soon only scattered ones remain in the neighborhood of the fertilized ova. That this change is due to the fertilization of the ovum and not to loss of activity on the part of the spermia may be demonstrated readily by adding unfertilized ova to the dish and watching the process repeat itself.

Only the head of the spermium, which is almost entirely condensed nuclear material, and the neck containing the centrosomal apparatus enter the ovum. The tail is dropped off (Fig. 11, B). Once within the ovum, the nuclear material contained in the head of the spermium loses its condensed form and begins to show its chromosomal content. It is now known as the *male pronucleus* (Fig. 11, C). In most mammals the second maturation division of the ovum is not completed until after ovulation. Not uncommonly it is delayed until a spermium has penetrated the egg cell. Always, however, by the time the male pronucleus has been formed the second maturation division has occurred. The reduced nucleus of the ovum is then known as the *female pronucleus*. *Fertilization* is said to have occurred only when the male and female pronuclei have merged with each other (Fig. 11, E). As each pronucleus contributes half the species number of chromosomes, the full species number of chromosomes is reëstablished in the fertilized ovum.

During the period between the penetration of the ovum by the spermium and the actual fusion of the pronuclei, a mitotic spindle has been forming from the centrosomal apparatus brought in by the spermium (Fig. 11, C, D). On this spindle the chromosomes from both the male and female pronuclei become arranged preparatory to the first mitotic division in the development of the new individual.

CHAPTER 4

The Process of Cleavage and the Formation and Early Differentiation of the Germ Layers

Cleavage. The series of cell divisions which, immediately after fertilization, follow upon one another in close succession are known as the *cleavage* or *segmentation divisions*. In the various groups of animals cleavage differs much in detail, but these differences are not as fundamental as at first glance they might appear. They are due almost entirely to the varying amount of yolk stored in the egg cells.

A mitotic division, whether it be one of the cleavage divisions of an ovum or the division of some other cell, is carried out by the active protoplasm of the cell. The food material stored in the cytoplasm of an egg cell is non-living and inert. This deutoplasm, as it is called, plays no part in mitosis except as it exerts a local retarding effect by the mechanical impediment it offers to the process.

Figure 12 shows diagrammatically how the first cleavage division is carried out in three types of eggs having different relative amounts of yolk, and different distributions of the yolk in the cytoplasm. In the egg of such forms as Amphioxus the yolk is relatively meager in amount and fairly uniformly distributed throughout the cytoplasm. An ovum with such a yolk distribution is termed *isolecithal* (*homolecithal*). An isolecithal egg undergoes a type of cleavage which is essentially an unmodified mitosis. The yolk is not sufficient in amount, nor sufficiently localized, to alter the usual mode of cell division.

In Amphibia the ovum contains a considerable amount of yolk and the accumulation of the yolk at one pole has crowded the nucleus and the active cytoplasm of the ovum toward the opposite pole. An egg in which the yolk is thus concentrated is termed *telolecithal*. The region of the egg in which the yolk is accumulated is designated as its *vegetative pole* and the opposite region where the nucleus and most of the active cytoplasm are situated is called its *animal pole*.

Cleavage in all types of eggs is initiated by the mitotic division of the nucleus. Division of the cytoplasm follows nuclear division. In

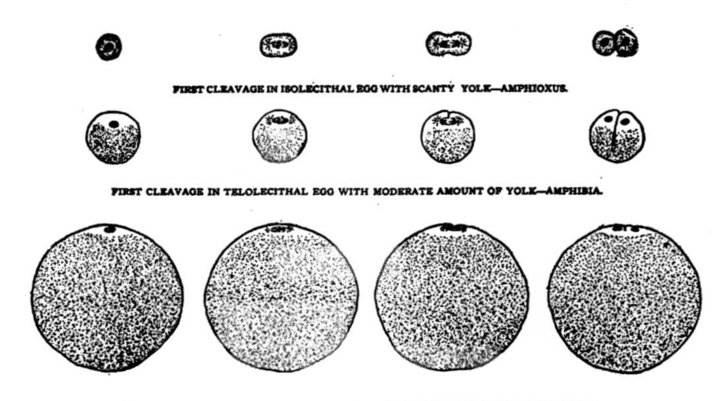

FIG. 12. Schematic diagrams to indicate the effect of yolk on the first cleavage division.

telolecithal eggs the cytoplasm at the animal pole where there is little or no yolk divides promptly, but where the yolk mass is encountered the process is greatly retarded. So slowly, in fact, is the division of the yolk accomplished that succeeding cell divisions begin at the animal pole of the egg before the first cleavage is completed at the vegetative pole.

The eggs of birds are also telolecithal, but the amount of yolk which they contain is both relatively and actually much greater than that in Amphibian eggs. Cleavage in birds' eggs begins as it does in the eggs of Amphibia, but the mass of the inert yolk material in them is so great that the yolk is not divided. The process of segmentation is limited to the small disk of protoplasm lying on the surface of the yolk at the animal pole, and is for this reason referred to as discoidal cleavage. The fact that the whole egg is not divided is indicated by designating the process as partial (meroblastic) cleavage in distinction to the complete (holoblastic) cleavage seen in eggs containing less yolk. The cells formed in the process of segmentation are known as blastomeres whether they are completely separated, as is the case in holoblastic cleavage, or only partially separated, as in meroblastic cleavage.

In mammalian ova the deutoplasmic content is exceedingly small, correlated with the fact that the embryo at an early stage in its development draws upon the uterine circulation of the mother for its nutrition. For this reason the cleavage of mammalian ova reverts to the simple type seen in primitive forms with a scanty and uniformly distributed yolk content. It is practically an unmodified mitosis. To put it into the usual technical terms of comparative embryology, we are dealing with the equal holoblastic cleavage of an isolecithal egg.

Through the pioneer work of Assheton and the subsequent extensive work of Streeter and Heuser the pig is one of the few mammals from which anything like a complete series of cleavage stages has been secured. Like other mammalian ova, that of the pig shows but a scanty amount of deutoplasm. This deutoplasm, in the pig ovum, consists chiefly of fat droplets scattered through the peripheral part of the cytoplasm. In sectioned specimens the fat has been dissolved by the alcohols through which the material is passed preparatory to its embedding but it leaves in the fixed cytoplasm numerous vacuoles as a record of its presence (Fig. 14, A). The uniformity of the distribution of the deutoplasm gives no clear-cut differentiation of a "vegetative pole" such as is afforded by the yolk in the case of Amphibian and Sauropsidan ova. The point at which the polar bodies are given off, however, establishes the animal pole and gives us thereby a basis of

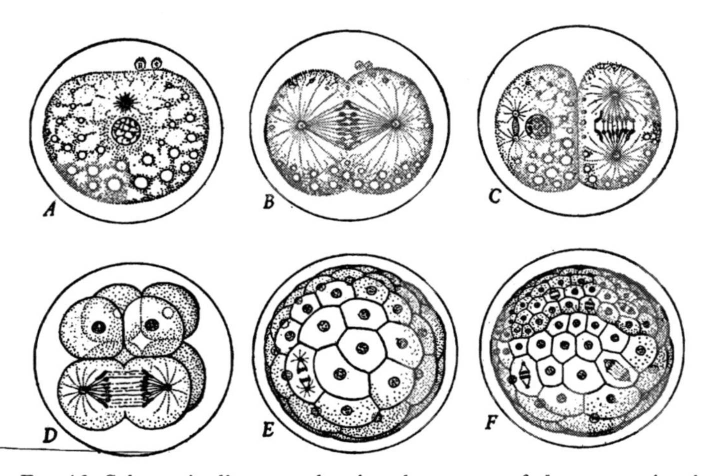

FIG. 13. Schematic diagrams showing the process of cleavage as it takes place in ova having only a small amount of yolk in their cytoplasm. (After William Patten.) Note that the blastomeres are separated from one another promptly and completely. (Patten: "Early Embryology of the Chick," The Blakiston Company.)

orientation. As is the general case, the first cleavage spindle forms at right angles to an imaginary axis drawn through the ovum from animal to vegetative pole (Fig. 13, B). When the first cleavage division

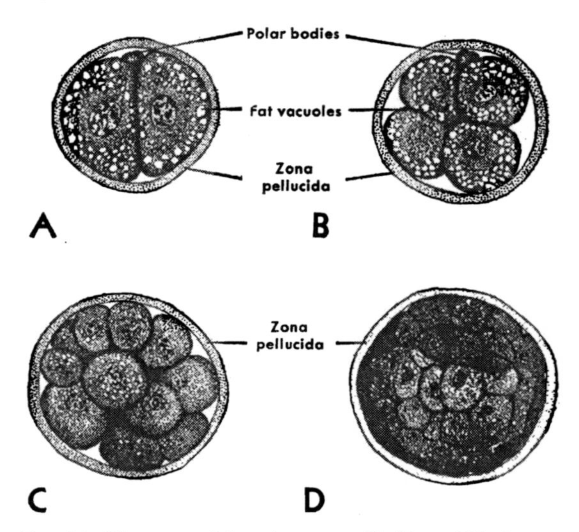

FIG. 14. Cleavage of the pig ovum. (A, B, and D, drawn from preparations loaned by Streeter and Heuser; C, after Assheton.) (All figures × 300.)

A, Two-cell stage in section. Specimens secured from the oviduct of a sow killed 2 days, 3½ hours after copulation.

B, Four-cell stage in section. Probable age about 2½ days.

C, A morula of about 16 cells. Drawn from unsectioned specimen, probable age about 3½ days.

D, Blastula stage, drawn from an unsectioned specimen secured from the uterus of a sow killed 4¾ days after copulation. Note the lighter central area indicating the beginning of the formation of the segmentation cavity (blastocoele) by cell rearrangement (cf. Fig. 15).

of the ovum is completed the plane of separation between the resulting blastomeres, being in the equator of the spindle, lies in the imaginary animal-vegetative axis of the ovum (cf. Figs. 13, C, and 14, A).

A second cleavage spindle is formed in each of the first two blastomeres almost as soon as they are established. The second spindles form at right angles to the first (Fig. 13, C), and the second cleavage planes are consequently perpendicular to the first cleavage plane.

Further cleavage divisions follow one another in such rapid succession that the growth interval usually intervening between succeeding mitoses is curtailed. In consequence, the individual blastomeres in each succeeding generation become smaller and smaller (cf. Fig. 14, A–D). Since the zona pellucida persists intact throughout the period of cleavage, the blastomeres are forced to dispose themselves within its spheroidal cavity. After several segmentation divisions have taken place, the resultant group of blastomeres appears much like a solid ball of cells suggestive of a mulberry or a blackberry. The embryo in this condition is said to be in the *morula* (translated = little mulberry) stage (Fig. 14, C).

After a characteristic morula has been formed, the term cleavage is not ordinarily applied to the cell divisions which occur. The inference should not be drawn that active cell division ceases or even that it is retarded in rate. On the contrary it continues with unabated rapidity. But processes of segregation and differentiation even thus early begin to make their appearance and the term cleavage, which implies merely increase in cell numbers through repeated cell divisions, ceases adequately to characterize the phenomena.

The Blastula Stage. Terms designating "stages of development" are convenient in discussing the progress of events, and the relative uniformity which has gradually been established in their usage is a great aid to mutual understanding. It should be borne in mind, however, that the delimitation of "stages" is purely arbitrary, for development is a continuous process and one phase merges into another without any real point of demarcation.

When the blastomeres of a morula begin to be rearranged and organized about a central cavity (Fig. 14, D), we say that the morula is becoming a *blastula* or that the embryo is entering the blastula stage. The newly formed cavity within it is called the segmentation cavity or *blastocoele* (Fig. 15). Because of the large size attained by the blastocoele in mammals, mammalian embryos in the blastula stage are commonly called blastocysts, or blastodermic vesicles.

In the formation of the blastocyst an internal cluster of cells is established at one pole. This, for want of a better term, has been called the *inner cell mass* (Fig. 15). Although it cannot be carried through in all details, the general distinction may be made that the

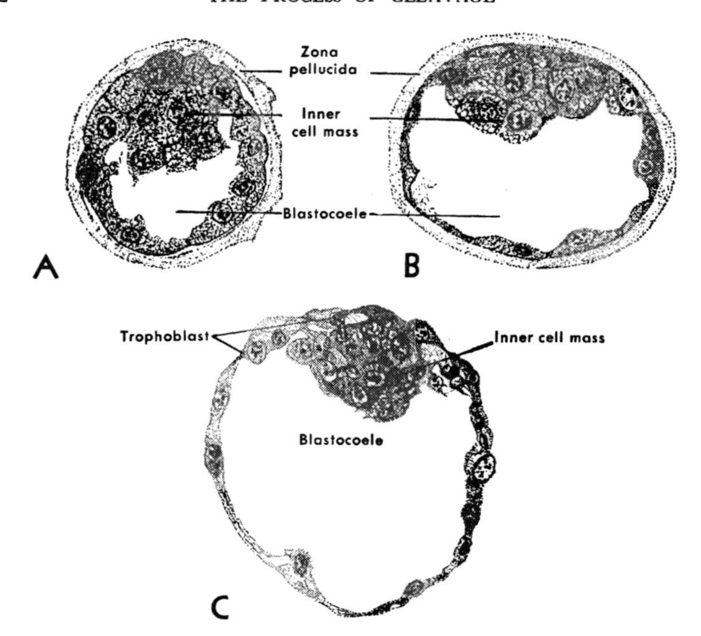

FIG. 15. Three stages of the blastodermic vesicle (blastocyst) of the pig, drawn from sections to show the formation of the inner cell mass. (A, B, from embryos in the Carnegie Collection; C, after Corner—all × 375.) A, Removed from uterus of sow 4¾ days after copulation (cf. Fig. 14, D). B, Copulation age, 6 days, 1¾ hours. C, Copulation age, 6 days, 20 hours.

inner cell mass is destined to be concerned primarily with the formation of the embryonic body, whereas the thin outer wall of the blastocyst contributes, not to the make-up of the embryo, but to the formation of certain membranes which acquire intimate relations with the uterus of the mother and are concerned with the absorption of food. For this reason the thin layer of cells which constitutes the blastocyst wall is called the *trophoblast* (Fig. 15).

The Formation of the Entoderm. The blastocoele, once established, enlarges rapidly so that the inner cell mass appears proportionately smaller because it occupies relatively less and less of the lumen of the blastocyst (Fig. 15). Actually, however, its cells are not only proliferating rapidly in situ but some of them apparently push out of the mass into the blastocoele. These are the first of the entoderm cells (Fig. 16, A). They are increased in numbers very rapidly after their first appearance and soon come to constitute a second complete layer inside the original outer layer of the blastocyst (Fig. 16, B, C). The internal lumen bounded by the entoderm is known as the primitive gut (*archenteron*).

While the entoderm is being established another important change is taking place in the blastocyst. Originally the cell cluster constituting the inner cell mass was, as its name implies, completely inside the trophoblast. During the period of entoderm formation the overlying trophoblast degenerates and the original inner cell mass comes to lie

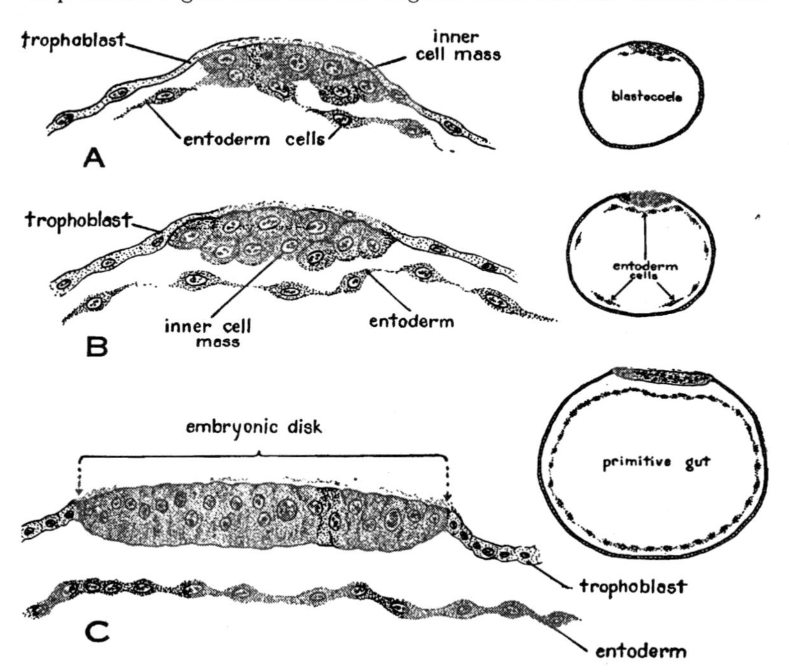

Fig. 16. Sections of pig blastocysts showing the first appearance and subsequent rapid extension of the entoderm. (From embryos in the Carnegie Collection.) Left, detailed drawings of inner cell mass (× 375). Right, sketches of same sections entire. The approximate age of the embryos represented ranges from 7 to 8 days.

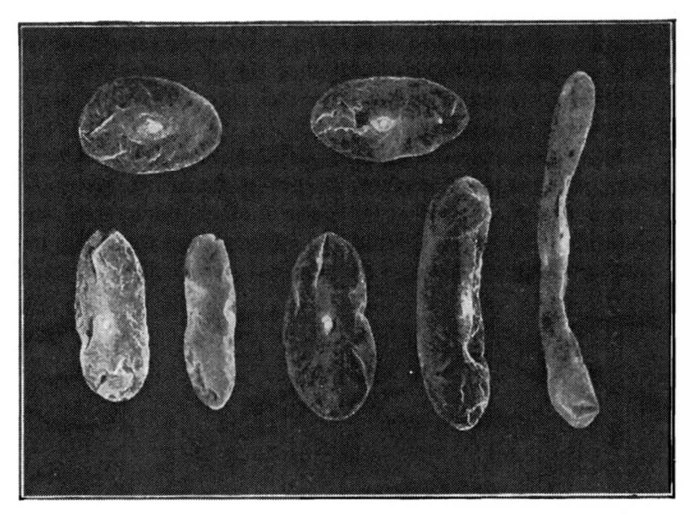

FIG. 17. Blastodermic vesicles (blastocysts) of the pig at the beginning of elongation (about 8 days old). All seven embryos were removed from the same uterus. (Carnegie Collection, C264, after a photograph (× 3) by Dr. Heuser.)

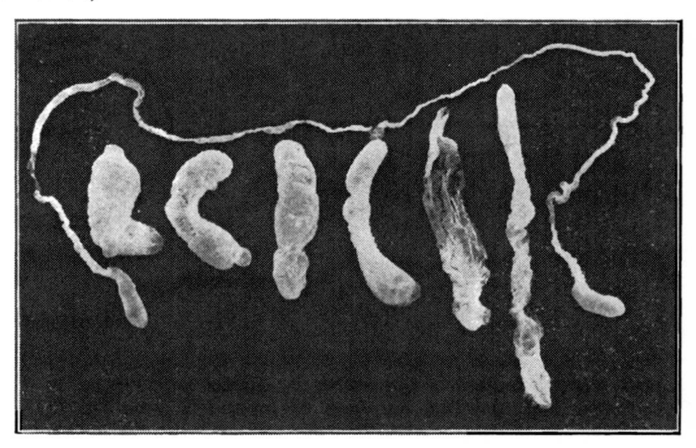

FIG. 18. A litter of embryos slightly older than those of the preceding figure, showing the exceedingly rapid elongation of the blastocyst which occurs at this stage (about the ninth day) of development. (Carnegie Collection, C213, after a photograph (× 2½) by Dr. Heuser.)

at the surface, constituting part of the outer wall of the blastocyst (Fig. 16). At the same time its cells proliferate rapidly and become aggregated into a disk-shaped, thickened area sharply differentiated from the adjoining trophoblast. We now call this differentiated area arising from the inner cell mass the embryonic disk (Fig. 16, C). In fresh embryos viewed under a dissecting microscope the embryonic disk appears as a whitish area of markedly greater density than adjoining portions of the blastocyst where the wall consists only of the trophoblast and the even thinner entodermal layer within it (Fig. 17).

The Elongation of the Blastocyst. Shortly after the entoderm has been established as a definite layer, the blastocyst as a whole undergoes a striking change in shape. From an approximately spherical vesicle it becomes converted into a tubular sac. The rapidity with which this alteration in form takes place is astonishing. The embryos shown in figure 17 are all from the same litter and consequently very nearly of the same age. The least developed embryo in the group departs but slightly from its originally spherical shape, yet the most developed has already attained a length twelve times its diameter.

Figure 18 shows a slightly older litter. Even the most elongated embryo in this group has by no means attained its maximum length. In another day or two of development all these blastocysts would have become of thread-like thinness and approximately a meter long. A small region near the middle of the thread remains somewhat less attenuated and there the embryonic disk is located. The disk itself is practically unaffected in this process of elongation which involves the trophoblast and to a somewhat lesser extent the layer of entoderm lying within it. The significance of this increase in extent of the extra-embryonic portion of the blastocyst will be appreciated when we see the part it plays in the formation of the extensive membranes through which the embryo draws upon the uterine circulation of the mother for its food supply (Chap. 6).

The Formation of the Primitive Streak. At about the same time that the blastocyst begins elongation, a local differentiation occurs in the embryonic disk which presages the formation of the primitive streak. Sections through the disk show at one part of its margin a heightened rate of cell proliferation accompanied by a definite increase in thickness (Fig. 19, A). Interpreting this thickening in the light of later developments it is possible for us to say that its appear-

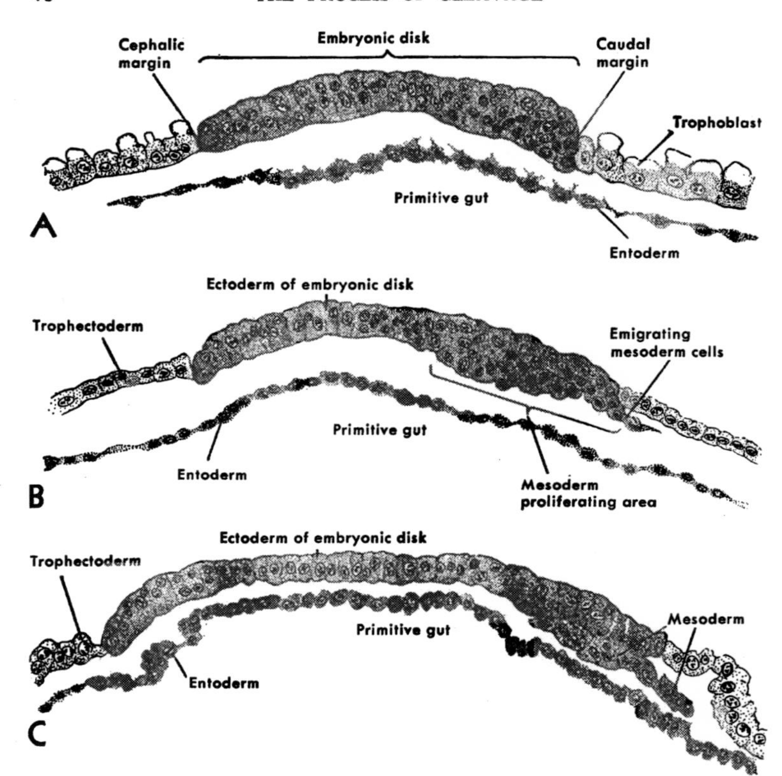

FIG. 19. Longitudinal sections of the embryonic disk of the pig during the ninth day of development, showing three stages in the origin of the mesoderm. (Projection drawings (× 200) from sections of embryos in the Carnegie Collection.)

ance definitely establishes the longitudinal axis of the embryo. The thickening occurs at the part of the disk which is destined to become differentiated into the caudal end of the embryo.

In dorsal views of an entire embryo the thickened area when it first appears is crescentic in shape, with its convexity indicating the caudal extremity of the embryonic disk, and its horns spreading out over the greater part of the caudal half of the margin of the disk (Fig. 20, A). At this stage of development the embryonic disk apparently undergoes rapid concrescence caudally. That is, while the anterior margin of the disk is spreading out radially in a manner typical of uniform rate and unspecialized direction of growth, the posterior margins grow at an accelerated rate toward a point of convergence at the caudal extremity of the disk. (See arrows in Fig. 20, B.) This tends at the same time to lengthen the disk itself cephalocaudally, and to crowd the horns of the crescentic thickened area toward the mid-line. Further progress of this convergent differential growth changes the originally crescentic thickened area of the embryonic disk to an oval (Fig. 20, D) and then pulls it out into a band lying in the long axis of the embryo (Fig. 20, E–G). This thickened longitudinal band is known as the *primitive streak.*

The Primitive Streak as a Growth Center. The change in shape and position undergone by the originally crescentic area of the embryonic disk in no way retards its activity as a growth center. We find this area, throughout its transformation and later when it has become the primitive streak, still a region of rapid proliferation from which newly formed cells are constantly being pushed forth to take their part in the formation of the rapidly expanding body of the embryo. It seems not unlikely that the formation of the groove in the primitive streak (Fig. 23) is a local structural modification entailed by the rapid emigration of cells from this region (Fig. 22, E).

The Formation of the Mesoderm. The most striking manifestation of the activity of the cells in the primitive streak region is the formation of the mesoderm. As its name implies, the mesoderm is developed between the original outer cell layer of the blastocyst and the subsequently formed inner cell layer or entoderm. Taking their origin from the growth center at the primitive streak (Fig. 19, B, C), the mesodermal cells proliferate with astonishing rapidity and establish a third definite and coherent cell layer in the embryonic body (Figs. 20, 21, and 22).

Intra- and Extra-embryonic Mesoderm. In its peripheral growth the mesoderm soon extends well beyond the boundaries of the embryonic disk. We may distinguish that part of the mesoderm underlying the embryonic disk as intra-embryonic mesoderm and that part of it which extends peripherally between the trophoblast and the entoderm of the blastocyst as extra-embryonic mesoderm (Fig. 22, A). The distinction is one of convenience in description, but of course purely arbitrary, for there is as yet no line of demarcation between the two areas. It is helpful, however, to realize at the outset that much of the peripheral mesoderm, together with the trophoblast and some

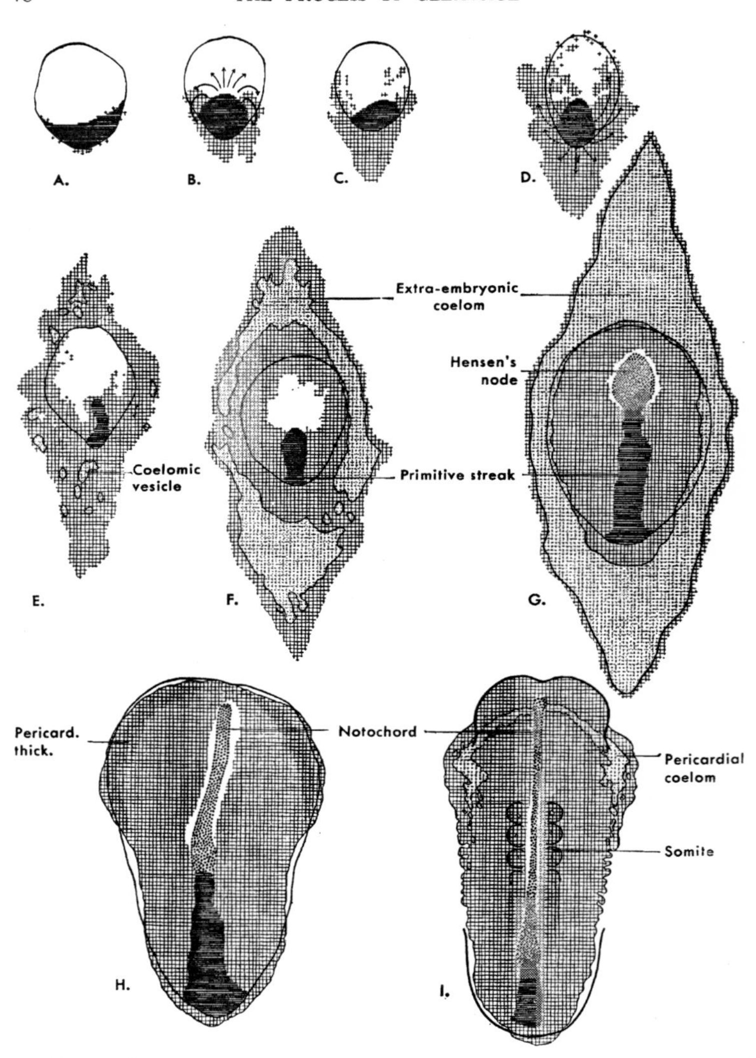

FIG. 20. Diagrams showing origin, extension, and early differentiation of the mesoderm in a series of pig embryos ranging from about the ninth to the fifteenth day of development. (After Streeter, slightly modified.)

In each figure the embryo is supposed to be viewed in dorsal aspect as a

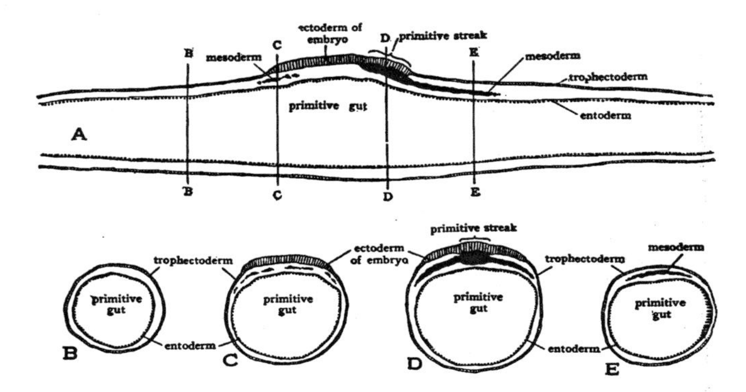

FIG. 21. Diagrams (× 50) of sections of pig embryos in the early primitive streak stage. (From series in the Carnegie Collection.) A is a longitudinal section; B–E are transverse sections at the levels indicated by the correspondingly lettered lines on the longitudinal section. See also the mesoderm plot, figure 20, D, made from an embryo of about the same age (approximately 10 days).

of the entoderm, goes into the fabrication of protective and trophic membranes which later completely envelop the growing embryo. The fact that these membranes are not incorporated in the embryonic body but are discarded at the time of birth is implied in their designation as extra-embryonic membranes.

The Formation of the Notochord. The notochord both phylogenetically and ontogenetically is of great morphological importance. In the most primitive of the vertebrate group it is a well developed

---

transparent object. Except for outlining the embryonic area, only mesodermal structures are represented. The area indicated by heavy horizontal hatching in A, is the thickened part of the embryonic disk from which mesoderm is first proliferated (cf. Fig. 19). Concrescent growth at this region (see arrows in B) gives rise to the primitive streak (cf. shape of horizontally hatched area, A–I). Mesoderm is indicated by crosshatching and the notochord by stippling. The regions in which the coelom has been established by splitting of the lateral mesoderm are indicated by the outlined areas shown in fainter crosshatching. Local thickenings in the mesoderm are indicated by strengthening of the crosshatching. The extra-embryonic mesoderm is omitted in H and I.

Abbreviation: Pericard. thick., thickened mesoderm in future pericardial region.

fibro-cellular cord lying directly ventral to the central nervous system and constituting the chief axial supporting structure of the body. In fishes such as those of the shark family (elasmobranchs) ring-like cartilaginous vertebrae are formed about the notochord. Although somewhat compressed where the vertebrae encircle it, the notochord persists in such forms as a well-defined continuous structure extending throughout the length of the vertebral column. When, in the progress of evolution, cartilaginous vertebrae are replaced by highly developed bony vertebrae the notochord is still more compressed. But even in mammals a minute canal persists for a time in the center of the developing vertebrae marking the position of the notochord. In the early stages of development the notochord of a mammalian embryo is a conspicuous structure, at once a record of evolutionary history, and an advance indication of the location of the vertebral column.

The notochord is established immediately cephalic to the primitive streak. The cells of which it is composed take their origin from the thickened area (*Hensen's node*) at the anterior end of the primitive streak. These cells grow out in the form of a rod-shaped mass lying medially in the body. Meanwhile the mesoderm in its peripheral growth has spread out more or less uniformly except in this region,

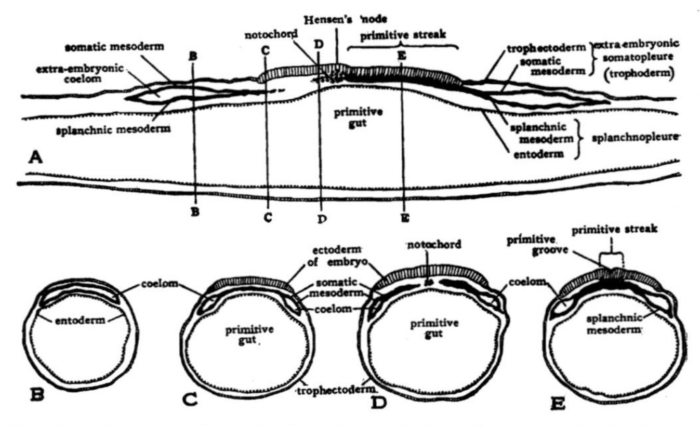

FIG. 22. Diagrams (× 50) of sections of pig embryos at the beginning of notochord formation. (From series in the Carnegie Collection.) A is a longitudinal section; B–E are transverse sections at the levels indicated by the correspondingly lettered lines on the longitudinal section. See also the mesoderm plot, figure 20, G, and drawing of an entire embryo, figure 23, made from specimens of the same age (approximately 12 days).

where it has left a temporarily free area (Fig. 20, E, F). It is into this unoccupied space between the ectoderm and entoderm that the notochord grows (Figs. 20, G, H, and 22, A, D).

It will be recalled that the primitive streak is nothing else than an elongation of the original proliferation center of the embryonic disk from which we traced the origin of the mesoderm. The source and manner of origin of the notochord are therefore similar to that

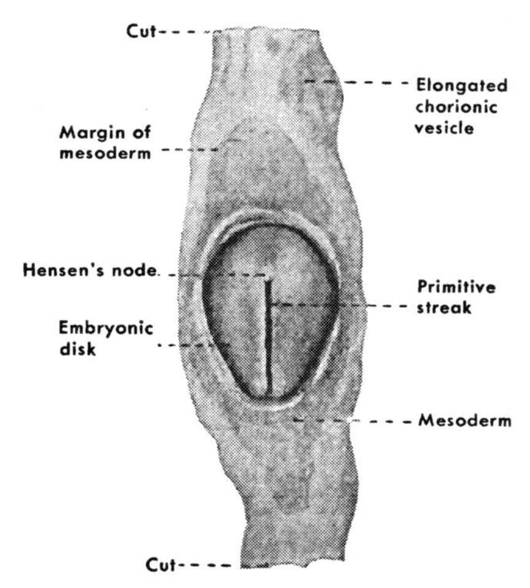

FIG. 23. Drawing (× 15) of pig embryo about 12 days old. (Carnegie Collection, C181–12.) The ends of the long thread-like blastocyst (cf. Fig. 18) have been cut off.

described for the previously established mesoderm. The only differences are that the notochord appears somewhat later, and that it grows as a rod-shaped mass of closely packed cells instead of spreading out freely as does the mesoderm (Figs. 20, G–I and 22, A, D).

The Coelom. The mesoderm which spreads out peripherally from the primitive streak is at first a fairly uniform sheet of cells (Fig. 21, A, D). It does not, however, remain long undifferentiated. Sections of slightly older embryos show the lateral portions of the mesoderm splitting into two layers (Fig. 22). The outer layer is called the *somatic mesoderm* and the inner layer the *splanchnic mesoderm*. The cavity between somatic and splanchnic mesoderm is the *coelom*. Because the somatic mesoderm and the ectoderm are closely associated and undergo many foldings in common, it is frequently convenient to designate the two layers together by the term somatopleure. For the same reasons splanchnic mesoderm and entoderm together are designated as splanchnopleure.

The splitting of the lateral mesoderm does not occur simultaneously throughout its extent. The earliest indications of the process

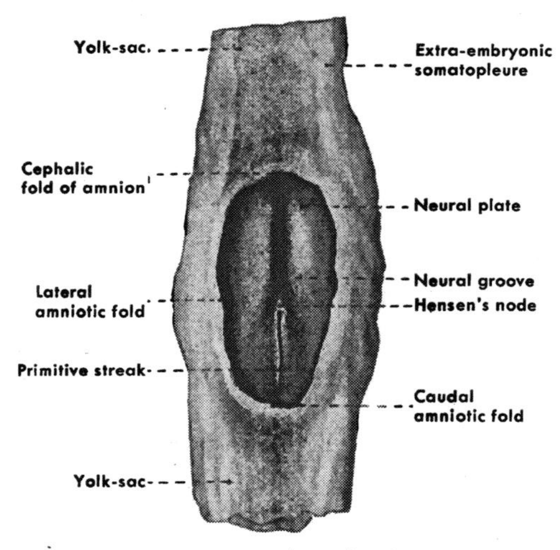

FIG. 24. Drawing (× 15) of pig embryo showing the first appearance of the neural groove. (Carnegie Collection, C160–68.) Compare with mesoderm plot from embryo of same age (approximately 13 days) shown in Fig. 20, H.

appear here and there in the more peripheral parts of the mesoderm. The first small local areas of separation give rise to isolated vesicles (Fig. 20, E) which rapidly extend and become confluent to establish the coelom (Figs. 20, E–G, and 22). A definite coelom is thus first established in the extra-embryonic portions of the mesoderm, and is correspondingly designated as the extra-embryonic part of the coelom or, more briefly, *exocoelom*.

As development progresses, the splitting initiated peripherally continues to extend toward the embryo and soon involves the intra-embryonic portion of the mesoderm (Fig. 20, G–I). Thus an intra-embryonic portion of the coelom is established which is at first directly continuous with the exocoelom (Fig. 26, C). Later in development, as the growing embryo is more definitely separated from its surrounding membranes, we shall see the demarcation between intra- and extra-embryonic coelom quite sharply established. The part of

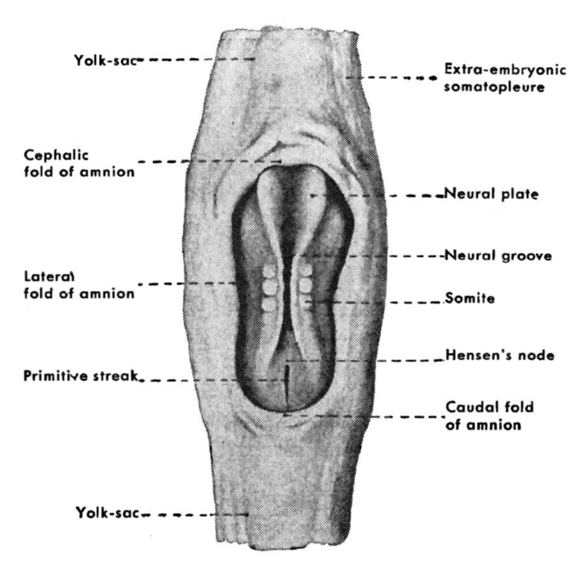

FIG. 25. Drawing (× 15) of pig embryo at the time of the appearance of the first somites. (Carnegie Collection, C190–2 and C196–1.) Compare with the section diagrams in figure 26, and the mesoderm plot in figure 20, I, made from embryos of the same age (about 14 days).

the coelom then included within the embryo gives rise to its body cavities (pericardial, pleural, and peritoneal cavities).

It is of interest to note in passing that the first part of the intra-embryonic coelom to be established is the region where the heart will develop (pericardial region of coelom, Fig. 20, I). This precocious formation of the *pericardial coelom* presages the early appearance of the cardiovascular system as a whole. Another condition of interest is the extensive development of the extra-embryonic layers which foreshadows an early differentiation of the membranes derived from them. The accelerated differentiation of these two systems is a very striking feature of mammalian development, and would seem to be quite definitely correlated with the paucity of yolk in the mammalian ovum. In the absence of a readily available supply of stored food material, membranes capable of establishing metabolic interchange with the maternal circulation, and a fetal circulation capable of transporting and distributing the food material absorbed through these membranes, are both indispensable factors for the growth of the embryo.

The Mesodermic Somites. The earliest manifestations of metamerism to appear in the body of the mammalian embryo are the segmentally arranged thickenings of the mesoderm called *somites*. Before the somites themselves are distinguishable there is a clearly marked differentiation of the mesoderm from which they are derived. On either side of the notochord the immediately adjacent mesoderm becomes markedly thickened (Figs. 20, H, and 26, C). In distinction to the lateral mesoderm with which they are continuous, these paired zones of thickened mesoderm constitute the *dorsal mesoderm*. The narrow region which forms the transition from dorsal to lateral mesoderm is called the *intermediate mesoderm*. (Fig. 26, C.)

Differentiation of the dorsal mesoderm first becomes apparent about midway between the cephalic and the caudal end of the embryo (Fig. 20, H). Somites are formed from these thickened zones of the mesoderm by a series of transverse divisions brought about by cell rearrangement and resulting in the establishment of block-like masses of more or less radially arranged cells (Figs. 20, I, and 42, A, B). The first pair of somites is formed from the cephalic part of the dorsal mesoderm. As continued growth from the primitive streak region progressively increases the length of the embryo, the first somites formed are carried cephalad in the general expansion of the embryonic body. Keeping pace with the increase in cephalo-caudal elongation, more and more dorsal mesoderm becomes differentiated caudally and new pairs of somites are added behind those previously established.

Caudal Growth and Cephalic Precocity. The fact that at this stage of development the most active growth is taking place from the caudal part of the embryo about the primitive streak, and that the newly formed tissue is being forced thence toward the head, is indi-

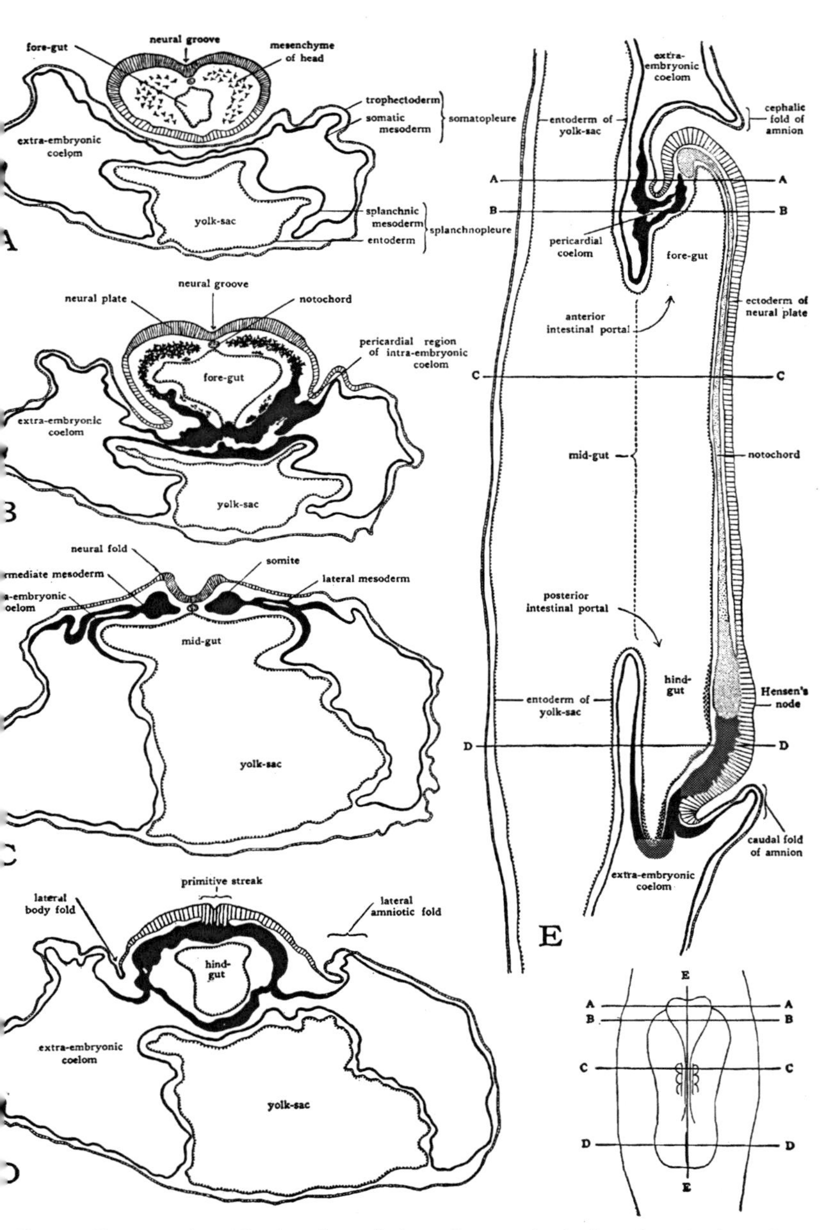

FIG. 26. Diagrams (× 50) of sections of pig embryos at beginning of somite formation. rom series in Carnegie Collection.) Location of sections indicated on small key diagram, which ould be compared with figure 25 drawn from embryo of same age (about 14 days).

cated by other conditions besides the succession in which new somites are added. The size of the primitive streak and its position in the body offer additional evidence concerning the nature of the growth processes going on at this time. In spite of the very rapid cell division which can be seen to occur in it, the primitive streak does not increase in size. Nor does it move cephalad in the growing body as do the somites. It becomes, on the contrary, a relatively less and less conspicuous structure and retains its original caudal position in the body. The reason for this is the fact that the cells proliferated in the primitive streak region do not remain there but push forth as fast as they are formed. The majority of these new cells are crowded in between the primitive streak and the already established part of the embryo cephalic to the streak. This results in rapid expansion of the body cephalic to the primitive streak. One is very likely, in observing a series of embryos in which the progress of elongation in the cephalic region is so striking, to attribute it entirely to especially active growth in this region itself. In reality it is due rather to rapid growth from behind, which pushes the cephalic region ahead.

The fact that the growth of a young embryo is taking place chiefly from its caudal end has a bearing also on the relative progress of differentiation in different regions of the body. It is a striking fact that the cephalic end of an embryo will always be found precocious in differentiation as compared with the more posterior portion of the embryo. This much-commented-on condition seems but natural when we consider that the head is actually older in development. For the structures posterior to the head are laid down by cells which were proliferated from the growth center at the primitive streak, subsequently to the establishment of the head itself. Differentiation does occur exceedingly rapidly in the head. Were this not so, other regions would pass it in developmental progress. But we cannot, in taking cognizance of this condition, afford to overlook the fact that the head is given a considerable lead at the outset by its earlier establishment.

The Neural Folds. During the period in which the early differentiation of the mesoderm occurs, there have appeared in the embryonic disk indications of the formation of the central nervous system. A broad mesial zone of the ectoderm cephalic to Hensen's node becomes markedly thickened as compared with the rest of the ectoderm. This thickened part of the ectoderm is called the *neural plate* (Fig. 26, B). Almost as soon as it is differentiated, the neural plate becomes folded so that its medial portion is depressed and its lateral portions elevated.

The longitudinal groove thus formed in the mid-dorsal surface of the embryo is called the *neural groove* and the ectodermal folds flanking it are known as the *neural folds* (Figs. 24, 25, and 26, C). The folding of the neural plate is at this time most pronounced at the level of the first somites, that is, in what is destined to become the hind-brain region of the embryo. Cephalically the broad, well-developed neural plate marks the future fore-brain region. Caudally the neural folds diverge and flatten out. At the primitive streak region even the ectodermal thickening which constitutes the neural plate has disappeared. The central nervous system caudal to the hind-brain region has yet to be laid down even in primordial form. Thus we see in the central nervous system the same conditions of cephalic precocity that were evident in somite formation. The brain region is definitely established before the caudal end of the spinal cord is even foreshadowed by a primordial cell aggregation.

The Embryological Importance of the Germ Layers. In looking back over the development thus far undergone by the embryo, perhaps the most conspicuous thing, at first glance, is the multitude of cells formed from the single fertilized egg cell by repeated mitoses. Of more significance, however, is the fact that even during the early phases of rapid proliferation the cells thus formed do not remain as an unorganized mass. Almost at once they become definitely arranged as a hollow sphere which is called the blastocyst. Scarcely is the blastocyst established as a single-walled vesicle when, from the knot of cells which constitute the inner cell mass, certain cells migrate out and become arranged as a second layer inside the first. This second layer, because of its position inside the original layer, is called the entoderm. Shortly a third cell layer makes its appearance between the first two, being called, appropriately enough, the mesoderm. That part of the original wall of the blastocyst which still constitutes its outer covering after the entoderm and mesoderm have been established is now properly called the ectoderm. These three cell layers are spoken of as the *germ layers* of the embryo.

The germ layers are of interest to the embryologist from several angles. The very simple organization of the embryo when it consists of first a single, then two, and finally three primary structural layers is reminiscent of ancestral adult conditions carrying far back into the invertebrate series. From the standpoint of probable ontogenetic recapitulations of remote phylogenetic history several facts are quite suggestive. The nervous system of the vertebrate embryo arises from

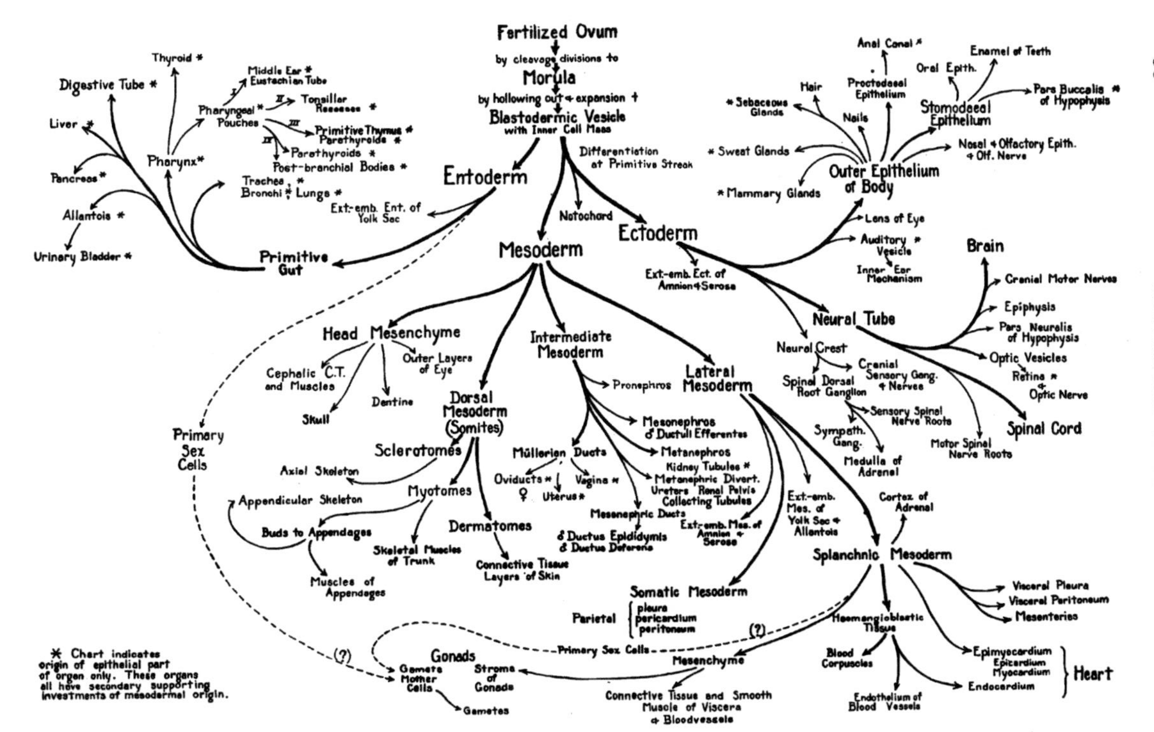

FIG. 27. Chart showing the derivation of the various structures of the body by progressive differentiation and divergent specialization. Note especially how the origin of all the organs can be traced back to the three primary germ layers, ectoderm, entoderm, and mesoderm.

the ectoderm—the layer through which a primitive organism which has not as yet evolved a central nervous system is in touch with its environment. The lining of the vertebrate digestive tube is formed from the entoderm—the layer which in very primitive forms lines a gastrula-like enteric cavity. The chief vertebrate skeletal and circulatory structures are derived from the mesoderm—the layer which in small, lowly organized forms is relatively inconspicuous but which constitutes a progressively greater proportion of the total bulk of animals as they increase in size and complexity and consequently need more elaborate supporting and transporting systems.

Interesting as are the possibilities of interpreting the germ layers from the standpoint of their phylogenetic significance, our chief concern with them centers about the part they play in the development of the individual. Their establishment marks the first segregation of cell groups which are clearly distinct by reason of their definite positional relations to each other, and within the embryo. This positional relationship, moreover, is fundamentally the same for the germ layers of all vertebrate embryos—a fact which speaks forcefully of the common ancestry of all the members of this great group.

Of more importance still is the fact that the different germ layers contain cells with different developmental potentialities. As development progresses, cell groups of given potentialities are, so to speak, sorted out of the germ layers, perhaps by being folded off from the parent layer, perhaps by migrating out as individual cells and later becoming re-aggregated elsewhere. The story of the embryological origin of the various parts of the body is the history of the growth, subdivision, and differentiation of the germ layers. From cell groups thus derived the organs with which we are familiar in the adult gradually take shape.

The idea of repeated regrouping and progressive differentiation and specialization is expressed graphically in figure 27. This chart at the present stage of our study will serve as a means of pointing out in a general way where the early processes we have been dealing with are leading. As we follow the phenomena of development farther we shall find each natural division of the subject centers more or less sharply on a certain branch of this genealogical tree of the germ layers.

CHAPTER 5

The Early Development of the Body Form and the Establishment of the Organ Systems

I. Body Form

The formation of the body is initiated by the same growth processes which establish the germ layers. But even after the germ layers have been laid down and have begun to show considerable differentiation, the configuration of the young embryo is so unlike that of the adult that except to one familiar with embryology there are no readily identified landmarks. There is no distinct head, no neck, no trunk; there are no appendages;—in short there are none of the conspicuous structural features by which we are accustomed to orient ourselves in dealing with adult anatomy (Figs. 23–25). It therefore seems wise to interrupt tracing in detail the processes which give rise to the internal organs of the embryo and follow very briefly the changes in the outer form of its body. Such a preliminary survey should make clear the significance of the unfamiliar topography of the young embryo and give us a perspective from which to follow out the internal changes occurring in the various regions.

The Embryonic Disk. Very early in development we noted the differentiation of a region of the blastocyst which was designated as the embryonic disk (Figs. 16–19). Although the embryonic disk has no semblance of adult anatomical form it is, nevertheless, the beginning of the body of the embryo. The remaining regions of the blastocyst will take part only in the formation of membranes accessory to the body.

Establishment of Body Axis. In the preceding chapter it was stated that with the establishment of the primitive streak in the embryonic disk the embryo had definitely differentiated its future body axis. As far as any anatomical features then exhibited by the embryo were concerned, this statement had to be taken on faith. If, however, the position of the primitive streak is followed in a series of progressively older embryos (Figs. 23, 24, 25, and 28) and if then the place of its final disappearance in the floor of the sinus rhomboidalis is observed (Fig. 29), one readily comprehends why the primitive streak is said to mark at once the longitudinal axis, and the caudal end of the body.

Shortly after the primitive streak becomes clearly defined, the thickened ectoderm of the neural plate is folded to form the neural groove (Figs. 24, 25, and 26). This folding coincides in direction with the primitive streak and further emphasizes the longitudinal axis of the growing embryo.

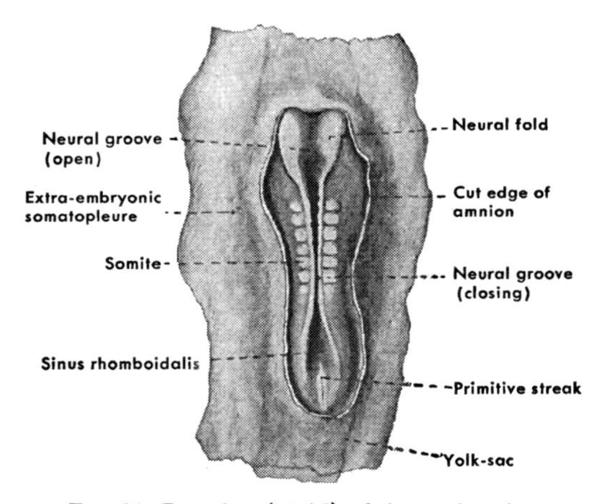

FIG. 28. Drawing (× 15) of six-somite pig embryo. (Carnegie Collection, C195–1 and C162.) Age approximately 14½ days.

Differentiation of the Cephalic Region. With the establishment of the neural groove, body landmarks begin to appear with rapidly increasing clearness. The neural folds in the cephalic region are of much greater size than they are farther caudally (Figs. 25 and 28). This condition foreshadows the differentiation of the neural tube into a conspicuously enlarged cephalic portion, the brain, and a more attenuated caudal portion, the spinal cord. The cephalic region of the embryo is, therefore, already indicated by this enlargement of the fore part of the neural plate (Fig. 28).

At first the topography of the head is vaguely defined, but it soon becomes more precisely marked out by the appearance of characteristic local structures. On either side of the head, in the future oral and cervical region, a series of circumscribed elevations appear. These are *gill arches* (*branchial arches*), homologous with the gill arches of ancestral water-living forms. In the mammalian embryo the most cephalic of these gill arches appears just caudal to the primitive mouth opening. Because it is involved in the formation of the lower jaw, it is called the mandibular arch. The arches caudal to the mandibular arch become less conspicuous and are incorporated into the neck (Figs. 30–34).

Coincidently the primordia of both the ear and the eye become

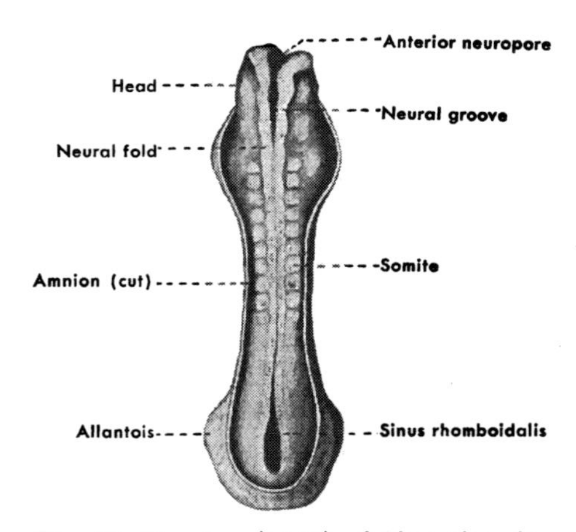

FIG. 29. Drawing (× 15) of 10-somite pig embryo. (Carnegie Collection, C176–6.) Age approximately 15 days.

recognizable. The ears arise as a pair of locally thickened areas (*auditory placodes*) in the superficial ectoderm at the level of the more posterior part of the brain. The auditory placodes soon sink below the surface to form the *auditory pits*. As the pits are deepened and enlarged they are designated as the *auditory* or *otic vesicles*. For a time they maintain an opening to the surface (Fig. 30), but this opening soon closes and the vesicles are not prominent externally (Figs. 31, and 32). They can, however, be seen readily in cleared specimens (Fig. 39) or in sections (Figs. 41, A, and 47). The auditory vesicles are destined to form the inner portion of the auditory mechanism. Somewhat later the beginnings of the external ear can be made out not far from the site of the original invagination of the internal ear primordium (Figs. 33 and 34).

The eyes arise as local outgrowths from the lateral walls of the rostral part of the brain. Long before the *optic vesicles*, as the early outgrowths are called, bear any resemblance to adult eyes, their position can be seen because of the prominence they make in the overlying ectoderm (Optic vesicle, Fig. 30). Specializations of the super-

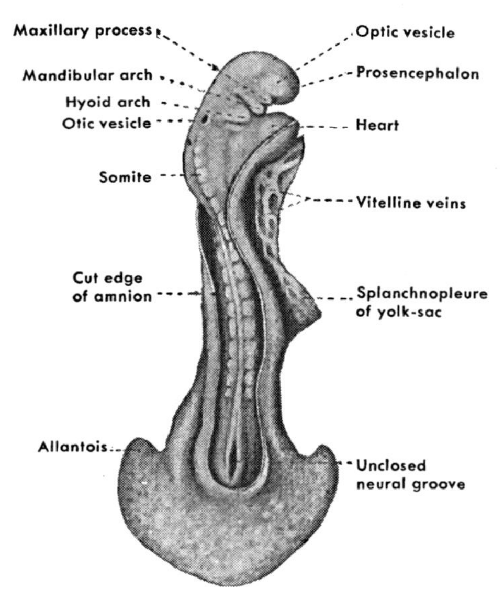

FIG. 30. Drawing (× 15) of 17-somite pig embryo. (Modified from Keibel.) Age approximately 16 days.

ficial tissues about the optic vesicles soon make the developing eye readily identifiable (Figs. 32–34). With the establishment of the mouth parts, the ear, and the eye there is no longer any difficulty in recognizing the general topography of the cephalic region.

The Trunk. Like their invertebrate ancestors, all vertebrates have a segmentally organized body. In adult mammals the underlying metamerism is largely masked by local fusions and specializations. But even so, unmistakable evidences of this fundamental plan of structure persist in the segmentally arranged spinal nerves and ganglia, in the series of vertebrae constituting the "back bone," and in the arrangement of the ribs and of the intercostal musculature. In the young mammalian embryo metamerism is much more obvious. One of its most conspicuous superficial markings is the series of paired prominences which indicate the location of the *mesodermic somites* (Figs. 25 and 26, C). These masses of mesodermal tissue are clearly metameric in arrangement. In fact, it is through them that we trace the origin of the segmental arrangement of the axial skeleton and the

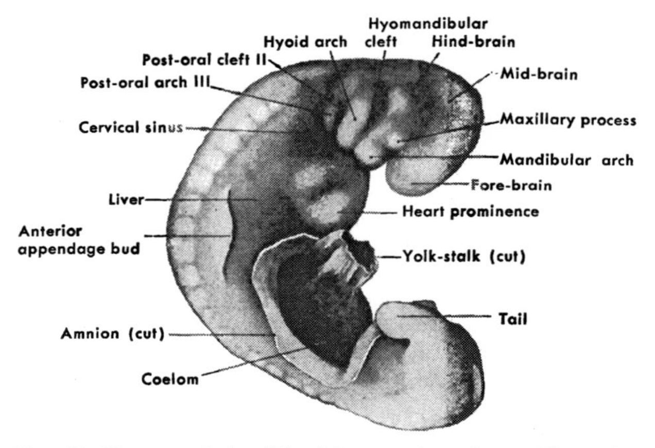

FIG. 31. Photograph (× 12) of 5 mm. pig embryo. (Carnegie Collection, C266–1.) Age approximately 17–18 days.

thoracic musculature just alluded to as one of the characteristic evidences of metamerism in adult mammalian anatomy.

The external prominence made by the developing heart appears in mammals at a strikingly early stage of development. The heart at first lies far toward the head from its definitive position. Bearing in mind that the mandibular arch will form the lower jaw, we can say that the heart originates "under the chin" (Fig. 30). As growth proceeds there is rapid elongation of the embryo between its head and trunk, which results in the establishment of the neck region. In this process the heart is carried caudad to lie in its characteristic position in the anterior part of the trunk (Figs. 30–34). A slight depression may be seen between the external prominences due to the heart and

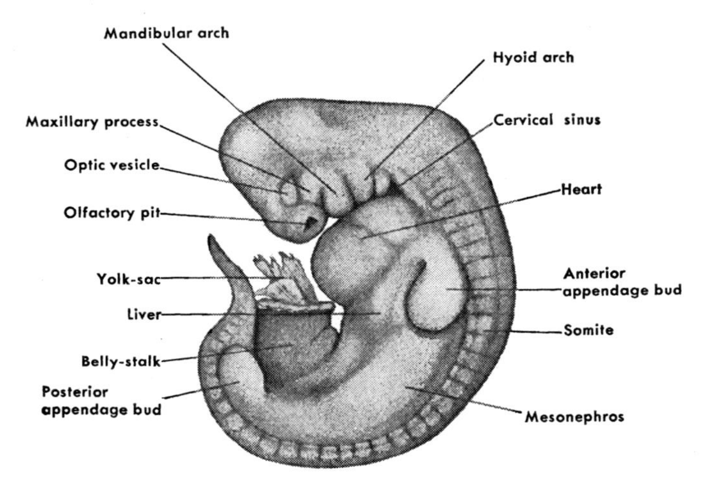

FIG. 32. Drawing (× 8) of 7.5 mm. pig embryo. (Modified from Minot.) Age approximately 18–19 days.

the liver. This depression indicates the position at which the diaphragm develops, and with its appearance we can differentiate the thoracic from the abdominal region of the trunk (Figs. 32–34).

Caudal to the hepatic prominence is the conspicuous *belly-stalk*. Over this stalk the tissues of the embryo are continuous with the extra-embryonic membranes. In it are embedded the large blood vessels by way of which the embryo receives its food and oxygen supply from the uterus of the mother.

The Appendages. From the sides of the body two pairs of outgrowths are formed. The first pair makes its appearance at the level of the heart. These bud-like tissue masses (Fig. 32) become differentiated into the anterior appendages. A second pair, located caudal to the belly-stalk, appears somewhat later and gives rise to the posterior appendages. The bulk of the tissue which goes to make up the appendage buds arises by outgrowth from the mesodermic somites at that level.

Flexion. The embryos of all the higher vertebrates develop within a confined space. The growing body must conform itself to the limitations imposed by the egg shell, as in birds and reptiles, or the uterine cavity, as in mammals. It is not at all surprising, therefore, that young embryos show a marked tendency to become curled, head to tail. This process by which an embryo at first straight (Fig. 29) becomes bent into more or less the shape of a letter C (Fig. 32) is called flexion. Flexion becomes apparent first in the cephalic region (Fig. 30) but soon thereafter involves the entire body (Fig. 32). At certain points flexion is especially strongly marked. This has led to speaking, for convenience in description, of the cranial flexure, the cervical flexure, the dorsal flexure, and the lumbo-sacral flexure. These so-called regional flexures in reality grade into one another and are nothing more than local accentuations of a process which involves the entire body.

Torsion. Preceding or accompanying flexion the body tends to show some twisting about its long axis (Fig. 30). This torsion, as it is called, is by no means as conspicuous or constant in mammalian embryos as it is in certain other forms, as for example the chick. In embryos with a large yolk, the body at first lies prone on the surface of the yolk sphere. Flexion would be impeded by the yolk except for the fact that the embryo, preceding flexion, twists its body through 90 degrees so that instead of lying prone on the yolk it lies on its side. The torsion which occurs in mammalian embryos is probably to be

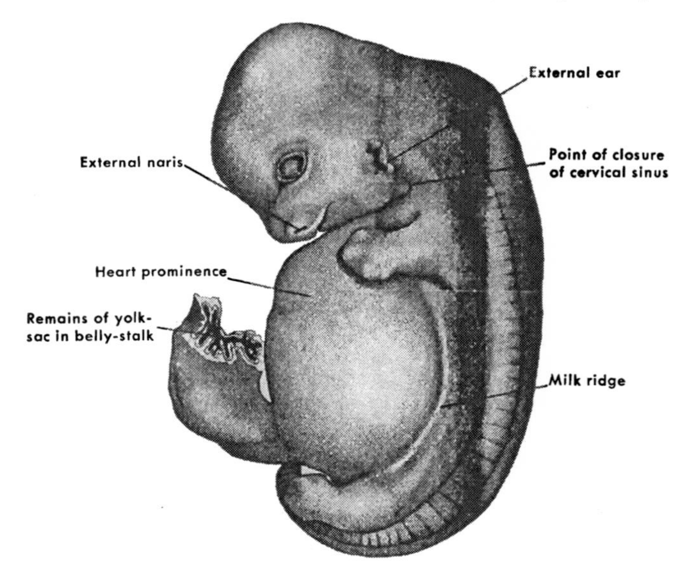

Fig. 33. Drawing (× 6) of 15 mm. pig embryo. (Modified from Minot.) Age approximately 24 days.

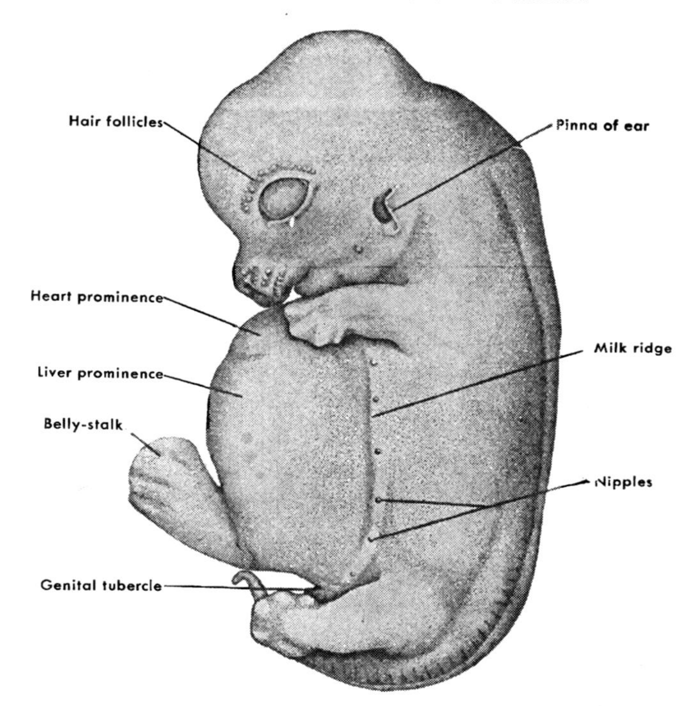

FIG. 34. Drawing (× 5.5) of 20 mm. pig embryo. (After Minot.) Age, about 4 weeks, i.e., approximately one-fourth of the total period of gestation.

regarded as a tendency inherited from their large-yolked ancestors. Like certain vestigial structures this change in body shape tends to make its appearance even after the conditions in connection with which it was developed have ceased to exist. Like vestigial structures, also, it is strikingly variable in the degree to which it develops. It may be hardly noticeable (Fig. 31) or it may be very strongly marked (Fig. 50). In either case it is but transitory, and the body soon loses all evidence of its torsion (Figs. 32–34).

II. The Establishment of the Organ Systems

The Nervous System

Formation of the Neural Tube. In dealing with the early differentiation of the germ layers, attention was called to a thickened area of the ectoderm, known as the neural plate, which constituted the

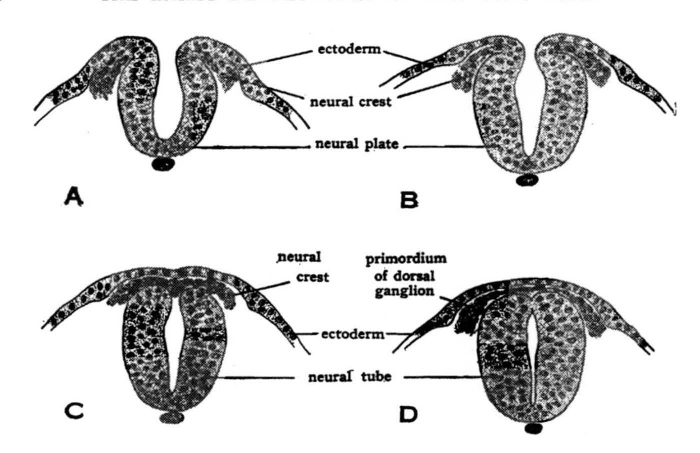

Fig. 35. Drawing (× 135) showing closure of the neural tube and formation of the neural crest. From pig embryos of: A, 8 somites; B, 10 somites; C, 11 somites; D, 13 somites.

primordium of the central nervous system. The first step in the formation of the brain and cord from this primordial mass of cells is its transformation from a superficial plate into a tubular mass of cells lying beneath the ectoderm.

As is the case with so many early embryological phenomena, the formation of the neural tube from the neural plate is brought about by a process which can most conveniently be described as "folding." Due to differential growth the neural plate is depressed centrally and elevated laterally. It is said to have become folded to form the neural groove (cf. Figs. 24 and 25 for its appearance in dorsal views of entire embryos and Figs. 22, C, and 26, C, for its configuration in transverse sections). The elevated lateral margins of the neural plate are now called the neural folds (Fig. 26, C).

As the neural folds become more elevated they grow toward each other, tending to close over the neural groove (Figs. 35, A, B). Up to this time the neural plate has remained directly continuous laterally with the superficial ectoderm, but when the neural folds meet in the mid-dorsal line this continuity ceases. A double fusion takes place. The mesial or neural plate components of the two folds fuse with each other and the lateral limbs consisting of unmodified ectoderm also fuse with each other (Fig. 35, C, D). Thus in the same process the original neural plate becomes the wall of the neural tube and the superficial ectoderm closes over the place formerly occupied by the open neural groove. Shortly after this fusion the neural tube and the superficial ectoderm become somewhat separated from each other leaving no trace of their former continuity.

The Neural Crests. There are cells located near the apices of the neural folds which are not involved in the fusion of either the superficial ectoderm or the neural plate. These cells form a pair of longitudinal aggregations extending one on either side of the mid-line in the angles between the superficial ectoderm and the neural tube (Fig. 35, A, B). With the fusion which closes the neural tube these two cell masses become, for a time, confluent in the mid-line (Fig. 35, C, D). But because this aggregation of cells arises from paired primordia and soon again separates into right and left components, it should be regarded as a paired structure. On account of its position dorsal to the neural tube it is called the neural crest.

When first established the neural crest is continuous anteroposteriorly. As development proceeds, its cells migrate ventrolaterally on either side of the spinal cord and at the same time become segmentally clustered. The metamerically arranged cell groups thus derived from the neural crest give rise to the dorsal root ganglia of the spinal nerves, and, in the cephalic region, to the ganglia of the sensory cranial nerves (Figs. 59 and 60).

Early Differentiation of the Brain. In dealing with the establishment of the cephalic region we noted the marked enlargement of the anterior portion of the neural plate and commented on the fact that this enlargement presaged the establishment of the brain. When the neural tube is formed by the folding of the neural plate, the cephalic part of the tube is of larger diameter corresponding to the greater size of the original neural plate in the future brain region (see Fig. 36).

The configuration of the primordial brain shows several things of interest. Not the least significant of these is the series of local enlargements, called *neuromeres*, which indicate its underlying metameric organization. Concerning the precise homologies of individual enlargements in the brain of a mammalian embryo with specific neuromeres of ancestral forms there is by no means complete agreement. The controversies center about the fusion of neuromeres in the more rostral parts of the brain. There are at least 11 enlargements recognizable in the embryonic brain, but only the more caudal ones

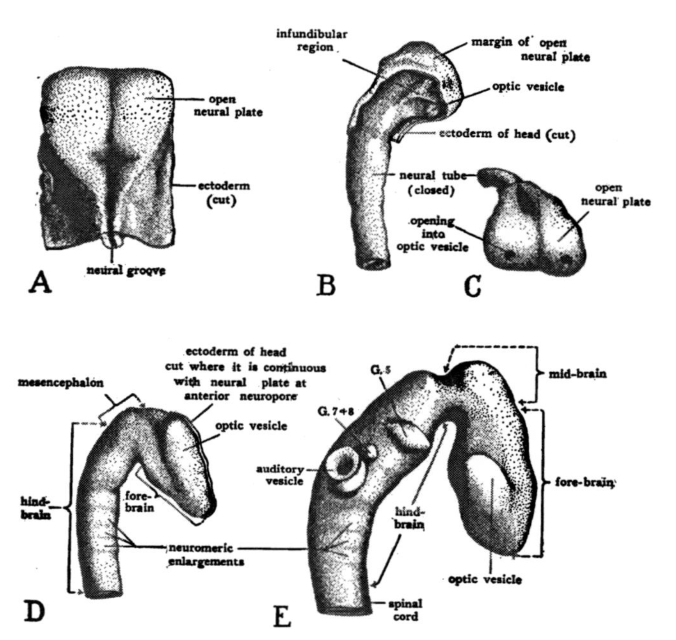

FIG. 36. Drawings (× 35) showing early stages in the formation of the brain of the pig. (Based on reconstructions made from series loaned by the Carnegie Institute.) A, Neural plate of 7-somite embryo, dorsal view. B, Neural plate of 13-somite embryo, lateral view. C, Same, anterior aspect. D, Brain of 17-somite embryo, lateral view. E, Brain of 24-somite embryo, lateral view.

show their individuality clearly. Some of the more rostral enlargements undoubtedly represent several neuromeres. In all probability there are as many as 15 neuromeres represented in the vertebrate brain. However this may be, for the beginning student the fact that metamerism is unmistakably present is to be emphasized rather than the controversies concerning the homologies of neuromeres.

Almost from its first appearance the brain shows certain indications of regional differentiation. In early stages we can recognize three fairly definite regions which later become subdivided to form the five divisions characteristic of the adult brain. The three primary regional divisions are known as the fore-brain, mid-brain, and hind-brain, or more technically, as *prosencephalon*, *mesencephalon*, and *rhombencephalon*. The prosencephalon is the broadest of the three divisions because of the presence of evaginations from its lateral walls which are the first indications of the formation of the eye. These evaginations are known as the primary optic vesicles (Fig. 36).

In the extreme rostral portion of the prosencephalon complete closure of the neural folds is somewhat delayed. There remains here, for a time, an opening known as the *anterior neuropore* (Fig. 29).

The mesencephalon is marked off by slight constrictions in the walls of the neural tube from the prosencephalon anteriorly, and somewhat less distinctly, from the rhombencephalon which lies posterior to it. In this early stage the mesencephalon shows no indication of local specialization presaging the formation of specific structures.

The most interesting feature of the hind-brain or rhombencephalon at this stage is the definite indication of neuromeric enlargements already mentioned (Fig. 36). Posteriorly the rhombencephalon grades without abrupt transition into the more slender part of the neural tube which will become the spinal cord.

At the extreme posterior end of the developing spinal cord, closure of the neural folds is delayed, just as it was anteriorly. The opening which thus persists for a time at the posterior end of the neural tube is known as the *sinus rhomboidalis* (Fig. 29).

The Digestive System

The Primitive Gut. Even before the body of the embryo takes shape, the formation of the digestive system has been initiated by the establishment of the entodermal layer of the spherical blastocyst (Fig. 16). At the time the blastocyst undergoes elongation (Figs. 17 and 18), the primitive gut, as the space within the entoderm is called, becomes correspondingly elongated. In fact it occupies nearly all the space within the blastocyst (Fig. 21).

When the mesoderm has been formed and split into somatic and splanchnic layers, the splanchnic mesoderm becomes closely associated with the entoderm as the splanchnopleure. Thus the primitive gut, very early in development, acquires a double-layered wall (Figs. 22 and 26). The entodermal component of the splanchnopleure gives rise to the epithelial lining of the gut tract and to its glands. The associated layer of splanchnic mesoderm becomes differentiated into the muscular and connective-tissue layers of the gut wall.

The Delimitation of the Embryonic Gut. By the time the wall of the primitive gut has received its mesodermal reinforcement the region which is to become the embryonic body is beginning to be more clearly defined. Formerly merely a disk-shaped area of the blastocyst distinguishable from the extra-embryonic portion of the germ layers by reason of its greater thickness, the body of the embryo now begins to be bounded by definite folds. These folds increase in depth, undercut the embryo, and finally, except for the communicating belly-stalk, separate it from extra-embryonic structures. The folds which thus definitely establish the boundaries between intra- and extra-embryonic regions are known as the limiting body folds or simply the body folds.

The formation of the body folds plays an important part in determining the configuration and relations of the gut tract of a young embryo. The "folding off" of the embryo begins with a ventral bending of the margins of the embryonic area so that the developing body takes on a marked dorsal convexity. Then the undercutting of these depressed margins cephalically and caudally, together with rapid increase in the length of the embryonic body, cause the embryo to overhang the extra-embryonic layers. The part of the embryo that juts out from the blastocyst anteriorly is the head (Fig. 37, B) and the portion which, slightly later, comes to project in a similar manner posteriorly, is the tail (Fig. 37, C). The folds of somatopleure which undercut the head and tail are known respectively as the subcephalic and subcaudal folds.

Coincidently the down-foldings on either side of the embryo become more definite, emphasizing its lateral boundaries. These folds are known as the lateral body folds (*lateral limiting sulci*) (Fig. 26, D). Toward the head they are continuous with the subcephalic folds and toward the tail with the subcaudal folds. The progressive deepening of all these circumscribing folds and the continued growth of the body itself constrict the connection of the embryo with the extra-embryonic membranes, and initiate the formation of the belly-stalk. The same folding process establishes the lateral and ventral body-walls of the embryo.

The superficial foldings which thus establish the boundaries between the embryonic body and extra-embryonic portions of the germ layers have their counterparts in the deeper lying layers. The changes which take place in the configuration of the splanchnopleure during this process bring about the division of the primitive gut into

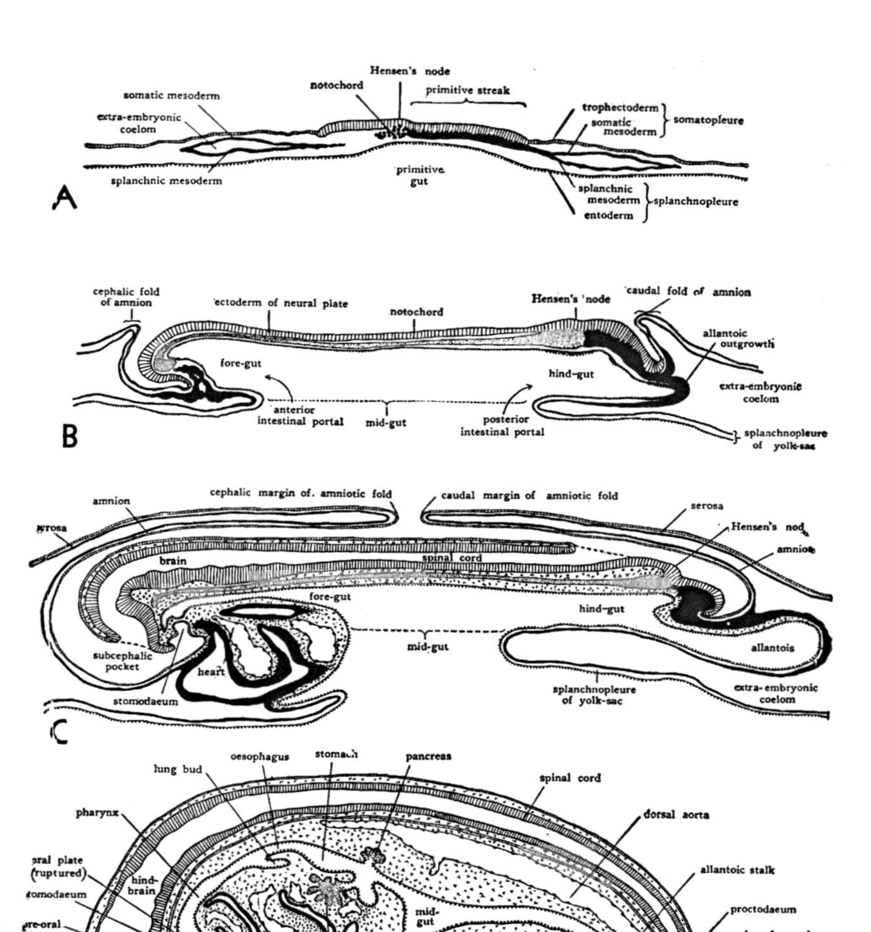

FIG. 37. Sagittal sections of pig embryos to show establishment and early regional differentiation of the gut. The drawings indicate schematically conditions: A, in the primitive streak stage; B, at the beginning of somite formation; C, in embryos having about 15 somites; D, in embryos having about 25 somites.

an intra-embryonic portion, and an extra-embryonic portion known as the *yolk-sac*.

The Fore-gut. The first part of the primitive gut to be definitely incorporated into the embryo is that portion lying beneath the head. With the forward growth of the head and its concomitant under-cutting by the subcephalic fold, an entodermally lined pocket is established in the cephalic region. This is the fore-gut (Fig. 37, B) Anteriorly the fore-gut ends blindly, posteriorly it remains in open communication with the rest of the primitive gut. The opening from the undifferentiated portion of the primitive gut into the fore-gut is known as the anterior intestinal portal.

The Hind-gut. In a similar manner a pocket of the primitive gut, known as the hind-gut, is formed beneath the caudal portion of the embryo (Fig. 37, B). The hind-gut ends blindly at its caudal extremity. Anteriorly it retains open communication with the rest of the primitive gut cavity by way of an opening termed the posterior intestinal portal.

The Mid-gut. Beneath the body of the embryo, between fore- and hind-gut, is a region of the primitive gut which is destined to be included within the body, but which as yet has no floor. This region is known as the mid-gut (Fig. 37, B). As the embryo is constricted off from the extra-embryonic layers by the progress of the subcephalic and subcaudal folds, the fore-gut and hind-gut are increased in extent at the expense of the mid-gut (cf. Fig. 37, B, C, D). The mid-gut is finally diminished until it opens ventrally by a very small aperture. This narrowed opening from the mid-gut to the yolk-sac is the yolk-stalk. When the extra-embryonic membranes are taken up we shall give further consideration to the fate of the yolk-sac.

The Stomodaeum and Proctodaeum. When first separated from the yolk-sac the embryonic gut ends blindly both cephalically and caudally. There are no indications of either oral or anal openings. Soon, however, there appear two depressions in the surface of the body which sink in to meet the gut. One of these depressions, the stomodaeum, is located on the ventral surface of the head in the future oral region. The other, the proctodaeum, is located caudally in the future anal region.

The stomodaeal depression gradually becomes deeper until its floor makes contact with the entoderm of the fore-gut (Fig. 37, C). The thin layer of tissue formed by the apposition of stomodaeal ectoderm to fore-gut entoderm is known as the *oral plate*. Not long

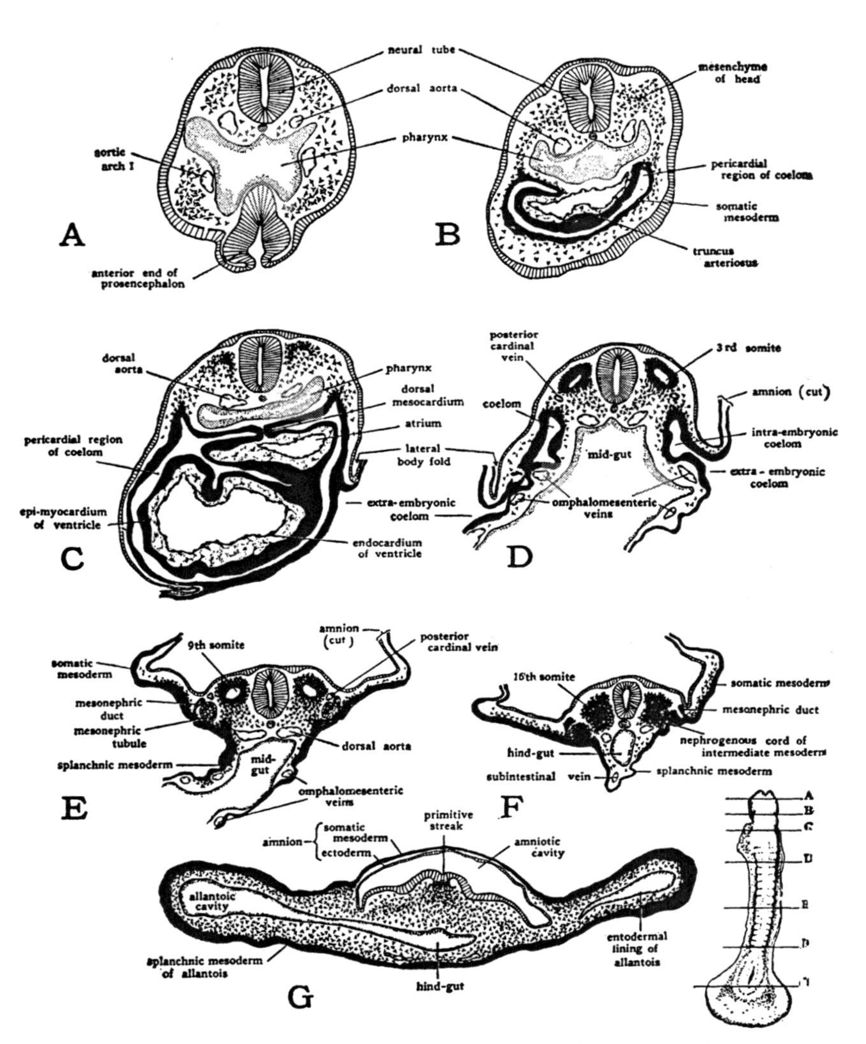

FIG. 38. Transverse sections (× 60) of pig embryo of 17 somites. (Carnegie Collection, C172–2.)

after the first appearance of the stomodaeum, the oral plate ruptures, establishing the anterior opening of the gut (Fig. 37, D). Growth of surrounding structures further deepens the original stomodaeal depression and it becomes the oral cavity. The region of the oral plate in the embryo becomes, in the adult, the region of transition from oral cavity to pharynx.

Somewhat later in development than the time at which the oral opening is established, the proctodaeum breaks through to the hind-gut, forming the cloacal opening. Subsequent differentiation in this region results in the separation of the originally single cloacal aperture, into anal and urogenital openings (see Chap. 10).

Pre-oral and Post-cloacal Gut. Neither the stomodaeum nor the proctodaeum break through into the gut at its extreme end. There

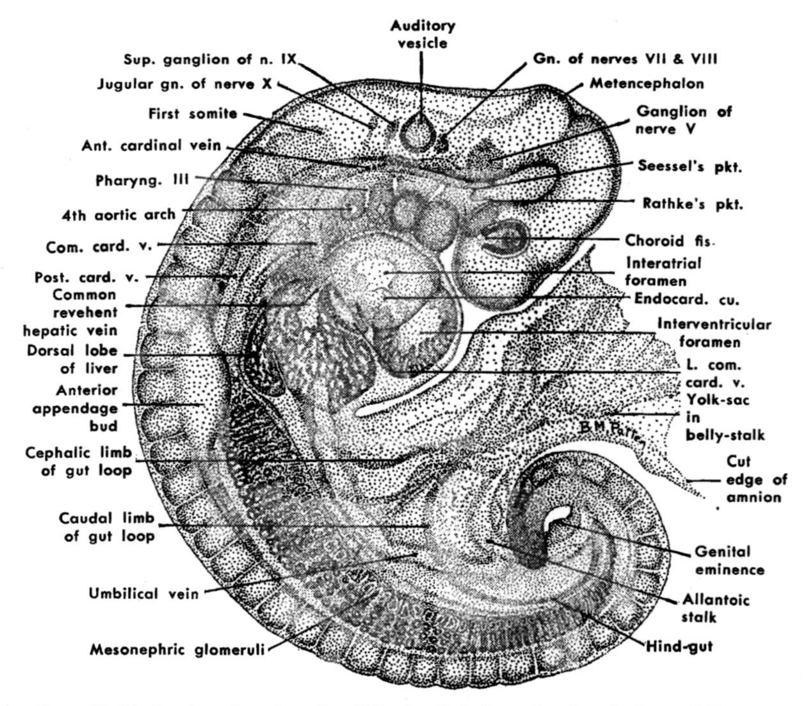

FIG. 39. Projection drawing (× 17) of a lightly stained and cleared 5 mm. pig embryo.

remains cephalic to the stomodaeal opening a small region of the fore-gut which is designated as the pre-oral gut. The comparable region of the hind-gut caudal to the proctodaeal opening is known as the post-cloacal gut (Fig. 37, D). Pre-oral and post-cloacal gut

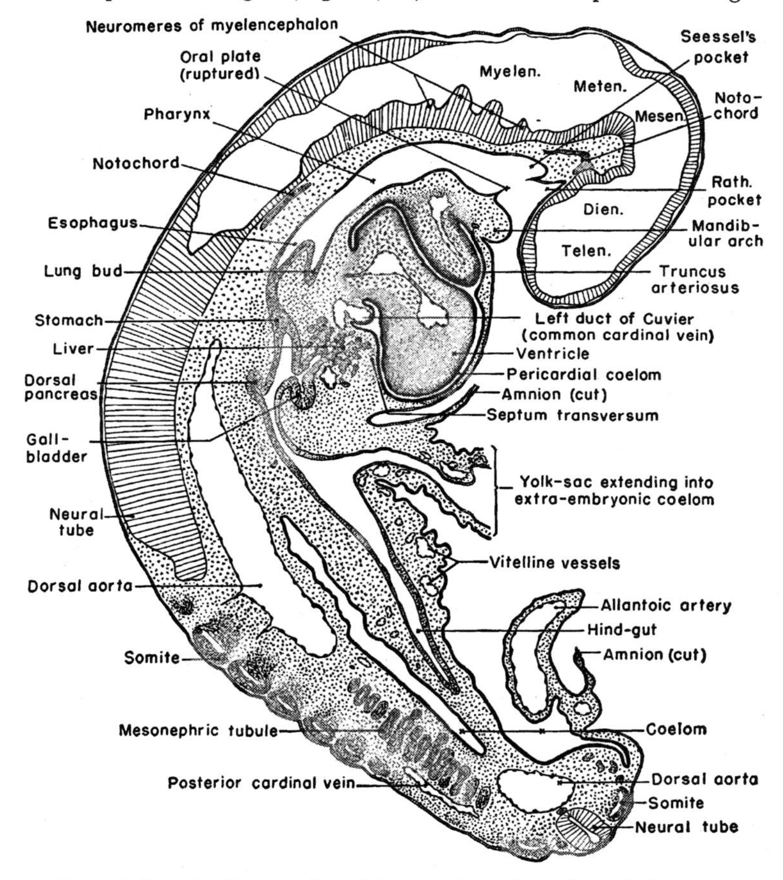

FIG. 40. Longitudinal section of 5 mm. pig embryo (× 25). The caudal end of an embryo in this stage of development is usually somewhat twisted to one side (see Fig. 31). For this reason sections which cut the cephalic region in the sagittal plane pass diagonally through the posterior part of the body. For a schematic plan of a completely sagittal section of an embryo of about this age see figure 37, D. Abbreviation: Rath., Rathke's.

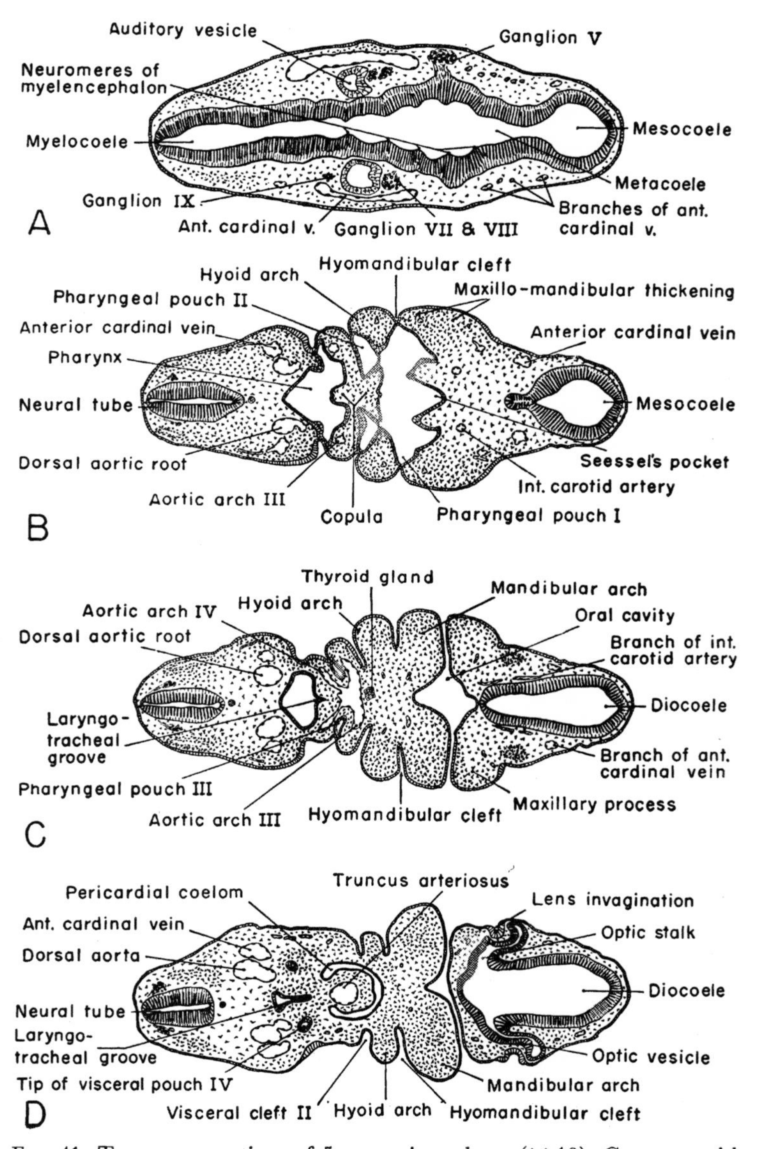

FIG. 41. Transverse sections of 5 mm. pig embryo (× 18). Compare with (Fig.

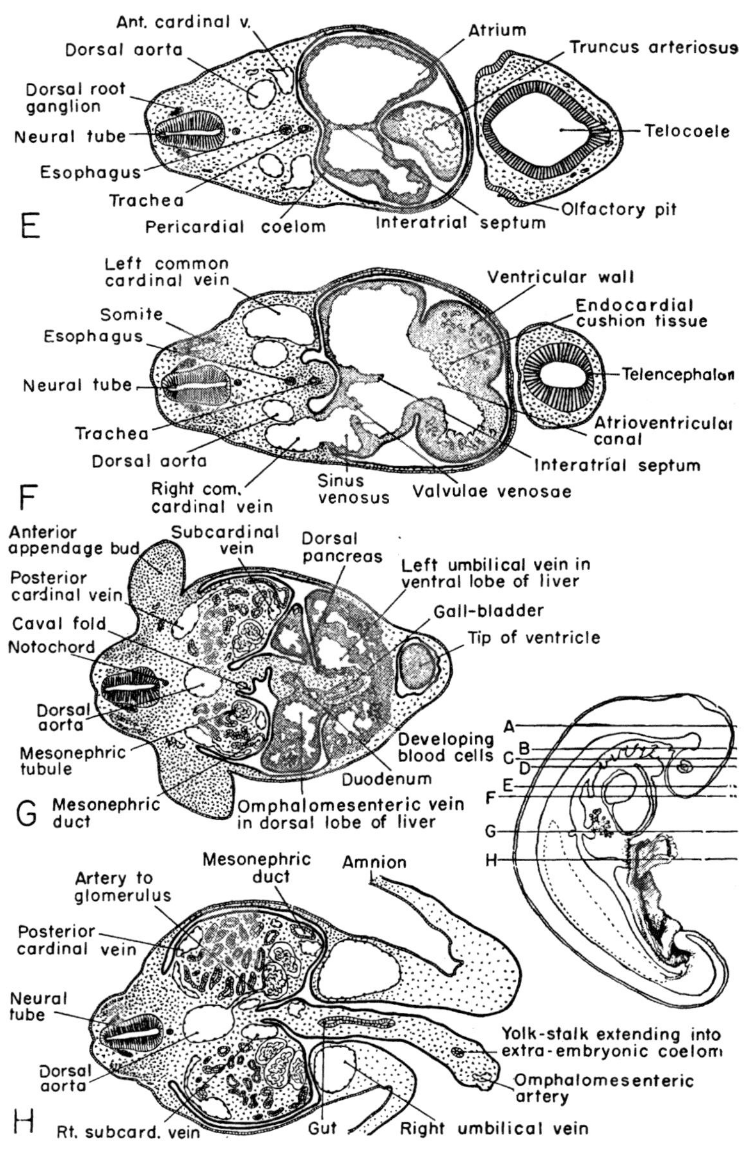

entire embryo (Fig. 39), and with longitudinal section of embryo of same age 40).

are of interest primarily as embryological landmarks helpful in following changing relations in the cephalic and caudal regions. They disappear later in development without giving rise to any definite structures.

Early Regional Differentiation of the Gut. The special structures and organs which arise from various parts of the gut tract will be considered subsequently in some detail. We cannot, however, ignore altogether the local differentiations which appear in young embryos indicating the regions where some of the more conspicuous organs will develop. The names of these primordial local differentiations clearly indicate their rôle in organ formation.

The fore-gut posterior to the stomodaeal opening becomes greatly expanded laterally and at the same time somewhat compressed dorso-ventrally. This region is the *pharynx* (Figs. 38, A–C, and 40). A series of lateral bays of the main pharyngeal lumen extend to either side. These are the *pharyngeal pouches.* Each pouch lies opposite one of the external grooves between adjacent, branchial arches (Fig. 41, B, C). The tissue between the floor of a branchial groove and the tip of its associated pharyngeal pouch is reduced to a thin membrane and sometimes may even break through transitorily to form an open gill cleft reminiscent of those in water-living ancestral forms.

From the extreme caudal part of the pharynx a medial ventral diverticulum arises which is destined to play an important part in the formation of the respiratory organs. Cephalically this diverticulum is little more than a furrow in the floor of the pharynx (*laryngo-tracheal groove*, Fig. 41, D). At its caudal extremity the outgrowth projects ventrad, entirely free of the gut tract. This blind end of the diverticulum (Fig. 40) soon bifurcates to form the *lung buds.*

Immediately caudal to the point at which the respiratory primordia appear, the gut narrows abruptly to form the esophagus and then dilates again in the region which will become the stomach. Just beyond the stomach is a conspicuous group of outgrowths from which are derived the pancreas, the liver and the gall-bladder (Figs. 40, 41, G, and 46). Caudal to this region there is as yet little local differentiation.

The Early Differentiation of the Mesoderm

The Somites. In the preceding chapter we traced the differentiation of the mesoderm into three primary divisions—dorsal, intermediate, and lateral. Comment was made, also, on the formation of the segmentally arranged somites in the dorsal mesoderm. The cells in a somite are not destined to a common fate. In fact these cells as a group have a wider diversity of developmental potentialities than any sharply localized aggregation of cells with which we have to deal. It is, therefore, a matter of especial interest to see the various steps by which they become, so to speak, sorted out, grouped according to their potentialities, and finally highly specialized in various ways (see Fig. 27).

When first formed a somite appears as a solid mass of cells without any definite organization—a mere local thickening of the dorsal mesoderm (Fig. 42, A). This initial mass grows rapidly in bulk and its cells take on a definite radial arrangement (Fig. 42, B). At the same time its boundaries become more clear cut and a small but quite definite lumen appears in its center (Fig. 42, B). This lumen, known as the *myocoele*, increases in size until the somite appears as a hollow vesicle with thick outer walls (Fig. 42, C).

By this time local differences within the somite are becoming apparent. Three regions are recognized and named on the basis of their later history. The dorso-mesial part of a somite is composed of cells which will form the skeletal muscles developing at that segmental level of the body. For this reason it is called the *myotome* (Fig. 42, B, C).

The ventro-lateral portion of the somite is made up of cells which have been believed to migrate out, become aggregated close under the ectoderm and give rise to the connective-tissue layer (dermis) which underlies the epidermis. Accordingly it has been called the *dermatome* (cutis plate). While some cells from this region of the somite undoubtedly are contributed to the formation of the deep layers of the skin, the conviction has been gaining ground that many, perhaps most, of them take part in the formation of muscle. Furthermore the connective-tissue layer of the skin is known to receive many cells from the somatic mesoderm generally, and from the diffuse mesenchyme in the cephalic region where there are no somites. The term dermatome is so firmly fixed that it is probably unwise to attempt to discard it, but we should bear in mind that while it does contribute to the dermis it probably does not do so any more extensively than other regions of the mesoderm which lie in close proximity to the ectoderm.

The third region of the somite is the so-called *sclerotome*, consisting of cells which migrate ventro-mesially from the original compact mass

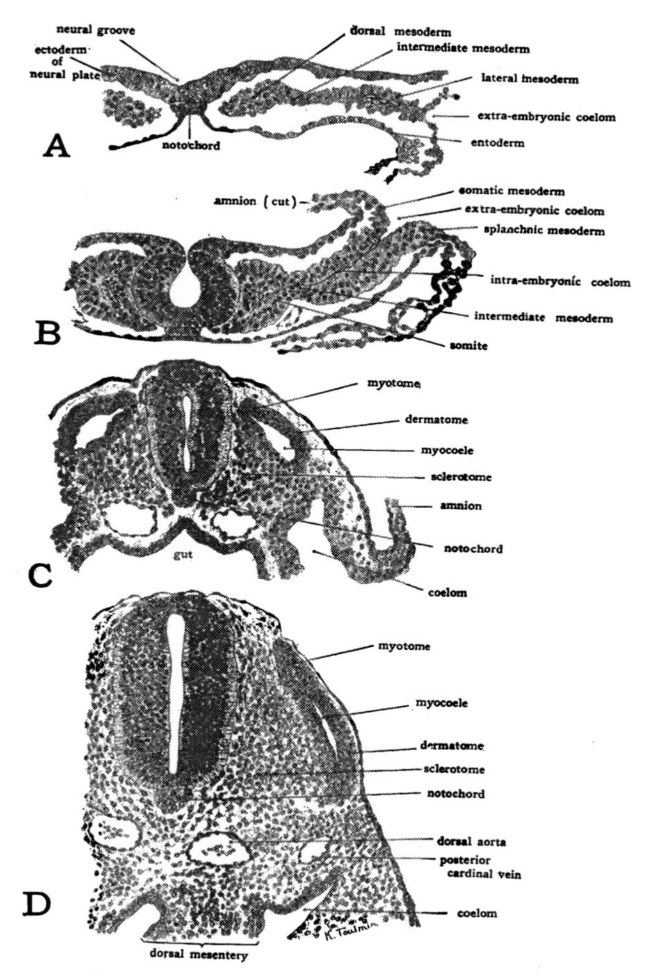

FIG. 42. Drawings (× 120) of transverse sections of pig embryos of various ages to show formation and early differentiation of somites. (From series in the Carnegie Collection.) A, Beginning of somite formation. B, 7-somite embryo. C, 16-somite embryo. D, 30-somite embryo.

(Fig. 42, C, D). These cells become concentrated about the neural tube and notochord, eventually giving rise to the vertebrae.

The Intermediate Mesoderm. There are, even at this early stage of development, changes becoming apparent in the intermediate mesoderm which foreshadow the formation of the embryonic excretory organs known as the mesonephroi. But the differentiation of other closely related parts of the urogenital system is as yet very slight. It therefore seems advisable to postpone consideration of the system as a whole and dismiss these early changes with a word of comment.

The mesonephroi are paired excretory organs conspicuous in young embryos but becoming rudimentary later in development. The name mesonephros (middle kidney) implies the existence in the vertebrate group of a more cephalic and a more caudal kidney. The cephalic kidney (pronephros) is vestigial in mammalian embryos. The caudal kidney (metanephros) becomes the permanent kidney of the adult.

Transverse sections of pig embryos with 16 or 17 somites show the mesonephric ducts as cords of cells arising on either side of the body where somatic and splanchnic mesoderm merge into the intermediate mesoderm (Fig. 38, F). Just mesial to the mesonephric duct, cells from the intermediate mesoderm become aggregated into a solid mass known as the nephrogenic cord (Fig. 38, F). Where development has progressed somewhat farther the nephrogenous cord can be seen to have given rise to a series of hollow vesicles, the primordial mesonephric tubules (Figs. 38, E, 39, and 40). These tubules soon make connection with the mesonephric duct, becoming at the same time much elongated and tortuous (Fig. 41, G). Meanwhile the mesonephric duct becomes patent and establishes an outlet into the hind-gut.

Somatic and Splanchnic Layers of the Lateral Mesoderm. Very early in development the lateral mesoderm becomes split into somatic and splanchnic layers with the coelom between (Figs. 22, 26, and 42). As long as the somites and the intermediate mesoderm remain undifferentiated the place of transition from somatic to splanchnic mesoderm is quite obvious and definite (Fig. 42, B, C). After the intermediate mesoderm has become organized into nephric tubules and no longer connects the somite with the lateral mesoderm, the line of demarcation between splanchnic and somatic layers is less readily determined. If the location of the mesonephric tubules is taken as a landmark, however, the point at which somatic and splanchnic mesoderm become continuous may be established with sufficient definiteness for all practical purposes (Fig. 38, E). Since the mesonephros grows primarily ventro-mesially, it is the splanchnic mesoderm which is pushed out to cover the mesonephros as it increases in size (cf. Fig. 38, E, with 41, G). At this stage the point of transition from somatic to splanchnic mesoderm can be approximated as the place at which the mesodermal lining of the coelom bends sharply from the inner face of the body-wall to be reflected over the lateral surface of the mesonephros.

Similarly in the cardiac region somatic mesoderm ends at the angles of the coelom on either side of the pharynx (Fig. 38, C). The splanchnic mesoderm covers the entodermal wall of the pharynx, forms the dorsal mesocardium, and is reflected to form the outer covering of the heart wall. It is this characteristic relationship of the two layers which accounts for their names. The somatic mesoderm lines the body-wall (somatic) face of the coelom. The splanchnic mesoderm forms the supporting membranes (denoted by the prefix meso-, e.g., meso-cardium, meso-gaster) of organs suspended in the coelom and covers the visceral (splanchnic) surfaces which project into the coelom.

The Circulatory System

The mammalian embryo, having practically no yolk available as food, is dependent for its survival and growth on the prompt establishment of relations with the circulation of the mother. This implies the necessity of a precocious development of the vascular system in the embryo, for the maternal circulation remains confined within the uterine walls, and the embryonic circulation must grow to it. Until this is accomplished the embryo is dependent on what food material it can obtain by direct absorption from the fluid within the uterine cavity—a method entirely inadequate to provide for the growth of the embryo except in its very early stages when its bulk is inconsiderable.

The Heart. In mammalian embryos the heart arises from paired primordia situated ventro-laterally beneath the pharynx. The fact that the heart, a median unpaired structure in the adult, arises from paired primordia which at first lie widely separated on either side of the mid-line is likely to be troublesome unless its significance is understood at the outset. The paired condition of the heart at the time of its origin is correlated with the fact that the embryonic body at first lies spread out flat on the surface of the blastocyst. The primordia

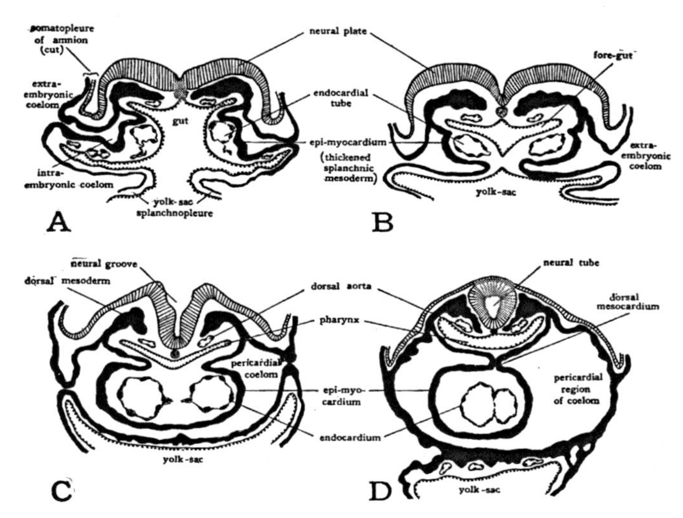

FIG. 43. Sections cut transversely through the cardiac region of pig embryos of various ages to show the origin of the heart from paired primordia. (Projection diagrams × 50, from series in the Carnegie Collection.) A, 5-somite embryo. B, 7-somite embryo. C, 10-somite embryo. D, 13-somite embryo.

of certain anatomically ventral structures arising at an early stage of development, therefore, first appear as separate halves lying on either side of the mid-line. With the folding under of the lateral margins of the embryonic area which brings the lateral walls of the body into their definitive position, the embryo is closed ventrally, and potentially median structures which arose as separate halves are established in the mid-line.

The primordial heart is double-layered, as well as paired right and left. The inner layer is called the *endocardium* because it is destined to form the internal lining of the heart. The outer layer is known as the *epi-myocardium* because it will give rise both to the heavy muscular layer of the heart wall (myocardium) and to its outer covering (epicardium).

The endocardium appears first in the form of irregular clusters and cords of mesenchymal cells lying between the splanchnic mesoderm and the entoderm. These cells become organized into two main strands lying one on either side of the gut. Soon after their establishment these strands acquire a lumen and are known as the endocardial tubes (Figs. 43, A, and 44, A). The endocardial tubes continue beyond the cardiac region as branching strands which will become, cephalically, the primitive efferent vessels and, caudally, the afferent vessels of the heart. Meanwhile the splanchnic mesoderm has become markedly thickened where it is reflected laterally over the endocardial tubes (Fig. 43, A). This thickened region of the splanchnic mesoderm constitutes the epi-myocardial layer of the heart.

While these changes have been occurring in the heart, folding off of the embryonic body has been going on with concomitant progress in the closure of the fore-gut at the level of the heart (cf. Fig. 43, A, B). As a result the paired endocardial tubes are brought progressively closer together. Finally they are approximated to each other and fused into a single tube lying in the mid-line (Figs. 43 and 44).

In the same process the epi-myocardial layers are bent toward the mid-line enwrapping the endocardium. Ventral to the endocardial tubes the epi-myocardial layers of opposite sides come into

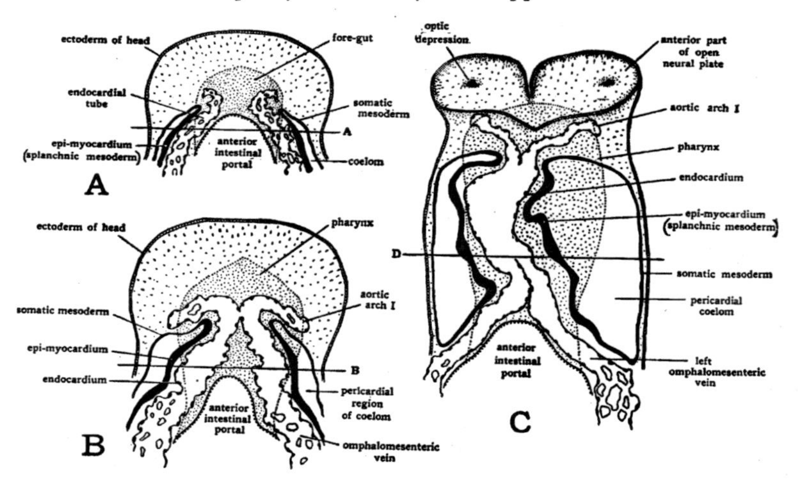

FIG. 44. Diagrams showing progress of fusion of cardiac primordia as seen in ventral views. A, 5-somite embryo. B, 7-somite embryo. C, 13-somite embryo.

The embryos are supposed to be viewed as transparent objects with the outlines of the cardiac primordia showing through. The lines A, B, and D indicate the levels of sections A, B, and D in figure 43.

contact and fuse with each other. This fusion occurs in such a manner that the limbs of the mesodermic folds next to the endocardium fuse with each other forming an outer layer of the heart no longer interrupted ventrally; and so that the limbs of the folds which line the pericardial coelom ventrally also fuse with each other to form an unbroken layer (Fig. 43, B, C). Thus the originally paired right and left coelomic chambers become confluent to form a median unpaired pericardial cavity in the same process which establishes the heart as a median structure. Dorsally the right and left epi-myocardial layers become contiguous, but here they do not fuse immediately as happens ventral to the heart. They persist for a time as a double-layered supporting membrane called the *dorsal mesocardium*. In this manner the heart is established as a nearly straight, double-walled tube suspended mesially in the most cephalic part of the coelom.

Blood Vessels. While these changes have been occurring in the cardiac region, the main vascular channels characteristic of young embryos are making their appearance. The cephalic prolongations of the endocardial tubes beyond the region in which the heart itself is formed constitute the start of the main efferent channels or *aortae*. The aortae are further extended by a process entirely similar to that involved in the formation of the endocardial tubes themselves. Cords and knots of cells of mesodermal origin[1] become aggregated along the course of the developing vessel. These strands of cells are then hollowed out to form tubes walled by a single layer of thin, flattened cells (endothelial cells). Where the main blood vessels are about to become established there is found first a meshwork of these small channels with their delicate endothelial lining. Gradually some of these primitive channels are enlarged and straightened to form the main vessels and their walls are reinforced by the addition of circularly disposed connective-tissue fibers and smooth muscle cells. In this manner the

[1] By some writers these cells are called mesenchymal because of their similarity to the rest of the mesenchyme in their mesodermal origin, in their sprawling shapes, and in their migratory proclivities. By others they are called angioblasts (collectively "the angioblast") because of the part they play in the formation of blood vessels. Unfortunately, because of a long-standing controversy, the very appropriate term angioblast carries with it the special connotation that all cells of this type originate from a common center and migrate thence to all places where vascular endothelium is formed. (This is the so-called "angioblast theory" of the origin of vascular endothelium.) Recent experimental evidence indicates that cells with vasculogenetic potentialities arise in many places by direct differentiation from local mesoderm.

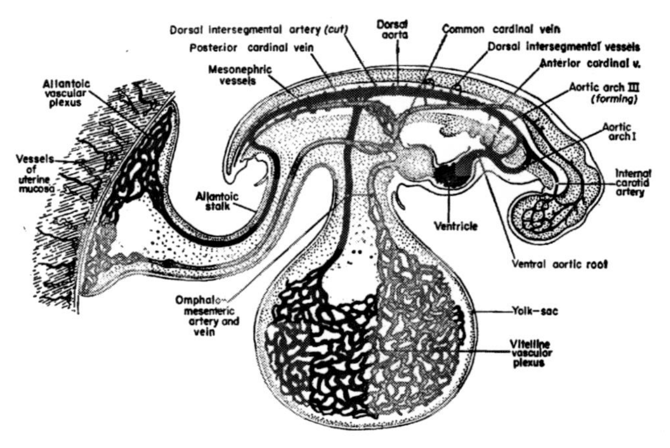

FIG. 45. Schematic plan of the circulation in a young pig embryo. At this stage all the blood vessels are paired (right and left), the entire circulatory system except for the heart being bilaterally symmetrical. Only the vessels on the side toward the observer are shown in the figure.

primitive efferent channels are prolonged from the heart cephalad beneath the pharynx as the *ventral aortae*. They then bend laterad and dorsad about the pharyngeal walls to form the *aortic arches*, and finally turn caudad to extend nearly the entire length of the embryo as the *dorsal aortae* (Fig. 45).

At first there is but a single pair of aortic arches which is located in the tissue of the first branchial arch (mandibular arch). Later in development five additional pairs of arches connecting the ventral and dorsal aortae are formed. Each of these aortic arches lies in one of the branchial arches caudal to the mandibular (Figs. 41, B, C, and 47). At present we are not interested in the history of individual aortic arches but merely in the fact that blood passes by way of one or more pairs of aortic arches around the pharynx from the ventrally located heart to the dorsally located aortae which form the main distributing trunks of the embryonic circulation (Fig. 45).

The vessels serving to collect the blood which is distributed to all parts of the embryo by branches from the aortae are called the *cardinal veins*. They arise somewhat later than the aortae but by an entirely similar process. There are two pairs of these vessels, the *anterior cardinal veins* draining the cephalic, and the *posterior cardinal veins* draining the caudal region of the body. Under favorable conditions the position of the cardinal veins can be made out in lightly stained and cleared entire embryos (Fig. 39). Their relations are best seen, however, in parasagittal sections (Fig. 47). At the level of the heart the anterior and the posterior cardinal veins on either side of the body unite as the *common cardinal veins* (*ducts of Cuvier*) (Figs. 45 and 47). The common cardinals are short trunks which at once turn ventro-mesiad and become confluent with the omphalomesenteric veins in the sinus venosus. The sinus venosus in turn discharges into the atrial part of the heart. The entrance of the sinus venosus into the atrium is guarded by a pair of delicate flaps called the *valvulae venosae* (Fig. 46).

In addition to the vessels limited in their distribution to the body of the embryo, there are conspicuous channels leading beyond the confines of the body to the yolk-sac and to the allantois. The main arteries from the aorta to the yolk-sac are called the omphalomesenterics and their terminal branches the vitellines. The main vessels leading to the allantois are known as the *allantoic arteries*, or, especially in mammals, as the *umbilical arteries*.

The afferent channels which lead from the yolk-sac to the heart are the *omphalomesenteric veins*. Near the heart these vessels arise as extra-cardiac continuations of the endocardial tubes (Fig. 44). The fusion which takes place in the heart does not at this stage as yet involve the omphalomesenteric veins and they enter the sinus venosus as paired channels.

Distally the omphalomesenteric veins are extended along the walls of the gut toward the yolk-sac. In the splanchnopleure of the yolk-sac the main omphalomesenteric vessels are continuous with a rich plexus of small tributaries, the vitelline vessels (Fig. 45). These smaller blood vessels can be traced into prevascular cords of mesodermal cells as yet not hollowed out. In these cellular cords are frequent knot-like enlargements, known as blood islands, containing not only cells which are destined to form the endothelium of blood vessels but also cells which will give rise to blood corpuscles.

Blood Islands. In the differentiation of a blood island the peripherally located cells become flattened and somewhat separated from the central cells (Fig. 48, B). Then they arrange themselves as a coherent layer a single cell in thickness, which invests the remaining cells of the island (Fig. 48, C). Meanwhile similar changes have been occurring in the neighboring islands and, with their progressive enlargement, adjacent blood islands coalesce (Fig. 48, C). Continuation of this process results in the transformation of the original endothelial vesicles into a plexus of anastomosing endothelial tubes, the primordial capillary bed of the yolk-sac.

Meanwhile fluid has accumulated within the endothelial vesicles and the central cells have become rounded, forming primitive blood corpuscles (Fig. 48, C). When these vesicles with their contained

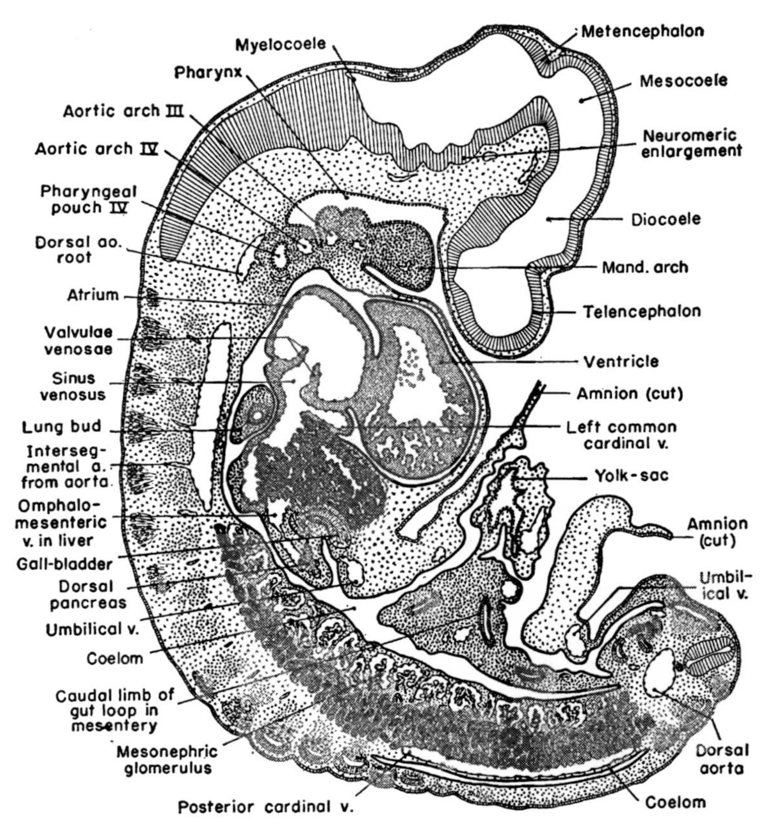

Fig. 46. Drawing of parasagittal section of 5.5 mm. pig embryo. (Projection outlines × 22.) The section was taken from the series to the right of the mid-line, at a plane favorable for showing the entrance of the sinus venosus into the right atrium.

corpuscles have become confluent to form capillaries and the capillaries have acquired open communication with the vitelline arteries on the one hand and the vitelline veins on the other, all the conditions necessary for active circulation of blood have been established. Under the pumping action of the heart which begins at this time, the fluid accumulated in the blood islands acts as a vehicle conveying the

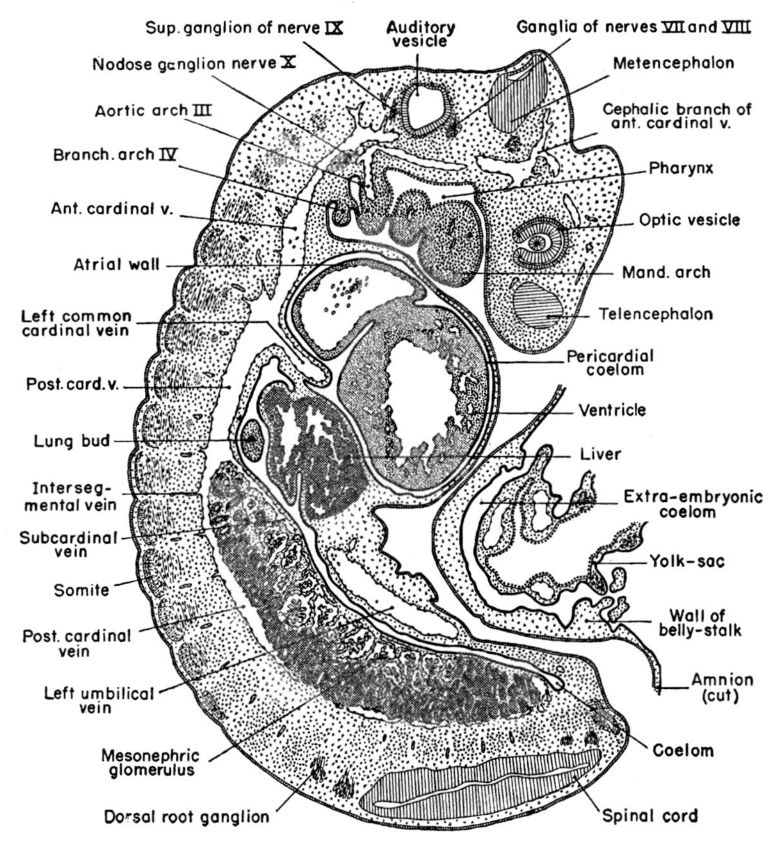

FIG. 47. Drawing of a parasagittal section of a 5.5 mm. pig embryo. (Projection outlines × 22.) The section was taken from the series to the left of the mid-line, at a plane especially favorable for showing the cardinal veins.

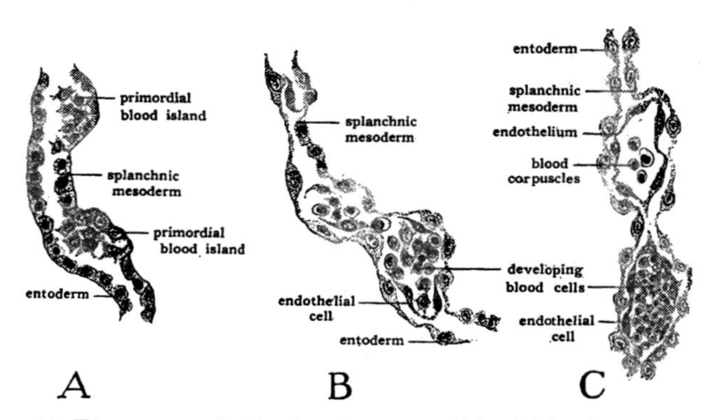

FIG. 48. Three stages in the development of blood islands.

A, The primordial blood island is merely a knot of mesodermal cells. (3-somite embryo.)

B, The beginning of specialization of peripheral cells to form endothelium and central cells to form corpuscles is clearly indicated. (13-somite embryo.)

C, The upper island has a complete endothelial wall which has become continuous with that of neighboring vascular channels as shown by the fact that most of the corpuscles formed within it have moved out into the blood stream. The lower island has not yet acquired a complete endothelial wall nor open communication with neighboring vessels, and the corpuscles formed within it still densely pack its lumen. (20-somite embryo.) All figures drawn from yolk-sac blood islands of pig embryos in the Carnegie Collection.

corpuscles formed in the blood islands of the yolk-sac to all parts of the body.

Changes similar to those described for the yolk-sac occur in the allantois so that, by the time the circulation of blood actually begins, there is a rich plexus of small vessels in the allantois. This plexus is fed by the allantoic branches of the aortae and blood is returned from it to the heart by way of a pair of large veins, the allantoic (umbilical) veins. These vessels traverse the allantoic stalk and the lateral body-wall, emptying into the posterior end of the heart along with the omphalomesenteric and the common cardinal veins (Fig. 45).

The circulatory system of a mammalian embryo at this stage in its development can be analyzed into three distinct sets of afferent and efferent channels. Each set of main channels with its interpolated capillary bed can be conveniently designated as a circulatory arc. One of these is known as the intra-embryonic arc because it consists of vessels which lie wholly within the body of the embryo. The blood, pumped by the heart, is distributed to all parts of the embryo over the aortae. Small branches from the aortae break up locally in various parts of the body into capillaries, thus bringing the blood into intimate relation with the developing tissues. The blood is then collected by the cardinal veins and returned by way of the ducts of Cuvier to the heart.

The other two arcs are, the vitelline which runs to the yolk-sac, and the allantoic or umbilical to the allantois (Fig. 45). Both these arcs start within the embryo, for the heart serves as a common receiving and pumping station, and the aortae as a common distributing main for all three of the circulatory arcs. But because their main vessels extend outside the body with their terminal ramifications in the extra-embryonic membranes, these latter arcs are ordinarily spoken of as extra-embryonic.

Later we shall trace the origin of the main systemic vessels of the adult from the primitive channels of the intra-embryonic circulation. We shall see the vitelline arc transformed into the hepatic-portal circulation, and the allantoic arc highly developed to become the placental circulation. It is, therefore, of prime importance that the primitive ground plan of the circulation as graphically summarized in figure 45 be fixed clearly in mind as a basis on which to build.

CHAPTER 6

The Extra-embryonic Membranes and the Relation of the Embryo to the Uterus

I. The Formation of the Extra-embryonic Membranes

In dealing with the establishment of the embryonic body we have already seen that the germ layers extend far beyond the region in which the embryo itself develops. These peripheral portions of the germ layers give rise to membranes which are of service as a means of protection and as a means of securing food from the blood of the mother. Because they are not incorporated in the body of the embryo but discarded at the time of birth, they are called extra-embryonic membranes. These membranes are the yolk-sac, the amnion, the serosa, and the allantois. Their designation as extra-embryonic should not lead us to lose sight of their vital importance to the embryo during its intra-uterine existence.

The Yolk-sac. In mammals, although there is virtually no yolk accumulated in the ovum, a large yolk-sac is formed just as if yolk were present. Such persistence of a structure, in spite of the loss of its original function, is not an uncommon phenomenon in evolution and has given rise to the biological aphorism that "morphology is more conservative than physiology." Not only does the yolk-sac itself persist, even the blood vessels characteristically associated with it in its functional condition appear in the empty yolk-sac of mammalian embryos (Fig. 45). When the mammalian yolk-sac is spoken of as a vestigial structure, therefore, we must bear in mind that such a statement has reference primarily to its function. Morphologically it is of considerable size in young embryos.

The yolk-sac may be defined as that part of the primitive gut which is not included within the body when the embryo is "folded off." In dealing with the formation of the digestive tube we have already sufficiently considered the processes which thus establish the boundaries between the yolk-sac and the intra-embryonic portion of the original gut cavity (Fig. 37). The peripheral extent of the yolk-

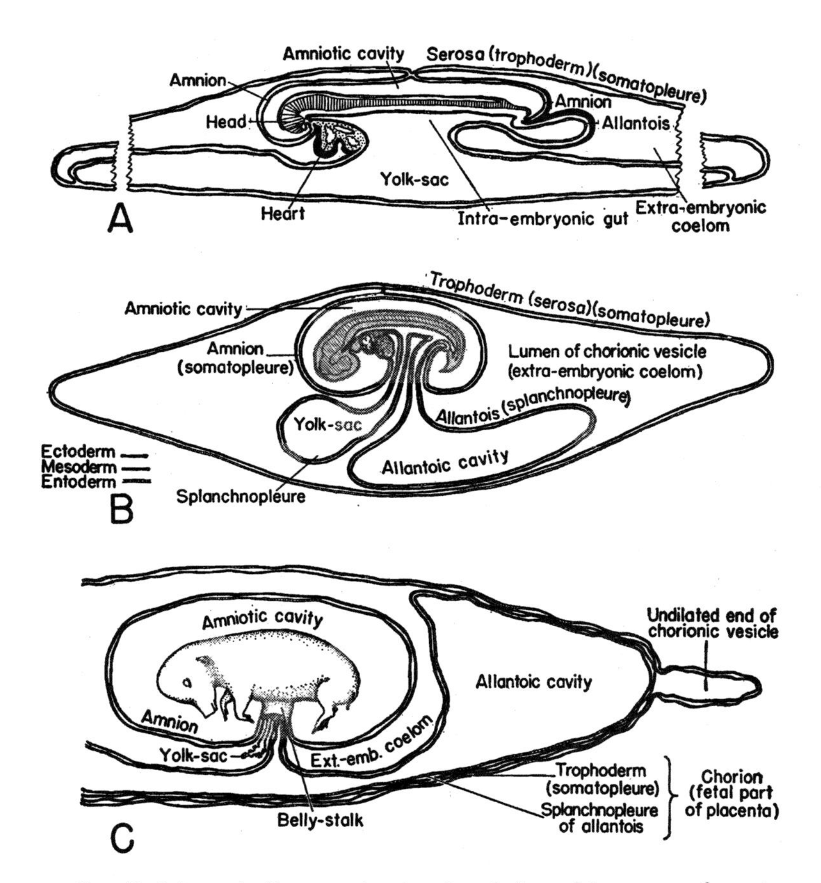

Fig. 49. Schematic diagrams showing the relations of the extra-embryonic membranes. The great length of the chorionic vesicle of the pig precludes showing the entire vesicle in correct proportions. In A, sections of the vesicle have been omitted. In B and C, the vesicle has been shortened from its true proportions.

A represents conditions in embryos of 15–20 somites; B, conditions in embryos of 4–6 mm. (after flexion); and C, conditions in embryos of 30 mm. or more.

sac is unusually great in young pig embryos correlated with enormous elongation of the blastocyst (Fig. 18). From the yolk-stalk, the yolk-sac extends beneath the embryo, beyond its head in one direction and beyond its tail in the other, nearly to the ends of the blastocyst (Fig. 49, A).

During this, the period of its greatest relative development, the yolk-sac serves as an organ purveying nutritive material to the embryo. Separated from the uterine wall only by the thin outer layer of the blastocyst, its abundant vessels are in a position which makes absorption from the uterus readily possible. The food material and oxygen thus absorbed are transported by way of the vitelline circulation (Fig. 45) to the growing embryo. When, somewhat later, the allantois becomes highly developed it takes over this function and the yolk-sac decreases rapidly in size (Fig. 49, B), finally becoming a shriveled sac buried in the belly-stalk (Figs. 49, C, and 65). The yolk-sac, therefore, exhibits another point of special interest in addition to being an organ which persists after the cessation of its original function. The persistent yolk-sac and its associated vessels are utilized by the embryo in a new manner. Containing no stored food materials, the mammalian yolk-sac turns about, so to speak, and absorbs from the uterus through its external surface. The fact that the function is temporary, later being taken over by the allantois, makes this physiological opportunism in seizing on a new source of supplies none the less striking.

The Amnion. The amnion arises as a layer of somatopleure which enfolds the developing embryo. When the amniotic sac is completed it becomes filled with a watery fluid in which the embryo is suspended. This suspension of the embryo in a fluid-filled sac, by equalizing the pressure about it, serves as a protection against mechanical injury. At the same time the soft tissues of the growing embryo, being bathed in fluid, do not tend to form adhesions with consequent malformations. The functional significance of the amnion is emphasized by the fact than an amnion appears only in the embryos of non-water-living forms. This has led to designating the embryos of fishes and Amphibia, which form no amnion, as "anamniotes," and those of birds, reptiles, and mammals, which do form an amniotic sac during development, as "amniotes."

In all mammalian embryos the amnion is formed at an exceedingly early stage of development. In some forms (e.g., man) the amniotic cavity appears even before the body of the embryo has taken definite shape. When the amnion is thus precociously established the processes in its formation are, as it were, hurried through and consequently they are difficult to analyze. In the pig the amnion is formed in more leisurely fashion, the process being quite similar to that in reptiles and birds.

The first indication of amnion formation in pig embryos becomes evident shortly after the primitive streak stage. The embryonic disk appears to settle into the blastocyst so that it is overhung on all sides by the extra-embryonic somatopleure (Figs. 25 and 26). As the embryo grows in length its head and tail push deeper and deeper into these folds of somatopleure (cf. Fig. 37, B, C). At the same time the folds themselves become more voluminous, growing centripetally over the embryo from the region of its head, its sides, and its tail. These circumferential folds of somatopleure are known as the amniotic folds. For convenience in description we recognize cephalic, caudal, and lateral regions of the amniotic folds, although in reality they are all directly continuous with each other.

Progress of the amniotic folds soon brings them together above the mid-dorsal region of the embryo to complete the amniotic sac (Figs. 37, C, and 49). Where the amniotic folds close, there persists for a time a cord-like mass of tissue between the amnion and the outer layer of the blastocyst (Fig. 49, B). In time even this trace of the fusion is obliterated and the embryo in its amnion lies free in the blastocyst (Fig. 49, C).

The amnion is attached to the body of the embryo where the body-wall opens ventrally in the region of the yolk-stalk (Fig. 50). As development advances this ventral opening becomes progressively smaller, its margins being known finally as the *umbilical ring*. Meanwhile the yolk-stalk and the allantoic stalk, in the same process of ventral closure, are brought close to each other, to constitute, with their associated vessels, the *belly-stalk* (Fig. 49, C). The amnion is now reflected from its point of continuity with the skin of the body at the umbilical ring, to constitute the covering of the belly-stalk, before it is recurved as a free membrane enclosing a fluid-filled space about the embryo.

The Serosa. The outer layer of the mammalian blastocyst has various names. In the inner-cell-mass stage, the outer layer of the blastocyst is most commonly called the trophoblast (Fig. 15, C). Since the primary germ layers are not at this time differentiated, this term is far more appropriate than ectoderm, which implies a layer having

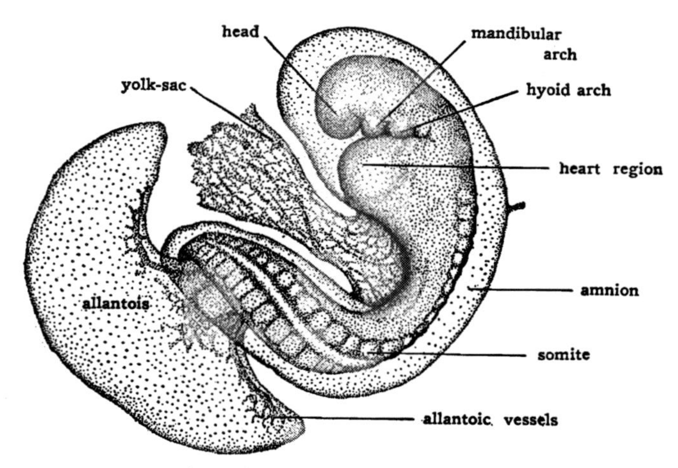

FIG. 50. Drawing (× 12) of pig embryo having about 28 pairs of somites (crown-rump length 4 mm.; age approximately 17 days). The embryo has been removed from the chorionic vesicle with the amnion and allantois intact but the distal portion of the yolk-sac has been cut off. (Cf. Fig. 49, B.)

relations which make it stand in contrast with entoderm, or with entoderm and mesoderm. The premature use of the term ectoderm would seem to have nothing to recommend it, and has been responsible for much confusion in thinking and much pointless discussion.

When the germ layers have been differentiated, the outer layer of the blastocyst is directly continuous with the ectoderm of the embryo (Fig. 19). There is then justification for calling it more specifically trophectoderm. The prefix troph- (Greek root, to nourish) suggests the part this layer plays in the acquisition of food materials from the uterus and at the same time sets it off from the ectoderm of the embryonic disk.

Still later in development when the mesoderm splits and its somatic layer becomes associated with the ectoderm, the name trophectoderm is no longer appropriate (Fig. 22). Three different names are in common use to cover this double layer. *Extra-embryonic somatopleure* is a term following the general embryological usage in designating by the suffix -pleure the double layers formed by the secondary association of somatic and splanchnic mesoderm with ectoderm and entoderm respectively. *Trophoderm* is widely used by those working especially in mammalian embryology. Those especially interested in comparative embryology are quite likely to use the term *serosa* for this layer, thus emphasizing its very evident homology with the part of the extra-embryonic somatopleure so designated in birds and reptiles. This multiplicity of names is not difficult to master if we understand they are all appropriate terms selected to emphasize somewhat different aspects of the same structure. Trophoblast, then, means the outer layer of the blastocyst before ectoderm, entoderm, and mesoderm are differentiated. Trophectoderm is the extra-embryonic part of the ectoderm when the germ layers have been differentiated, and trophoderm (serosa) (extra-embryonic somatopleure) is trophectoderm reinforced by a layer of somatic mesoderm.

In the formation of the amnion, the somatopleure is thrown into folds surrounding the embryo. The inner limb of the fold, as we have seen, becomes the amnion (Fig. 37, C). The outer limb of the fold is commonly called the serosa. It is, perhaps, in just this designation of the outer limb of the amniotic fold that the term serosa is most often encountered. After the amnion has been completed, interest in this outer layer centers about its rôle in food absorption and the term trophoderm is most commonly used to designate it.

The Allantois. Almost as soon as the hind-gut of the embryo is established, there arises from it a diverticulum known as the allantois (Fig. 37). The allantoic wall, because of its manner of origin, is necessarily composed of splanchnopleure. As the allantois increases in extent its original communication with the hind-gut is narrowed to a cylindrical stalk, while its distal portion becomes enormously dilated (Figs. 37 and 49).

Externally the developing allantois appears at first as a crescentic enlargement directly continuous with the caudal part of the embryo (Figs. 29 and 30). Its dilated distal portion continues to grow with great rapidity until it is a large semilunar sac free of the caudal end of the embryo except for the stalk by which it maintains its original connection with the hind-gut (Fig. 50).

The allantois early acquires an abundant blood supply by way of large branches from the caudal end of the aorta (Figs. 45 and 51). In the allantoic wall these arteries break up into a maze of thin-walled vessels. As the yolk-sac circulation undergoes retrogressive changes, this plexus of allantoic vessels becomes progressively more highly developed and soon takes over entirely the function of metabolic interchange between fetus and mother.

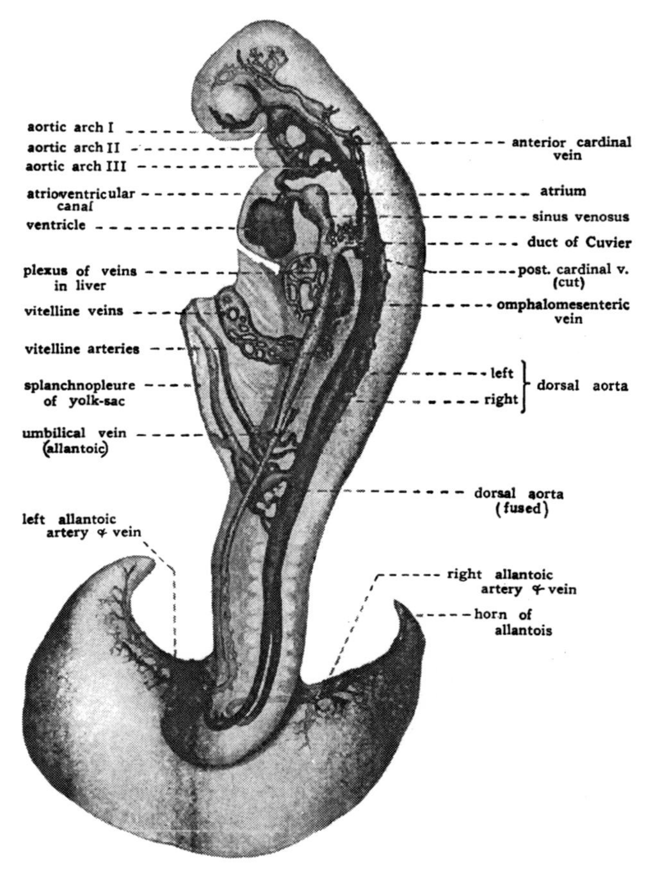

FIG. 51. Drawing showing the main vascular channels in a pig embryo of about 17 days (27 pairs of somites). Based on cleared preparations of injected embryos loaned by Dr. C. H. Heuser, and on Sabin's figures of similar material.

II. The Relation of the Embryo and Its Membranes to the Uterus

Having seen the manner of origin and the relations of the various extra-embryonic membranes to the embryo, we must now turn our attention to their relations to the uterus. It will be recalled that the ova

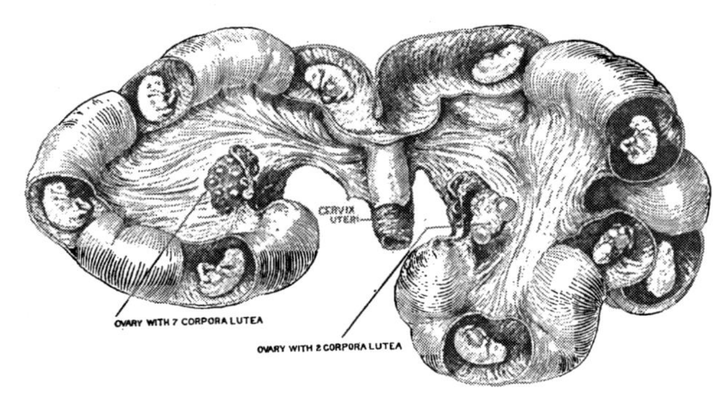

FIG. 52. Uterus of pregnant sow opened to show distribution of embryos. (After Corner.) Note the approximately uniform spacing of the embryos in the two horns of the uterus in spite of the fact that the corpora lutea indicate the origin of two ova from one ovary and seven from the other.

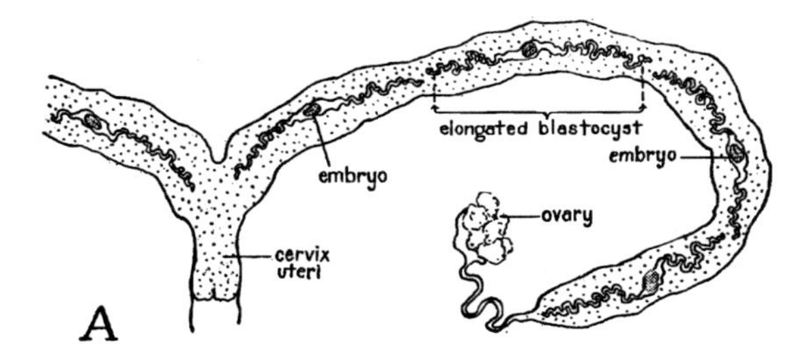

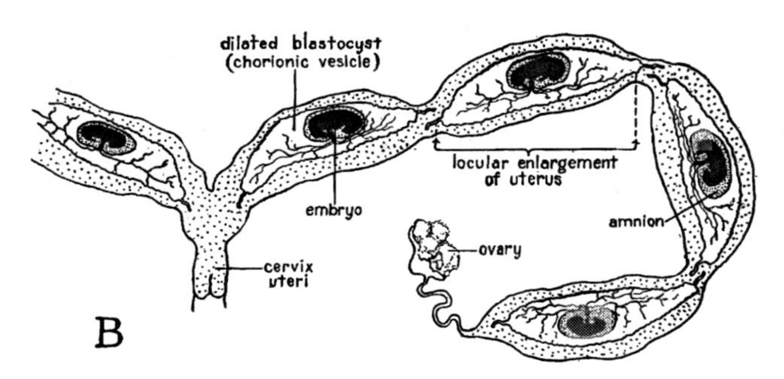

FIG. 53. Schematic diagrams indicating the intra-uterine relations of the embryos and their membranes: A, in the elongated blastocyst stage; B, when the blastocysts have been dilated to form the chorionic vesicles characteristic of the later stages of pregnancy.

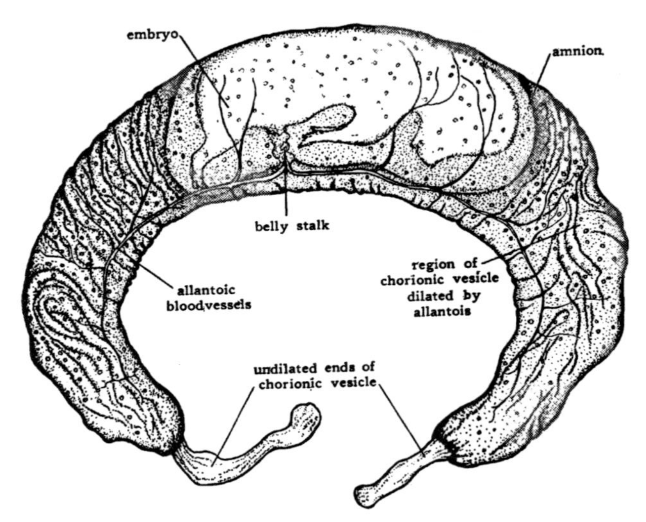

FIG. 54. Drawing of pig embryo in unruptured chorionic vesicle. (After Grosser.) Compare with figures 49, C, and 53, B.

are fertilized in the uterine tube shortly after their discharge from the ovary. It takes from three to four days for the ova to pass through the uterine tube so that by the time the embryos reach the uterine cavity the cleavage divisions are well under way (Figs. 14, C, D, and 15, A).

The Spacing of Embryos in the Uterus. Normally, of course, some of the ova fertilized were formed in one ovary and some in the other. The corpora lutea which develop from the ruptured ovarian follicles leave unmistakable evidence of the number of ova contributed by each ovary. Usually the numbers are approximately equal, but it not infrequently happens that, at a given estrus, one ovary is far more prolific than the other. If all the ova originating on one side remained in the corresponding horn of the uterus there would be crowding of the embryos on one side and unutilized space on the other. Such a condition does not ordinarily occur. The mechanism by which the embryos are arranged within the uterus is not known, but the careful observations of Corner show that they do tend to become uniformly distributed in the two horns of the uterus regardless of the side on which the ova were liberated (Fig. 52).

This spacing of the embryos apparently takes place before the elongation of the blastocysts occurs. Even when elongated, the blastocysts do not spread over as much space in the uterus as one might suppose. The lining of the uterus is extensively folded and the thread-like blastocysts follow along these folds, so that a blastocyst a meter long will occupy perhaps no more than 10–15 cm. of a uterine horn (Fig. 53, A).

The attenuated condition of the blastocyst does not persist long. With the increase in the extent of the allantois and growth of the embryo, the blastocyst becomes greatly dilated and somewhat shortened. Thus altered in shape and appearance it is commonly called a chorionic vesicle. The allantois never grows to quite the entire length of the chorionic vesicle and there remains an abrupt narrowing at either extremity of the vesicle where the allantois ends (Figs. 49, C, and 54).

Where these undeveloped terminal portions of neighboring chorionic vesicles lie close to each other the uterus remains undilated, sharply marking off the region where each embryo is located (Fig. 53, B). One of the local enlargements of the uterus containing an embryo and its membranes is known as a *loculus*.

The Chorion. While these changes have been occurring in the general relations of the embryo and its membranes to the uterus there have been further specializations of the fetal membranes themselves. In its peripheral growth the allantois comes into contact with the trophoderm (extra-embryonic somatopleure) (serosa) and becomes fused with it. The new layer thus formed by the fusion of allantoic splanchnopleure with extra-embryonic somatopleure is the *chorion* (Fig. 49, C). Lying in intimate contact with the uterine mucosa externally, and having in its double mesodermal layer the rich plexus of allantoic vessels communicating with the fetal circulation, it is well situated to carry on the functions of metabolic interchange between the fetus and the uterine circulation of the mother.

The Placenta. In many mammals the chorion becomes fused with the uterine mucosa. In the pig, although the chorion and the uterine mucosa lie in close contact with each other, they can always be peeled apart (Fig. 57, A). When the uterine mucosa and the chorion do not actually grow together and can readily be separated from one another without the tearing of either, they are said to constitute a contact placenta (semi-placenta). Most mammals show a more highly specialized condition in which the fetal (chorionic) and

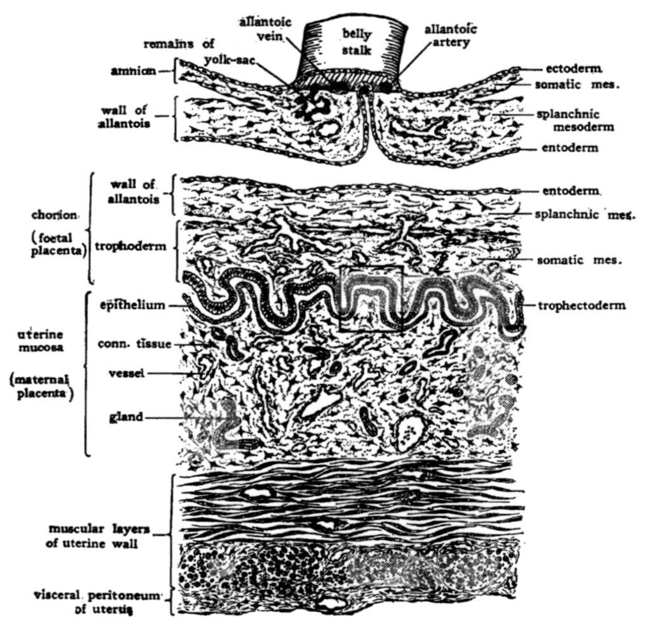

FIG. 55. Semischematic diagram showing the structure of the chorion and its relation to the uterine wall. Reference to figure 49, C, will show the general relations of the region here depicted. The area indicated in this figure by the rectangle is shown in detail in figure 56.

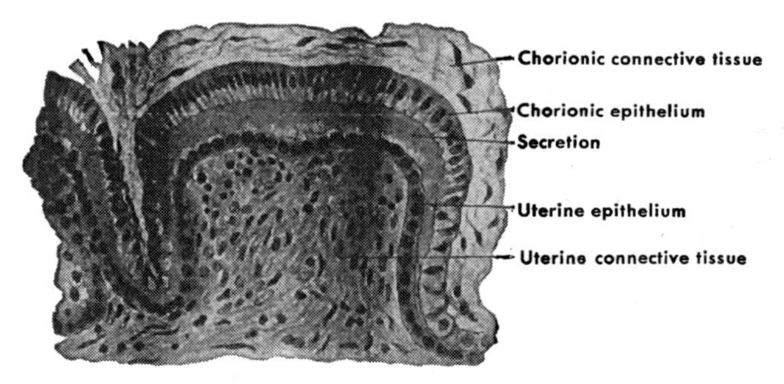

FIG. 56. Detailed drawing of small area of placenta of 27 mm. pig embryo showing the structure of its chorionic and uterine portions where they lie in contact. (After Grosser.) For relations of area drawn see figure 55.

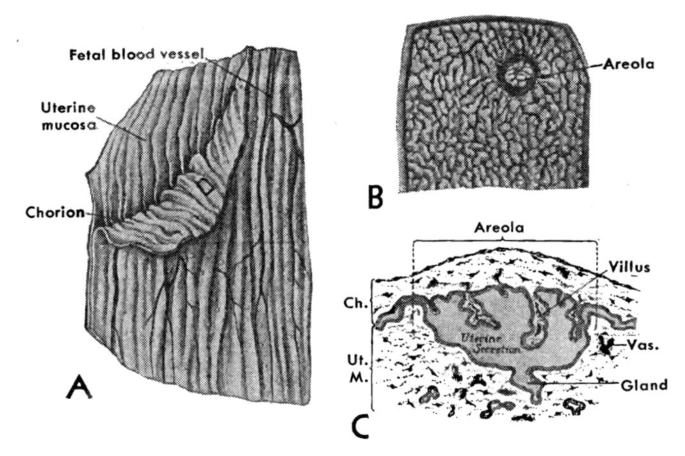

FIG. 57. A, Small portion of pig placenta drawn about natural size, to show the manner in which the chorion may readily be separated from the uterine mucosa without injury to either layer. (After Grosser.)

B, Small area of chorion of 19 cm. pig embryo showing its uterine surface magnified. (After Grosser.) For location of area depicted see rectangle in A. Note the minute irregular local elevations (chorionic villi of primitive type) which are superimposed on the major folds shown in A, further increasing the surface of contact between chorion and uterine mucosa.

C, Semischematic diagram of section in region of areola. (After Zeitzschmann.) The areolae fit into depressions in the uterine mucosa near the orifice of uterine glands. The villi of the areolae are somewhat more highly developed than those of the general chorionic surface and are believed to be especially active in absorbing nutritive materials from the uterine secretions.

maternal (mucosal) portions of the placenta actually grow together so they cannot be pulled apart without hemorrhage. So intimately do they become fused in fact, that when, shortly following the birth of the fetus itself, the extra-embryonic membranes are delivered as the so-called "after-birth," a large part of the uterine mucous membrane is pulled away with the chorion. In contrast with the primitive contact type such a placenta is called a burrowing or "true" placenta. An understanding of placental relationships in such a form as the pig where all the component parts are clearly identifiable furnishes the best possible background for interpreting the more complex types of placentae where there are extensive fusions between fetal and maternal portions.

Even in its fully developed condition the chorion of the pig is a relatively simple structure. The entodermal lining of the allantoic cavity becomes much reduced in thickness, forming a delicate single-layered epithelial covering on the internal face of the chorion. The mesoderm of the allantoic wall fuses with that of the overlying trophoderm and this combined layer becomes differentiated into a primitive type of gelatinous connective tissue (mucoid connective tissue) (Fig. 55). In this connective-tissue layer the blood vessels of the allantoic plexus are located. With the fusion of the allantoic mesoderm to the trophoderm these vessels invade the trophoderm and their terminal arborizations come to lie close to the trophectoderm (Fig. 55). The trophectoderm becomes a highly developed simple columnar epithelium with its surface exposure enormously increased by abundant foldings.

This plicated epithelial surface of the chorion follows the folds in the surface of the uterine mucosa so that the chorionic and the uterine epithelium lie in contact with each other except for a thin film of uterine secretion (Fig. 56). Under the uterine epithelium is a loosely woven layer of connective tissue richly supplied with maternal vessels. Thus there intervenes between fetal and maternal vessels only a scanty amount of connective tissue and the epithelial layers of the uterine mucosa and of the chorion.[1] Through these layers food materials and oxygen are passed on from maternal to fetal vessels and the waste products of metabolism are absorbed from the fetal blood stream by the maternal blood stream. What part the epithelial layers may play in this transfer is uncertain. It seems not unlikely that the chorionic epithelium is active in absorption as is the case with the yolk-sac epithelium in Sauropsida. Whether or not this is the case it certainly constitutes no serious impediment to the reciprocal transfer of materials between fetus and mother which is the function of the placenta.

---

[1] Because the fetal and maternal vessels both retain the integrity of their endothelium, and because both chorionic and uterine epithelium persist, a placenta of this type may be designated technically as *epitheliochorial*. The term is used by students of comparative placentation in contrast with *endotheliochorial* which designates a type of placenta such as that occurring in the cat or dog where the uterine epithelium is lost but the endothelium of the uterine vessels is retained; and in contrast also with the term *hemochorial* which designates the type of placenta occurring in man and a number of other mammals in which maternal blood comes in direct contact with chorionic tissue without the intervention of either uterine epithelium or the endothelium of maternal vessels.

CHAPTER 7

The Structure of Embryos from Nine to Twelve Millimeters in Length

Pig embryos of about 10 mm. crown-rump length are especially valuable as material for laboratory study. They are neither so small and delicate that procuring and preparing them involves unusually skillful manipulation, nor so large that complete series of sections are tedious to make and difficult to study. Developmentally they are not so far advanced that their structure is hard to comprehend if the student is familiar with the early stages of development in any of the vertebrates. Yet if their morphology is thoroughly mastered it forms a starting point from which the later developmental processes in mammals may readily be traced. It seems advantageous, therefore, to deal with this, the most commonly utilized stage of development, somewhat less in narrative fashion than has been done heretofore. We shall, as it were, pause for a while and look over carefully the conditions which thus far have been established so that we may follow the story of the later phases of development more understandingly.

I. External Features

Flexion. Like all forms which develop in the limited confines of an egg shell or within the uterine cavity, the body of a young pig embryo is curled up on itself. This flexion of the spinal axis is more marked in some regions than others. The points of conspicuous bending have received special designations from the region in which they appear. In a pig of 10 mm. (Fig. 58) the cranial flexure, cervical flexure, dorsal flexure, and lumbosacral flexure are all well developed.

The Head. In the cephalic region the thin skin leaves the contours of the brain clearly suggested (cf. Figs. 58 and 59). The nasal pits have appeared as definitely circumscribed depressions at the rostral end of the head and the bulging caused by the growing optic cups clearly marks the position of the eyes. Especially in fresh specimens the retinal pigment can be seen through the overlying skin.

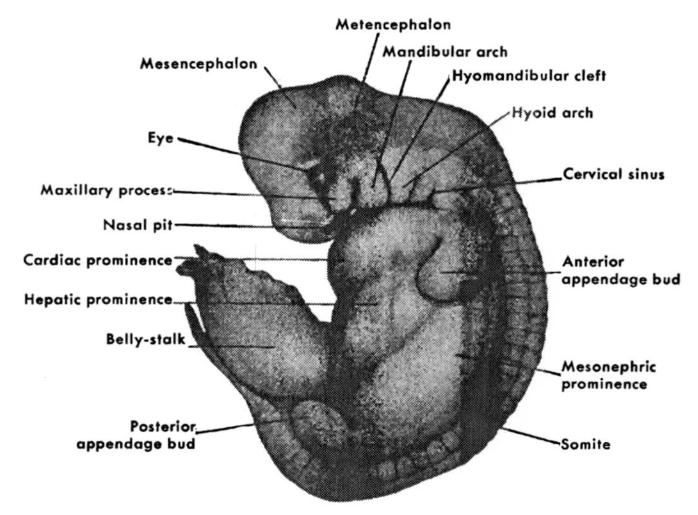

FIG. 58. Drawing (× 7) showing external appearance of 10 mm. pig embryo. (Modified from Minot.)

The Gill Arches. Flanking the oral cavity and caudal to it, in the region which will be under the chin, the gill (branchial) arches appear as strongly marked local elevations. This entire region is at this stage so compressed against the thorax that one gets a very incomplete view of its structure unless an embryo is decapitated and the head viewed in ventral aspect (Figs. 168 and 169). It is then seen that the maxillary processes form the lateral parts of the upper jaw, and that the two mandibular elevations meet each other in the mid-ventral line to form the arch (mandibular arch) of the lower jaw. Posterior to the mandibular are three similar arches, the hyoid and and the unnamed third and fourth post-oral arches. Later in development the arches posterior to the mandible become less conspicuous and are incorporated into the neck.

The Gill Clefts. Between the branchial arches are deep furrows which mark the position of ancestral gill clefts. Although in mammalian embryos these furrows do not ordinarily break through into the pharynx they are commonly called clefts because of their phylogenetic significance. Only the most cephalic of these clefts is named (hyomandibular cleft, Fig. 58); the others are designated by their post-oral numbers. The entire region about the third and fourth post-oral clefts becomes especially deeply depressed and is known as the cervical sinus.

The Appendage Buds. Both anterior and posterior appendage buds in 10 mm. pigs are still paddle-shaped (Fig. 58). Not until embryos have grown to a length of between 15 and 20 mm. do the five digits show as terminal enlargements (Figs. 33 and 34). It is significant that the ancestral five-digit type of appendage appears transitorily in the pig embryo, before its characteristic highly specialized hoof is even suggested.

The Trunk. At this stage the superficial bulges due to the somites are evident all the way from the cervical to the caudal region of the

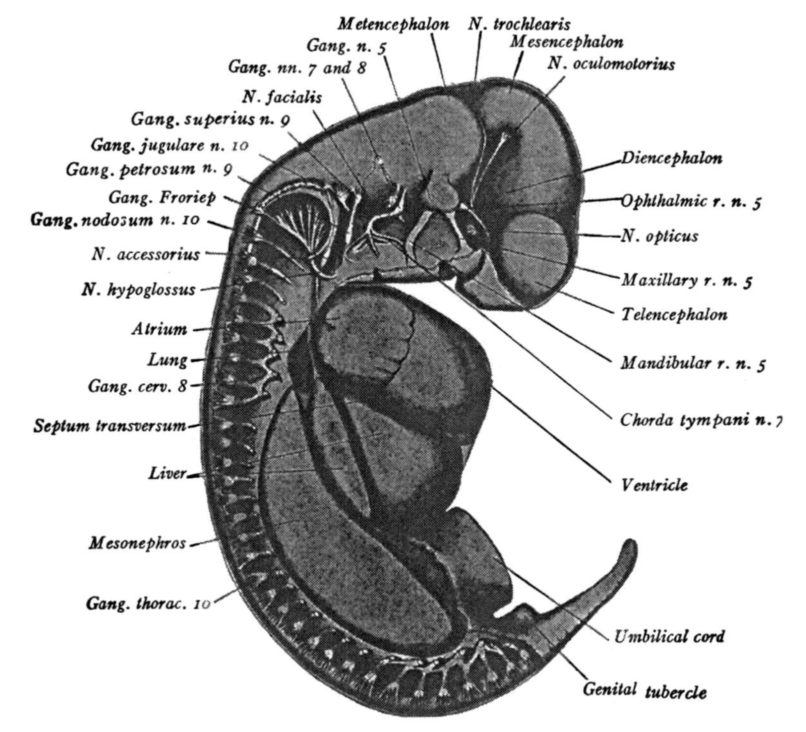

Fig. 59. Model of 10 mm. pig embryo showing the location and relations of the more conspicuous internal organs. (After Prentiss.)

body. Well-marked prominences in the lateral and ventral portions of the body-wall indicate the position of the heart, the liver, and the mesonephros (cf. Figs. 58 and 59).

The band of tissue ("milk ridge") which gives rise to the mammary glands is not ordinarily developed in embryos of the 9 to 12 mm. range. It usually becomes clearly marked by 15 mm. (Fig. 33), and by 20 mm. the nipples can generally be recognized (Fig. 34).

II. The Nervous System

The Brain. In younger embryos the brain consisted of three regions, fore-brain (prosencephalon), mid-brain (mesencephalon), and hind-brain (rhombencephalon) (Fig. 36, E). Now we find five regional divisions or vesicles, as they are called. The prosencephalon has divided to form the *telencephalon* and *diencephalon;* the mesencephalon has remained undivided; and the rhombencephalon has become differentiated into *metencephalon* and *myelencephalon* (Figs. 59, 60, 65, and 88).

The telencephalon consists of the most anterior median part of the brain and two lateral outgrowths from it called the *lateral telencephalic vesicles* (Figs. 60, 67, and 69). Its posterior boundary is conventionally established by drawing a line from a fold in the roof of the brain called the velum transversum (see Fig. 65 where the leader to Seessel's pocket crosses the dorsal wall of the neural tube) to the recessus opticus, a depression in the floor of the brain at the level of the optic stalks (Figs. 65 and 88).

The diencephalon is the hinder portion of the old prosencephalon. Its caudal boundary is conventionally established by drawing a line from an elevation in the floor of the neural tube called the tuberculum posterius (Fig. 88) to a depression in the roof of the neural tube which is just appearing at this stage of development, and may (Fig. 60) or may not (Fig. 65) be evident in the particular embryo under observation. The most conspicuous special features of the diencephalon are the paired lateral outgrowths from it which form the optic vesicles (Figs. 60 and 63) and the median ventral diverticulum which constitutes the infundibulum (Fig. 65). The median dorsal outgrowth of the diencephalon known as the epiphysis, which is so conspicuous a feature in chick embryos of the third and fourth day, appears relatively late in the pig. No suggestion of an epiphyseal evagination has appeared in 9 to 12 mm. embryos.

The mesencephalon shows little change from its condition in

younger embryos. Its demarcation from the metencephalon posteriorly is indicated by a conspicuous constriction in the neural tube (Figs. 59, 60, and 65).

The division of the primitive hind-brain (rhombencephalon) into metencephalon and myelencephalon is indicated at this stage, though not conspicuous or definite. The dorsal wall of the neural tube just caudal to the meso-rhombencephalic constriction is markedly thick, contrasting strikingly with the very thin roof of the more posterior part of the hind-brain (Fig. 65). The zone of the neural tube where the dorsal thickening exists is the metencephalon; the posterior, thin-roofed portion of the hind-brain is the myelencephalon. Although all external indications of individual neuromeres have by this time disappeared, the internal face of the myelencephalic wall still shows definite neuromeric markings (Figs. 61 and 67).

The Cranial Nerves. The peripheral relations of the cranial nerves to cephalic structures and their central relations to the brain are strikingly constant throughout the vertebrate series. In fishes we recognize 10 pairs of cranial nerves. In the mammals we encounter these same 10 cranial nerves with essentially similar relations both centrally and peripherally. But the mammalian brain in its progressive specialization has incorporated a part of the neural tube which in primitive fishes was unmodified spinal cord. One of the clearest evidences of this process is the fact that we find in the mammals 12 pairs of cranial nerves, the first 10 of which are homologous with the 10 cranial nerves of fishes and the last two of which represent a modification of nerves which in fishes were the most anterior of the spinal nerves.

The 12 cranial nerves of mammals are designated by numbers almost as commonly as they are referred to by name. Beginning with the most rostral they are: (I) olfactory, (II) optic, (III) oculomotor, (IV) trochlear, (V) trigeminal, (VI) abducens, (VII) facial, (VIII) acoustic, (IX) glossopharyngeal, (X) vagus, (XI) spinal accessory, and (XII) hypoglossal. In 9 to 12 mm. pig embryos all of the cranial nerves except the olfactory and optic are readily recognizable (Figs. 59–63, and 92). Those carrying sensory fibers show conspicuous ganglia near their point of connection with the brain (see nerves V, VII, VIII, IX, and X in Fig. 92). Except for the acoustic (VIII) all these ganglionated nerves carry also some motor fibers—that is, they are mixed nerves. The cranial nerves composed practically entirely of motor fibers have no external ganglia (nerves III, IV, VI, and XII, Fig. 92).

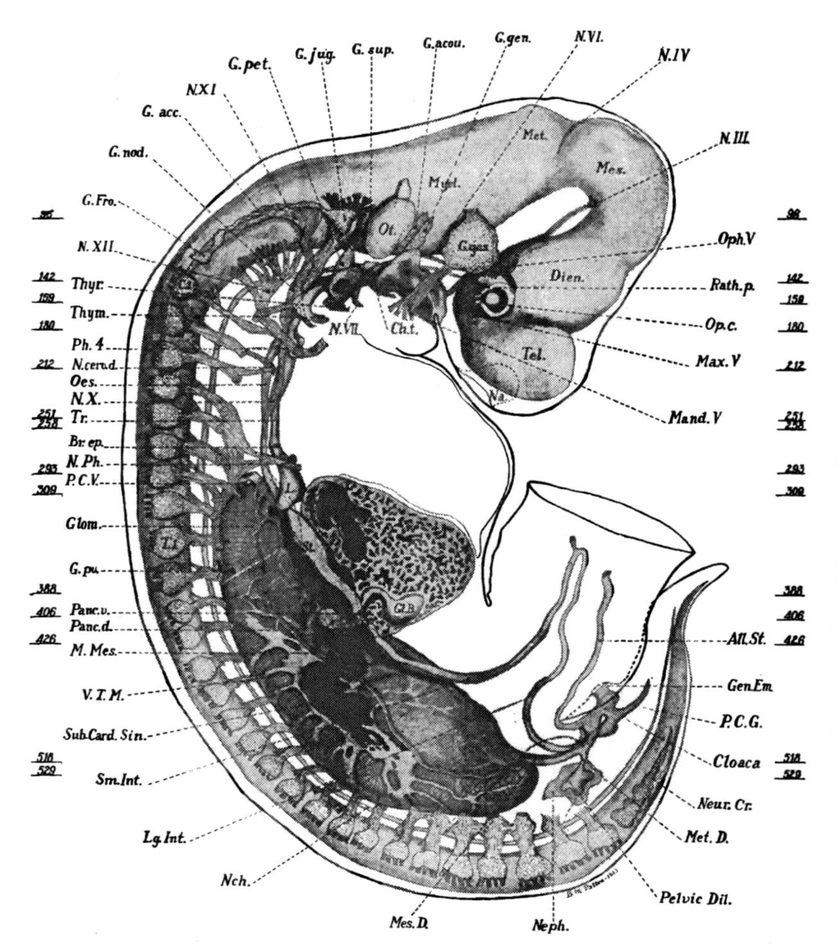

FIG. 60. Reconstruction (× 14) of the nervous, digestive, respiratory, and urinary systems of a 9.4 mm. pig embryo. Compare with Frontispiece and with figure 66. By laying a straight-edge across the numbered lines in the margin the locations of the accompanying cross-sections may be determined.

Abbreviations

All. St., allantoic stalk.

Br. ep., eparterial bronchus.

C.1., dorsal root ganglion of first cervical nerve.

FIG. 60—(*Continued*)

Abbreviations

Ch. t., chorda tympani branch of facial nerve.
Dien., diencephalon.
G.acc., accessory ganglion.
G.acou., acoustic ganglion of eighth nerve.
G.Fro., Froriep's ganglion.
G.gas., Gasserian (semilunar) ganglion of fifth nerve.
G.gen., geniculate ganglion of seventh nerve.
G.jug., jugular ganglion of tenth nerve.
G.nod., nodose ganglion of tenth nerve.
G.pet., petrosal ganglion of ninth nerve.
G.pv., prevertebral sympathetic ganglion.
G.sup., superior ganglion of ninth nerve.
Gen. Em. genital eminence.
Gl. B., gall bladder.
Glom., glomerulus.
L., lung.
Lg. Int., large intestine.
M. Mes., splanchnic mesoderm cut where reflected over mesonephros from mesentery.
Mand.V., mandibular branch of fifth (trigeminal) nerve.
Max.V., maxillary branch of fifth (trigeminal) nerve.
Mes., mesencephalon.
Mes. D., mesonephric duct.
Met., metencephalon.
Met.D., metanephric diverticulum.
Myel., myelencephalon.
N.III., third cranial (oculomotor) nerve.
N.IV., fourth cranial (trochlear) nerve.
N.VI., sixth cranial (abducens) nerve.
N.VII., seventh cranial (facial) nerve.
N.X., tenth cranial (vagus) nerve.
N.XI., eleventh cranial (accessory) nerve.
N.XII., twelfth cranial (hypoglossal) nerve.
N.cerv.d., descending cervical nerve.
N.Ph., phrenic nerve.
Na., location of nasal (olfactory) pit.
Nch., notochord.
Neur. Cr., neural crest.
Neph., nephrogenous tissue of metanephros.
Oes., esophagus.
Op.c., optic cup.
Oph.V., ophthalmic branch of fifth (trigeminal) nerve.
Ot., auditory vesicle (otocyst).
P.C.G., post-cloacal gut.
P.C.V., posterior cardinal vein.
Panc.d., dorsal pancreas.
Panc.v., ventral pancreas.
Pelvic Dil., pelvic dilation of metanephric diverticulum.
Ph.4., fourth pharyngeal pouch.
Rath. p., Rathke's pocket
Sm. Int., small intestine.
St., stomach.
Sub. Card. Sin., large venous sinus formed by the transverse anastomosis of the subcardinal veins (subcardinal sinus).
T.1., first thoracic spinal ganglion.
Tel., telencephalon (specifically the label is on the right lateral telencephalic vesicle).
Thym., thymus.
Thyr., thyroid.
Tr., trachea.
V.T.M., superficial mesonephric veins connecting post- and subcardinals (named venae transversales mediales by Sabin).

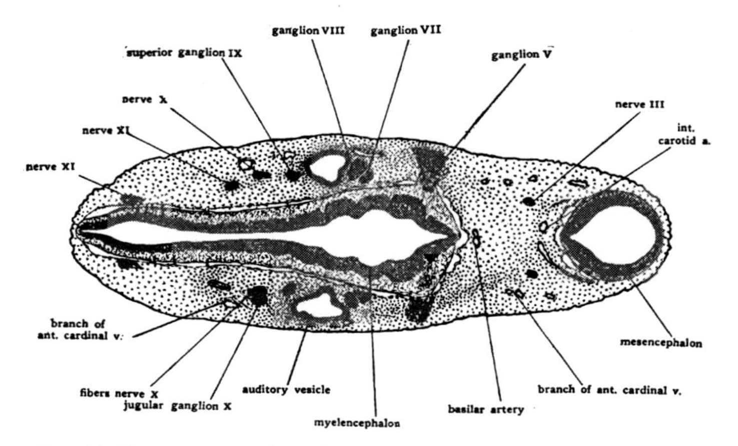

FIG. 61. Transverse section of 9.4 mm. pig embryo passing through the myelencephalic region (× 15). This and the following drawings of cross-sections were taken from the same series used in making the reconstructions which appear as figures 60 and 66. The serial number of this section is 96. Its location on the reconstructions may be determined by laying a straight-edge on the marginal lines numbered 96.

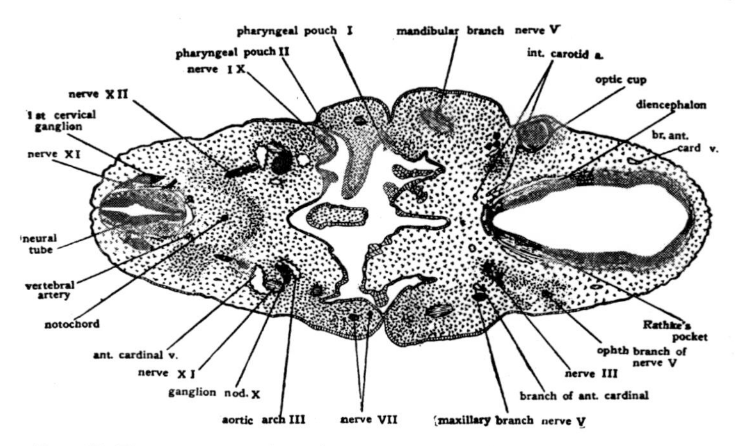

FIG. 62. Transverse section of 9.4 mm. pig embryo passing through the pharynx (× 15). (Its serial number on reconstructions appearing as figures 60 and 66 is 142.)

The Spinal Cord and Spinal Nerves. The caudal part of the myelencephalon merges without any definite line of demarcation into the spinal cord. The walls of the neural tube in the cord region have already begun to become differentiated. Dorsally and ventrally they remain thin, but laterally rapid proliferation of primordial nerve cells has caused them to increase greatly in thickness so that the originally oval lumen of the tube becomes slit-like (cf. Figs. 42, B–D, and 83).

The cells arising from the neural crests (Fig. 35) have become aggregated on either side of the cord into segmentally arranged clusters, the spinal ganglia (Fig. 59). From nerve cells in each of these ganglia, fibers grow both toward the cord and peripherally, establishing the *dorsal root* (afferent root) (sensory root) of a spinal nerve (Figs. 73, 84, and 85). The *ventral root* (efferent root) (motor root) of a spinal nerve is composed of fibers which grow out from cells lying in the wall of the neural tube (Fig. 84). Outside the cord the dorsal and the ventral root unite to form a spinal nerve trunk.

Immediately distal to the union of the dorsal and the ventral root of a spinal nerve, the nerve trunk breaks up into three main branches: a *dorsal ramus* carrying the fibers associated with the dorsal part of the body, a *ventral ramus* composed of fibers terminating in ventral parts

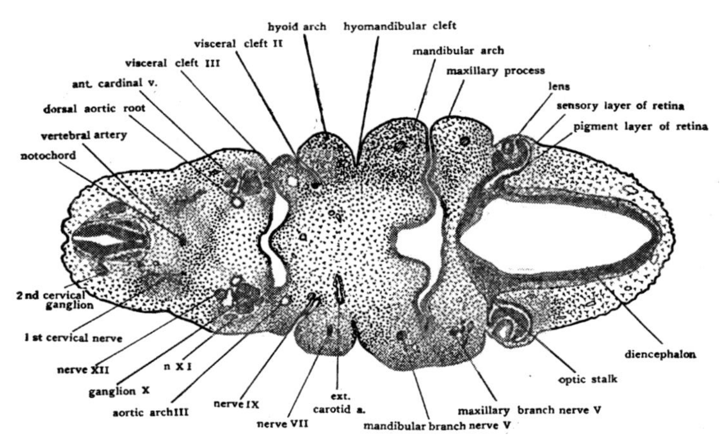

FIG. 63. Transverse section of 9.4 mm. pig at the level of the eyes (× 15). (Serial number on reconstructions, 159.)

of the body, and a *ramus communicans* containing the fibers which extend by way of the prevertebral sympathetic chain to the viscera (Fig. 85).

Primarily the spinal nerves are strictly metameric in arrangement, each nerve carrying the sensory fibers from, and the motor fibers to, that segment of the body in which it arises. But the underlying metamerism of the body in the adult is greatly modified, almost obliterated in many regions, by such processes as the fusion of primordial tissues from several metameres to form new, highly specialized structures, or by the migration of entire organs from their place of origin to new positions in the body. Since the spinal nerves arise very early in development and are at first associated with structures at their own metameric level, their final arrangement constitutes a valuable record of evolutionary and developmental history. The appendages, for example, arise by the coalescence and organization of primordial tissue from several adjacent metameres. The corresponding spinal nerves innervate these tissues. Originally entirely separate, these nerves merge peripherally to form the nerves to the appendages. But the story of polymetameric development is permanently recorded by the series of nerve roots which retain their independent origin from the spinal cord in spite of their peripheral fusion in the brachial and sacral plexuses.

Furthermore, the caudal migration of the appendages during development is clearly evidenced by the fact that their nerves arise from the cord at a more cephalic level than that occupied by the appendages themselves. (Note the involvement of cervical nerves in the formation of the brachial plexus, Fig. 60.) Similarly the caudal migration of the diaphragm from its place of first appearance at what is destined to be the level of the neck, is indicated by the cervical origin from the cord of the phrenic nerve to the diaphragm. In a pig embryo of about 10 mm. the phrenic nerve (Fig. 60) can be seen extending directly from its level of origin toward the septum transversum (Fig. 59) which is the beginning of the diaphragm. Later in development, as the diaphragm moves caudad, the terminal portion of the phrenic nerve will be pulled caudad with it, constituting a permanent record of the migration of the diaphragm.

The Sense Organs. Although the cranial nerves associated with the nose and the eye are not as yet well developed in 10 mm. pig embryos, the primordia of the sense organs themselves are established. The olfactory organs are represented by a pair of depressions situated at the rostral end of the head (Figs. 58, 168, and 169). From specialized cells in the ectodermal lining of these nasal pits, nerve fibers will later arise and grow into the telencephalon establishing the olfactory nerves.

The single-walled spheroidal optic vesicle seen in younger embryos (Fig. 36, E) has been converted by invagination of its distal portion into a double-walled cup (Figs. 41, D, and 63). The inner layer of the optic cup is already much thickened, foreshadowing its development into the highly specialized sensory layer of the retina. The outer layer remains thinner and becomes the pigmented layer of the retina. The invagination of the primary optic vesicle to form the optic cup takes place eccentrically. As a result the lip of the cup is not, at first, complete. It shows a ventral gap, called the *choroid fissure*, which does not close until much later in development (Figs. 47 and 60, choroid fissure shown but not labeled; Fig. 41, D, section cut somewhat on a slant so that it passes through the choroid fissure in one eye but not in the other). The choroid fissure is continued as a groove on the ventral surface of the optic stalk. When the fibers which constitute the optic nerve grow from cells in the sensory layer of the retina to the brain, they pass along this groove in the optic stalk.

While these changes have been occurring in the optic vesicle, the lens has been established by invagination of the superficial ectoderm overlying the optic cup (Fig. 41, D). By the time the embryo has attained a length of 10 mm., the lens has been completely separated from the parent ectoderm and appears as a spheroidal vesicle lying in the opening of the optic cup (Figs. 60 and 63).

The primordium of the internal ear mechanism is as yet very simple in form. It makes its first appearance as a local thickening of the superficial ectoderm overlying the hind-brain. This thickened area, the auditory placode, then sinks in to form a pit (Figs. 30 and 36, E) which soon becomes closed over to form the auditory (otic) vesicle. By 10 mm. the auditory vesicle has entirely lost all connection with the ectoderm from which it was derived. The only indication of its origin is a slender stalk, the endolymphatic duct, which extends dorsally toward the site of the original invagination.

Although the nerve connections of the auditory apparatus with the brain have not yet been definitely established, they are clearly indicated by the nerve fibers growing from cells in the acoustic ganglion to the brain on the one hand, and toward the otic vesicle on the other. Of significance, also, is the close proximity of the first pharyngeal (hyomandibular) pouch to the otic vesicle (Fig. 60).

This pouch is destined to give rise to the middle ear chamber and the Eustachian tube.

III. The Digestive and Respiratory Organs

The Oral Region. The digestive system of 10 mm. pigs still shows most of the primary landmarks seen in younger embryos. The stomodaeal depression has broken through into the fore-gut establishing the oral opening, but a small diverticulum called Seessel's pocket persists as a vestige of the pre-oral gut (cf. Figs. 37, 40, and 65). Seessel's pocket is of no especial interest in itself for it gives rise to no adult structure. In embryos of this age, however, it is a valuable landmark indicating as it does precisely the point at which stomodaeal ectoderm and fore-gut entoderm became continuous when the oral plate ruptured.

The part played by the growth of nasal, maxillary, and mandibular processes in deepening the original stomodaeal depression and in the formation of the face and jaws is discussed elsewhere (Chap. 13). But mention should be made here of another stomodaeal structure called Rathke's pocket. Rathke's pocket arises in the mid-line as a slender diverticulum of stomodaeal ectoderm growing toward the infundibulum (Fig. 65). Later in development it separates from the ectoderm and its deep portion becomes fused with the infundibulum to form an endocrine gland known as the hypophysis.

The Pharynx. Caudal to the oral opening, the fore-gut becomes very broad and considerably flattened dorso-ventrally to form the pharynx. A series of four pairs of pocket-like diverticula, the pharyngeal pouches, arise from it laterally (Fig. 60). Each pharyngeal pouch is situated opposite one of the external gill furrows, representing, as it were, an abortive attempt at establishing an open gill cleft. In mammalian embryos this process ordinarily stops just short of completion, the gill clefts remaining closed by a thin membrane (Figs. 41, B, and 62).

The Trachea and Lung Buds. In the floor of the pharynx at the level of the most posterior pair of pharyngeal pouches, a median ventral groove appears which is rapidly converted into a tubular outgrowth parallel to the digestive tract. This groove is the tracheal (laryngo-tracheal) groove, and the tubular outgrowth which is formed by its prolongation caudad is the trachea. In 10 mm. embryos the caudal end of the trachea has become enlarged and bifurcated to form the lung buds. Thus the original evagination from the pharynx is the

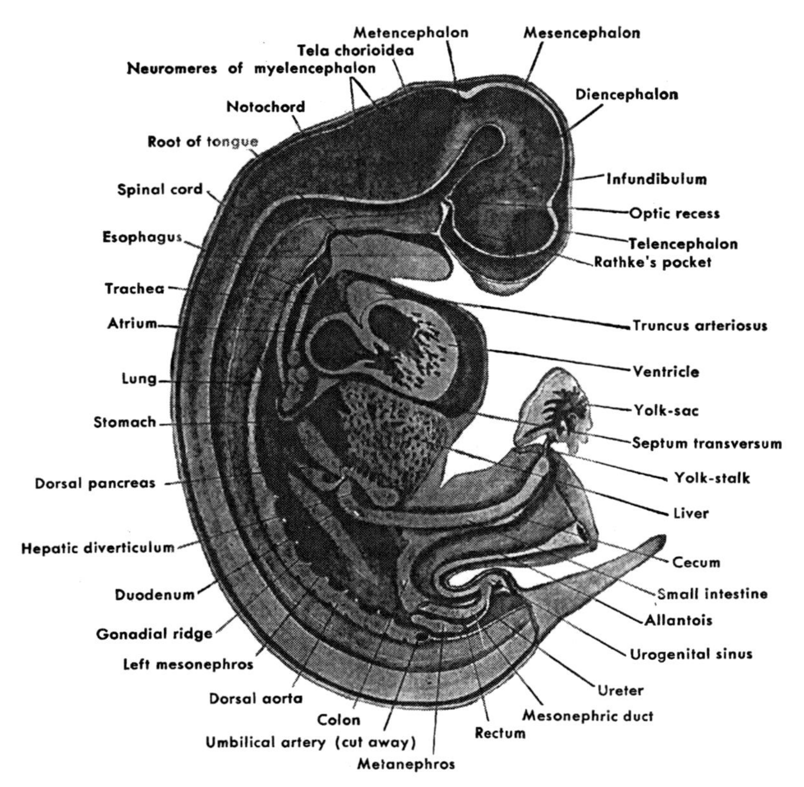

FIG. 64. Model of 10 mm. pig embryo dissected to the sagittal plane. (After Prentiss.)

primordium of larynx, trachea, bronchi, and lungs, but for the sake of brevity it is often referred to as "the lung bud" (Figs. 40, 60, 65, and 69–74).

The Esophagus and Stomach. Posterior to the pharynx the digestive tube is distinctly narrowed to form the esophagus (Figs. 65 and 69–74). A marked local dilation, already of suggestive shape, indicates the beginning of the stomach (Figs. 59, 60, 64, and 67).

The Liver and Pancreas. Immediately caudal to the stomach are the outgrowths of the gut which constitute the primordia of the pancreas and of the liver and gall-bladder. The pancreas at this stage consists of two independent parts, a large dorsal bud and a small ventral bud (Figs. 60, 76, and 104). The original hepatic diverticulum (Figs. 40 and 103, A) has given rise to a very extensive mass of glandular tissue which is crowded ventrally and cephalically from its point of origin to constitute the liver (Figs. 59, 64, and 65). The narrowed proximal portion of the original evagination from the gut persists as the duct draining the liver, and a diverticulum of it becomes enlarged to form the gall-bladder (Fig. 104).

The Intestines. The elongation of the intestines which later results in their characteristic coiling has just commenced in 10 mm. embryos. The gut has become relatively thinner than in earlier stages and protrudes into the belly-stalk in the form of a slender U-shaped loop (cf. Figs. 37, D, 40, 64, and 65). Communicating with the gut at the apex of the loop is the yolk-stalk. The yolk-stalk by this time has become greatly attenuated and the yolk-sac with which it communicates is reduced to a shriveled vesicle embedded in the belly-stalk (Figs. 64 and 65).

In some of the older embryos in the 9 to 12 mm. range the U-shaped bend of the gut has been twisted to form a loop and a slight enlargement of the gut just caudal to the yolk-stalk will be found indicating the beginning of the cecum (Fig. 64). Cephalic to this enlargement the gut will become small intestine, and caudal to it, large intestine.

The Cloaca. The dilated caudal end of the gut where the allantoic stalk and the mesonephric ducts enter is called the *cloaca* (Fig. 60). It is in this region that the proctodaeal depression ruptures into the gut establishing its posterior opening to the outside. In 10 mm. embryos the tissue intervening between the gut and the proctodaeum is never thick and usually shows definite signs of impending disintegration if not an actual rupture (Fig. 65). Caudal to the level of the proctodaeum a variable portion of the hind-gut persists for a time as the so-called post-cloacal (post-anal) gut (Fig. 60).

IV. The Coelomic Cavity

When first established the coelom consisted of paired cavities bounded by the splanchnic and somatic layers of the lateral mesoderm (Fig. 108, A). With the folding off of the body from the extra-embryonic membranes and the closure of the embryonic body ventrally, the primary right and left coelomic chambers are carried toward each other in the mid-line (Fig. 108, B–D). In this process the

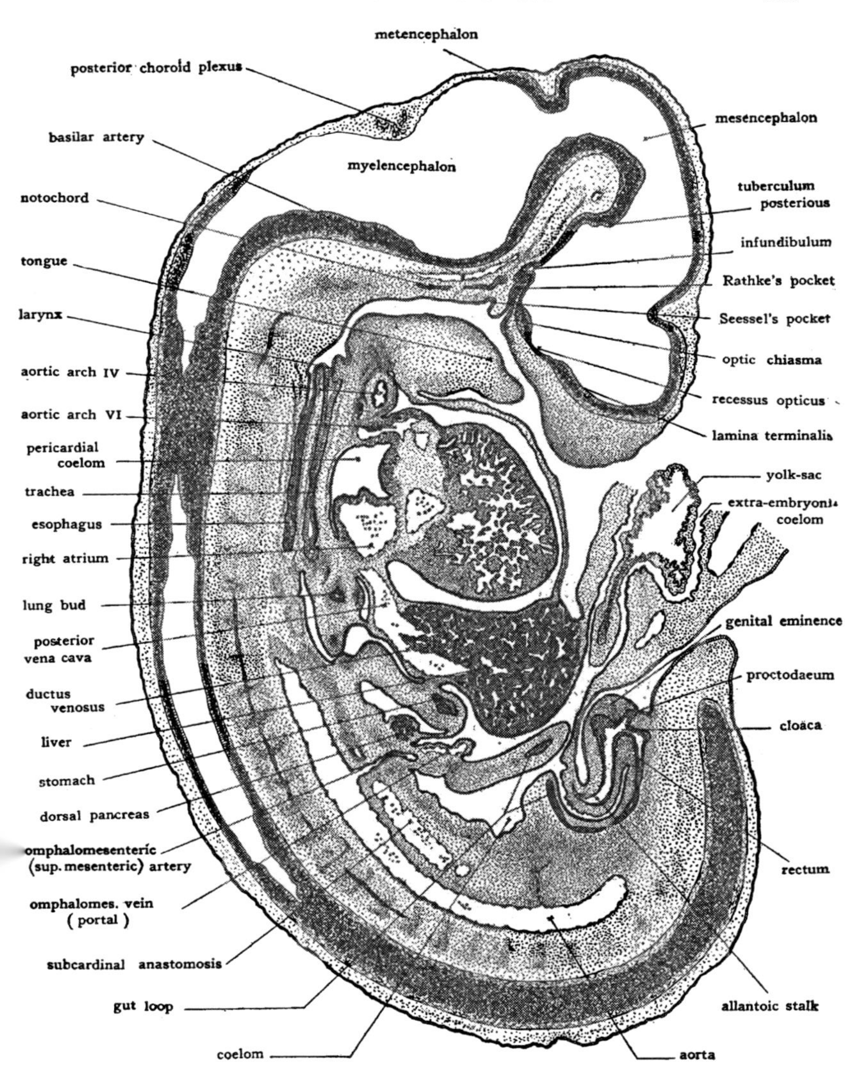

FIG. 65. Sagittal section of 10 mm. pig embryo (× 16).

gut tract is caught between the layers of splanchnic mesoderm which form the mesial boundaries of the coelomic chambers. The double layer of splanchnic mesoderm thus formed serves as a supporting membrane for the gut and is known as the mesentery. Shortly after its formation the part of the mesentery ventral to the gut breaks through, bringing the right and left coelomic chambers into confluence and thus establishing a single body cavity within the embryo (Fig. 108, F).

Only in the mid-body region do these changes take place exactly as described above. While they are essentially similar elsewhere there are certain local modifications of the process which are of special interest. At the level of the pancreas and liver the ventral mesentery persists, supporting the liver (Figs. 108, E, and 111). Where the extra-embryonic membranes are continuous with the embryo at the belly-stalk, the body cavity remains for a long time continuous with the extra-embryonic coelom (Figs. 108, C, and 111). In the cardiac region the digestive tract develops in the dorsal body-wall so that no dorsal mesentery is formed. Here the heart is formed ventral to the gut and suspended in the coelom by a double layer of splanchnic mesoderm in a manner quite suggestive of that in which the liver is suspended in the ventral mesentery farther caudally in the body (cf. Figs. 43, D, and 108, E).

In 10 mm. pig embryos the body cavity is not yet divided into separate pericardial, pleural, and peritoneal chambers. Between the liver and the heart, however, there has appeared a shelf-like structure which partially separates the thoracic region of the coelom from the abdominal. This incomplete partition is called the *septum transversum*. At this stage it is an ingrowth of ventral body-wall tissue fused to the cephalic face of the liver (Fig. 64). The septum transversum itself never extends all the way across the coelom. Later in development we shall see it supplemented by the pleuroperitoneal folds which arise from the dorsal body-wall, and complete the diaphragmatic partition across the coelom.

V. The Urinary System

The Pronephros. In mammalian embryos the pronephros is a vestigial organ appearing only transitorily in the form of a few rudimentary tubules. In 10 mm. pig embryos even these tubules have almost completely disappeared. The paired ducts which originally appeared in connection with the pronephric tubules persist, however, and are appropriated by the developing mesonephric tubules. After forming this new association they are called the mesonephric (Wolffian) ducts.

The Mesonephros. In younger embryos we saw the development of mesonephric tubules from the intermediate mesoderm and the manner in which they attained connection with the mesonephric duct (Figs. 38, E, F, 40, and 41, G, H). In pigs of 10 mm. mesonephric tubules have been formed in great numbers, and each tubule has become much elongated and exceedingly tortuous. As a result the mesonephros becomes an organ of great bulk—in fact the most conspicuous organ in the body of an embryo of this age (Figs. 59 and 60).

Sections passing through the mesonephros (Figs. 74–79) convey a very vivid impression of the interwoven mass of tubules of which it is composed. Details as to the shape of the tubules and the relations of the blood vessels to them can best be considered in connection with the development of the urogenital system (Chap. 10).

In the more cephalic part of the mesonephros it is not always easy to distinguish the duct from the tubules. Farther caudally the duct can easily be identified lying along the ventral border of the mesonephros. After leaving the substance of the mesonephroi, the mesonephric ducts curve ventro-mesially to enter the cloaca together with the allantoic stalk (Figs. 60 and 122).

The Metanephros. Just cephalic to the cloacal end of the mesonephric duct an outgrowth arises from it and extends antero-dorsad. This is the metanephric diverticulum (Figs. 60 and 122). The terminal portion of this diverticulum becomes dilated, presaging its ultimate fate as the pelvic cavity of the kidney. Its proximal portion remains slender as the ureter. About the enlarged pelvic end of the metanephric diverticulum a mass of mesoderm accumulates. Because it is destined to form the excretory tubules of the permanent kidney, this mass of mesoderm is known as the nephrogenous tissue of the metanephros (Figs. 60 and 79). Like the cell clusters which form the mesonephric tubules it arises from intermediate mesoderm.

VI. The Circulatory System

A functionally competent circulatory system is laid down before the developing pig has reached a length of 10 mm. Paired primordia fuse to establish the heart as a median tubular organ receiving the blood at its posterior end and pumping it out from its cephalic end (Figs. 43 and 44). The aortae arise as paired vessels which lead away from the heart, swing around to the dorsal side of the pharynx, and

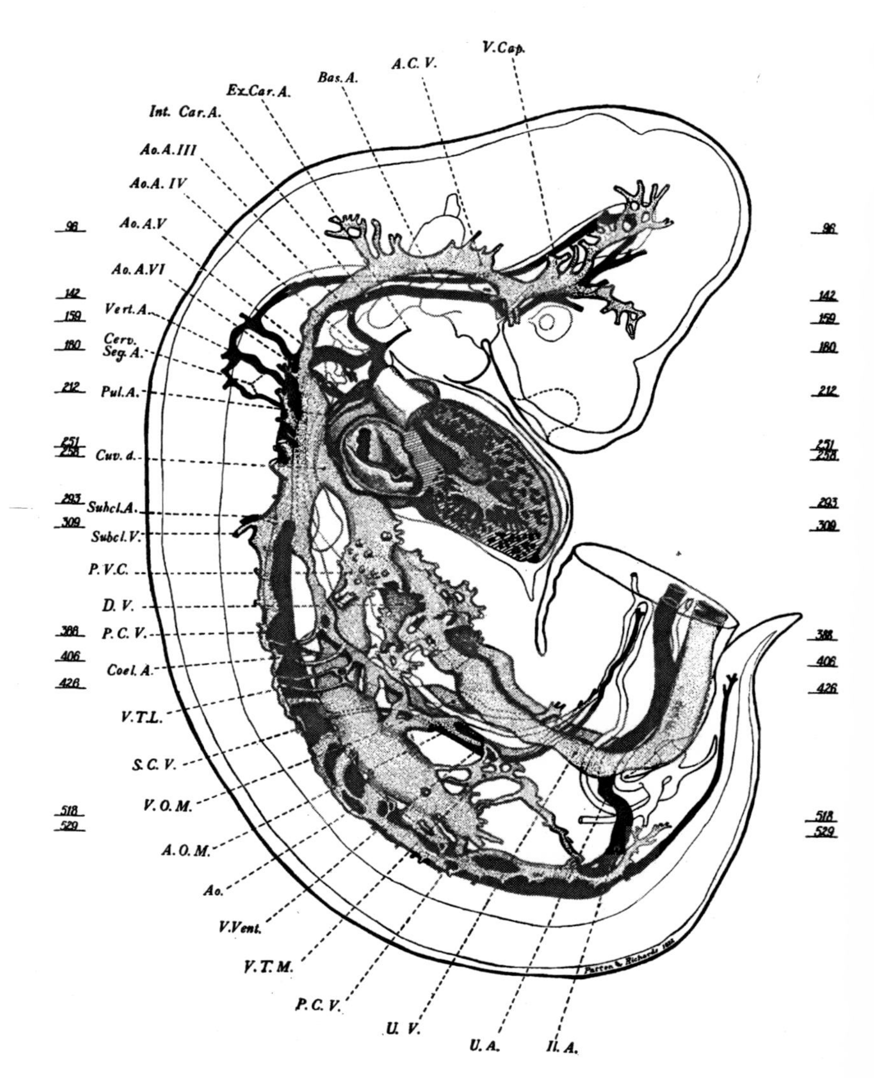

FIG. 66. Reconstruction (× 14) of the circulatory system of a 9.4 mm. pig embryo. By laying a straight-edge across the numbered lines in the margin, the locations of the accompanying cross-sections may be determined.

then extend caudad throughout the length of the embryo as the main distributing channels (Fig. 45). Large collecting vessels develop in the shape of the paired cardinal veins which receive the blood from the anterior and posterior regions of the body and return it over the ducts of Cuvier (common cardinal veins) to the sino-atrial part of the heart (Fig. 45). Pigs in the 9 to 12 mm. range still retain to a large extent this primary bilaterally symmetrical arrangement of the main vessels which is so characteristic of young embryos. There have been, however, many local elaborations and modifications which are of special interest because they initiate the series of changes which lead toward adult conditions.

The Heart. The chief factor in changing the configuration of the heart from its primitive straight tubular shape to the condition seen in 10 mm. embryos is its own rapid elongation and consequent bending. In this process the cephalic end of the heart remains anchored by the aortic roots and its caudal end by the omphalomesenteric veins. Thus the receiving and discharging ends of the heart suffer no radical change in position. The intervening portion of the heart is, however, bent into a loop which is carried first ventrally and then caudally to form the pumping part of the heart or ventricle (Figs. 142 and 143). The details involved in the bending of the heart tube and the special

---

FIG. 66—(*Continued*)

Abbreviations

A.C.V., anterior cardinal vein.
Ao., aorta.
Ao.A., aortic arch.
A.O.M., omphalomesenteric artery.
Bas.A., basilar artery.
Cerv. Seg. A., intersegmental branches of aorta in cervical region.
Coel. A., celiac artery.
Cuv.d., common cardinal vein (duct of Cuvier).
D.V., ductus venosus.
Ex. Car. A., external carotid artery.
Il.A., iliac artery.
Int. Car. A., internal carotid artery.
P.C.V., posterior cardinal vein.
Pul.A., pulmonary artery.
P.V.C., posterior vena cava.
S.C.V., subcardinal vein.
Subcl.A., subclavian artery.
Subcl.V., subclavian vein.
U.A., umbilical (allantoic) artery.
U.V., umbilical (allantoic) vein.
V.Cap., vena capitis (tributary of anterior cardinal vein).
V.O.M., omphalomesenteric (portal) vein.
V.T.L., lateral transverse veins of mesonephros.
V.T.M., medial transverse veins of mesonephros.
V. Vent., ventral vein of mesonephros.
Vert. A., vertebral artery.

For cardiac structures refer to figure 145.

features of its internal structure can best be considered in connection with later stages of development when their significance will be more apparent (see Chap. 11). At present we are chiefly interested in becoming acquainted with the more outstanding structural features of the embryonic heart.

The great veins converging to enter the heart become confluent in a thin-walled chamber called the sinus venosus (Fig. 144). The sinus venosus opens into the atrial portion of the heart by a slit-like orifice guarded against return flow by well-developed flaps known as the valvulae venosae (Figs. 46, 145, 146, and 147).

The atrial region has undergone extensive transverse enlargement so that it bulges out into pouch-like right and left chambers (Figs. 71, 72, 142, 144, and 147). Although the beginning of the separation of these chambers from each other is clearly indicated by the presence of an interatrial septum, this septum is not complete, and the atrial chambers remain in communication through a secondary perforation in the septum called the *interatrial foramen secundum* (*ostium II*) (Figs. 71, 145, and 148).

Leaving the atrium, the blood passes through a constricted region known as the *atrio-ventricular canal*. Previously a single channel (Fig. 147, A), in 10 mm. embryos this canal has, as was the case with the atrium, become more or less completely divided into right and left channels. The division is effected by a pair of plastic masses of mesenchymal tissue, the so-called *endocardial cushions of the atrio-ventricular canal*. Located one dorsally and one ventrally (Fig. 39), these cushions grow together and fuse to divide the atrio-ventricular canal (Figs. 147 and 148). Except for the least advanced specimens their fusion will ordinarily have occurred in embryos of the 9–12 mm. stage.

In the ventricle, also, there are indications of the impending separation of the heart into right and left sides. The interventricular septum has appeared as a well-marked median ridge extending from the apex of the ventricular loop toward the atrio-ventricular canal (Figs. 71, 72, 147, and 148). Above this septum the two parts of the ventricle are still in open communication.

Correlated with its activity in pumping, the ventricular wall has become greatly thickened. Irregular branching bands of developing muscle tissue protrude from the main part of the wall into the lumen. These *trabeculae carneae* already suggest the muscular bands which project so characteristically into the cavities of the adult ventricles.

From the ventricle the blood passes into the truncus arteriosus

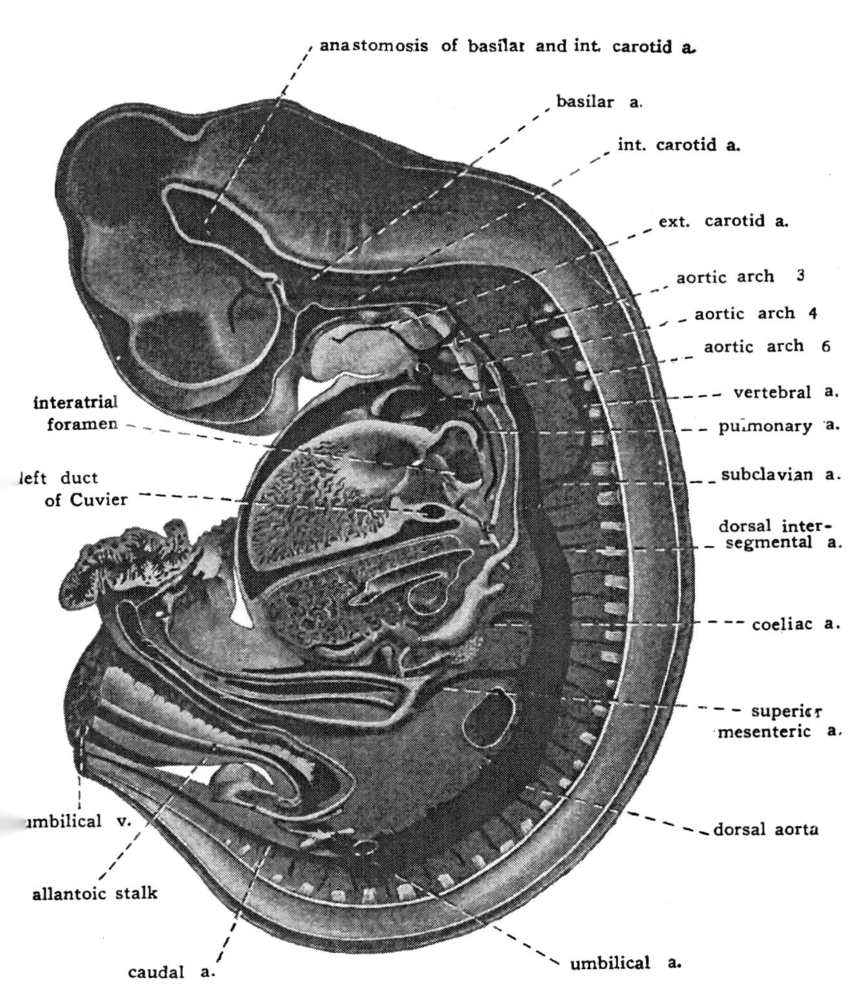

FIG. 67. Reconstruction of 12 mm. pig embryo showing the relation of the main arterial trunks to the viscera. (From Minot after Lewis.)

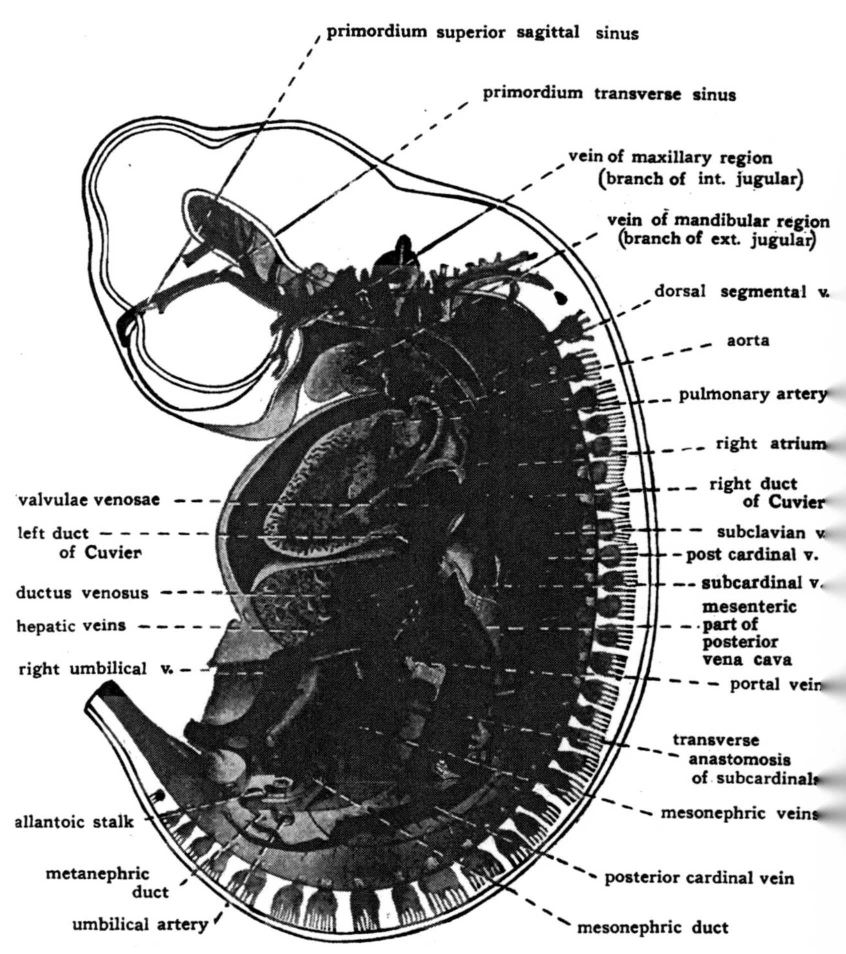

FIG. 68. Reconstruction of 12 mm. pig embryo showing the relation of the main venous channels to the viscera. (From Minot after Lewis.)

and thence out to the body by way of the ventral aortic roots. Aside from the marked thickening of its walls, the truncus arteriosus shows little change from its original condition as the anterior part of the heart tube. Its diameter remains small and the longitudinal division it is destined to undergo later in development is barely suggested by the irregular shape of its lumen seen in cross-sections (Fig. 70).

It should not be inferred from the modifications which have occurred in the different regions of the heart that it has as yet altered its primitive method of functioning. The heart tube has become bent and shows local dilations and constrictions which we name because of their future fate. Many internal conditions point toward its division into right and left sides. But the blood enters the heart posteriorly by way of the sinus venosus, is collected in the atria, and passes into the ventricle whence it is pumped out by way of the truncus arteriosus as an undivided stream, just as was the case in younger embryos where the heart was still a straight tube.

The Aortic Arches. In vertebrate embryos six pairs of aortic arches are formed extending around the pharynx from the ventral to the dorsal aorta. Of these the most cephalic are the first to appear and the other pairs are formed in sequence caudal to the first (Figs. 134 and 136). In pig embryos of 10 mm. the two most cephalic arches have already degenerated. The functional arches at this stage are the third, fourth, and sixth (Figs. 66, 67, 134, E, and Frontispiece). The fifth

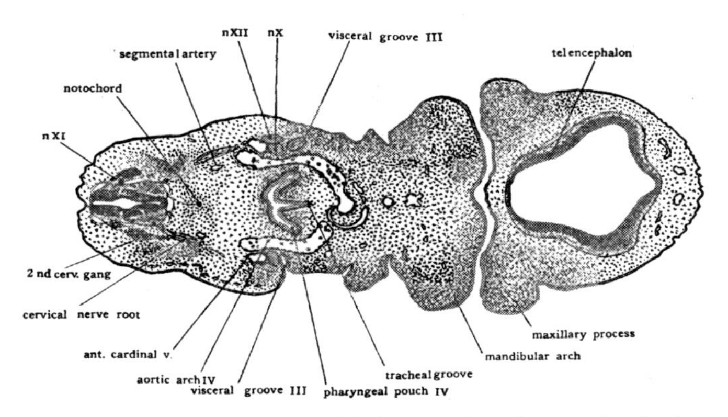

FIG. 69. Transverse section of 9.4 mm. pig embryo at level of posterior part of pharynx (× 15). (The serial number of this section on the reconstructions appearing as figures 60 and 66 is 180.)

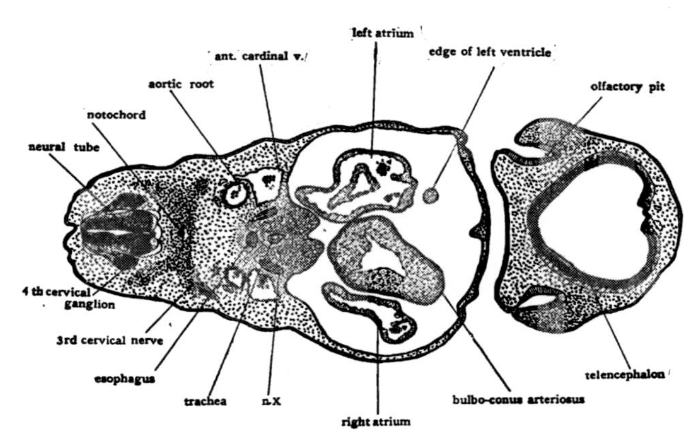

FIG. 70. Transverse section of 9.4 mm. pig embryo through telencephalon and cephalic part of pericardial chamber (× 15). (Serial No. 212.)

arch is always poorly developed in mammals and usually appears only as a small collateral channel appended sometimes to the fourth, but more commonly to the sixth arch (Fig. 66). From the sixth arches small branches (the pulmonary arteries) extend caudad to the developing lungs (Frontispiece).

Arteries of the Cephalic Region. The portions of the ventral aortic roots which led to the two anterior aortic arches do not disappear when the arches themselves degenerate but persist as the external carotid arteries (Figs. 66 and 67). The dorsal aortae are prolonged cephalad as the internal carotid arteries (Figs. 45, 66, and 67).

Throughout the length of the aorta small branches appear at regular intervals and extend dorsad on either side of the neural tube. Since these vessels are formed between adjacent somites they are known as the intersegmental arteries (Figs. 45, 67, and 159). In the cervical region the intersegmental vessels form a series of connections with each other which eventually result in the establishment of longitudinal vessels dorsal to, and parallel with, the aortae. These are the vertebral arteries (Fig. 67). This longitudinal anastomosing between the cervical intersegmental arteries appears cephalically first (Fig. 66) and then progresses caudad. As the vertebral arteries become established, the more cephalic intersegmental branches feeding them disappear. Only the most caudal of the intersegmentals concerned in the formation of the vertebral arteries persists. Since this intersegmental

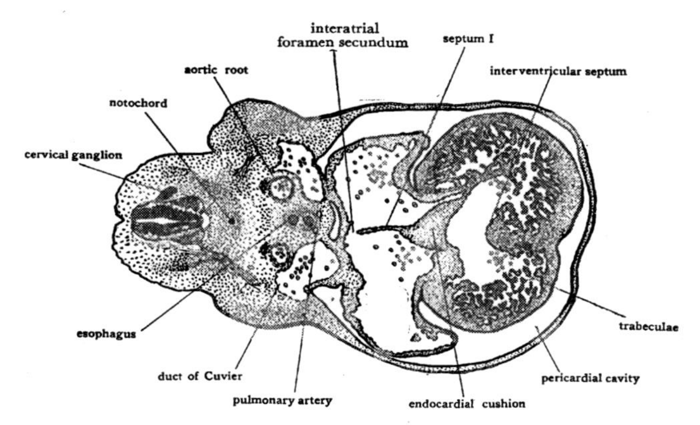

FIG. 71. Transverse section of 9.4 mm. pig embryo through heart at level of interatrial foramen secundum (× 15). (Serial No. 251.)

artery (seventh cervical) is at the same time the one situated at the level of the anterior appendage bud and consequently the vessel which is enlarged with the growth of the appendage to form the subclavian artery, the vertebral artery eventually appears as a branch of the subclavian (Figs. 67 and 133, B, C).

Cephalic to the cervical level the vertebral arteries bend toward the mid-line and unite with each other to form a median vessel lying ventral to the myelencephalon. This is the basilar artery (Figs. 65, 66, and 67). Ventral to the cephalic flexure in the neural tube, the internal carotid arteries send branches mesiad to unite with the basilar (Figs. 67 and 133). This anastomosis between the internal carotid and the basilar is the first step in the formation of the arterial circle (circle of Willis) which is such a conspicuous landmark in the adult anatomy of the hypophyseal region.

The Dorsal Aorta and Its Branches. When first formed the dorsal aorta is a paired vessel. This paired condition is retained in the branchial region, but posteriorly the two primitive aortae soon fuse with each other to form a median vessel. The fusion first occurs in the mid-body region (Fig. 51) and extends thence cephalad to about the level of the anterior appendage buds and caudad throughout the length of the aorta (Fig. 67).

In young embryos the most conspicuous vessels arising from the dorsal aorta are the omphalomesenteric trunks which are prolonged

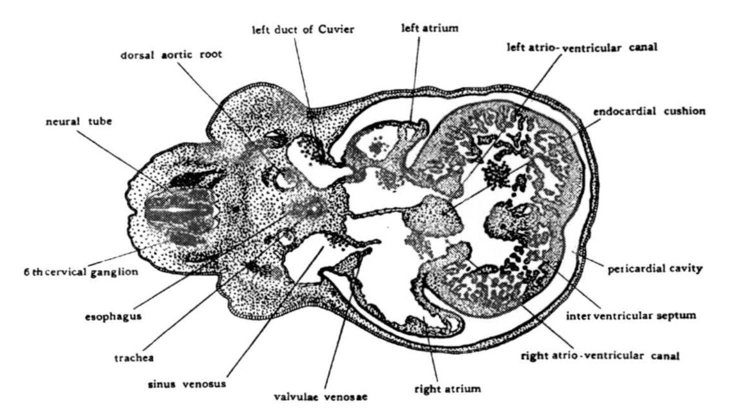

FIG. 72. Transverse section of 9.4 mm. pig embryo through heart at level of atrio-ventricular canals (× 15). (Serial No. 258.)

as the vitelline arteries to the yolk-sac, and the allantoic or umbilical arteries to the vascular plexus of the allantois (Figs. 45 and 51). Both these vessels arise from the aorta before its fusion and are themselves paired. The umbilical arteries retain their paired condition (Fig. 79). When the body is closed ventrally, the right and left omphalomesenteric roots are brought together in the mid-line and fuse with each other to form a median vessel running in the mesentery. With the early degeneration of the yolk-sac, this vessel becomes relatively less conspicuous and is known as the anterior (superior) mesenteric artery (Figs. 66 and 67). Its original relations are, nevertheless, apparent from its course along the intestinal loop into the belly-stalk to the place where the small yolk-sac still retains its attachment to the gut.

Somewhat cephalic to the anterior mesenteric artery, the celiac artery arises from the aorta and extends in the mesentery toward the gastric region of the gut tract (Fig. 67). In the adult, the celiac, anterior mesenteric, and posterior mesenteric arteries constitute a group of vessels which one naturally thinks of together because of their similar ventral origin from the aorta, their course through the mesenteries, and their termination in the gastro-intestinal tract. The third member of this enteric group, the posterior (inferior) mesenteric cannot ordinarily be found in pigs of 9 to 12 mm. It arises from the

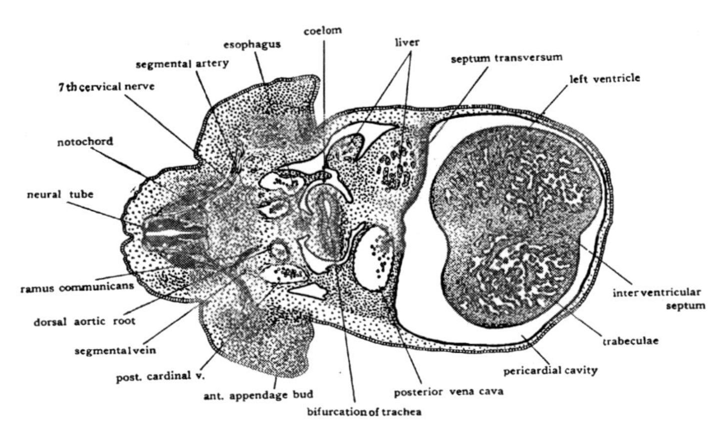

Fig. 73. Transverse section of 9.4 mm. pig embryo at level of tracheal bifurcation (× 15). (Serial No. 293.)

aorta, caudal to the other two vessels, at a slightly more advanced stage of development.

At the level of the mesonephros the aorta gives off, in addition to the series of dorsal intersegmental branches, many small branches which extend ventrally. These vessels feed the capillary plexuses (glomeruli) in the dilated ends of the mesonephric tubules and the network of capillaries which surround the tubules themselves (Fig. 117). Individually these branches are very small, but the volume of blood they handle collectively is surprisingly large as evidenced by the size of the veins (post- and subcardinals, Fig. 68) which drain the mesonephroi.

The Anterior Cardinal Veins. In 9 to 12 mm. embryos little alteration from primitive conditions (Fig. 47) has occurred in the veins of the cephalo-thoracic part of the body. Numerous large tributary vessels have appeared, especially in the cephalic region where they converge on either side of the head as the so-called venae capitis (Fig. 66). It is already possible to recognize in the larger of these branches the primordial vessels from which the main venous sinuses of the adult cranial region are derived (Fig. 68). Fundamentally, however, these veins are but an elaboration of the original anterior cardinal system. From them the blood passes caudad along the less modified portion of the anterior cardinals to enter the heart by way of

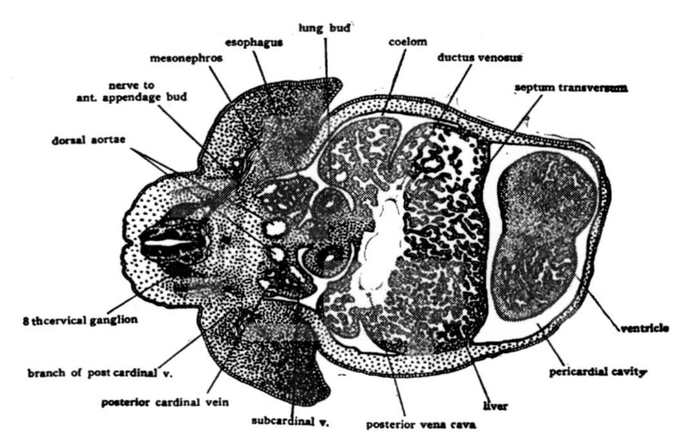

FIG. 74. Transverse section of 9.4 mm. pig embryo through cephalic part of liver (× 15). (Serial No. 309.)

the common cardinal veins (Fig. 66). Just before the anterior cardinal vein enters the common cardinal vein a series of small tributaries (dorsal segmental veins, Fig. 68) return to it the blood distributed by the intersegmental arteries to the cervical region. Near this point, also, a well-developed branch brings back the blood from the mandibular region. The vessel which thus returns the blood distributed by the external carotid artery is the beginning of the external jugular vein (Fig. 68). The anterior cardinal vein itself is later known as the internal jugular.

The Posterior Cardinal Veins. In very young embryos (Fig. 45) the posterior cardinal veins are the only conspicuous venous channels in the caudal half of the body. In 9 to 12 mm. pigs these veins have already begun to degenerate. Their relative position as vessels lying dorsal to the mesonephroi remains unchanged, but much of the blood formerly returned by them now reaches the heart over new channels (cf. Figs. 45, 66, and 139, A–F). As a result the posterior cardinal veins in the mid-mesonephric region have been interrupted. Toward the heart from this point the old channels persist, although they are much reduced in size. Caudal to their interruption the posterior cardinals rapidly degenerate as main channels (Figs. 66 and 139, D, F).

The Subcardinal Veins. The diversion of the blood from the posterior cardinals is brought about by the development of a col-

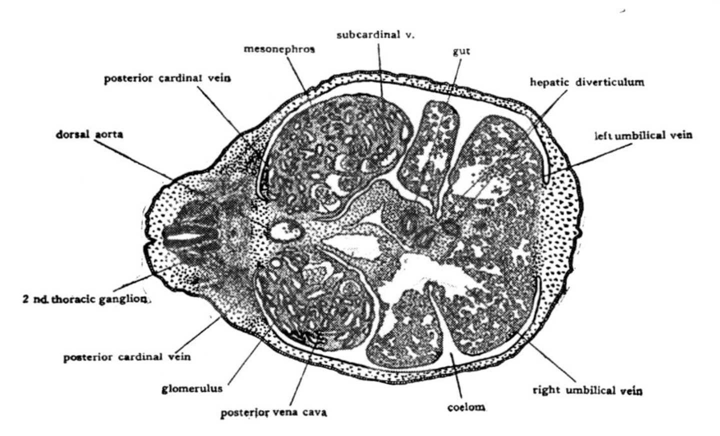

FIG. 75. Transverse section of 9.4 mm. pig embryo just cephalic to attachment of belly-stalk (× 15). The belly-stalk and the tip of the tail have not been included in the drawing. (Serial No. 388.)

lateral system of veins in the mesonephroi. When it first appears, this system of vessels is but an irregular plexus tributary to the postcardinals (Fig. 139, A). The organization of longitudinal channels in these plexuses establishes the main subcardinal veins as vessels extending cephalad in the ventro-mesial border of the mesonephroi, parallel with and ventral to the posterior cardinal veins. In the cephalic part of the mesonephros the newly established subcardinal blood stream enlarges some of the small channels already entering the posterior cardinal and discharges through them into the posterior cardinal vein (Figs. 47 and 139, C). Other vessels of the primitive plexus persist as a meshwork of small veins lying superficially in the mesonephros. These veins afford free intercommunication between the postcardinals dorsally and the subcardinal vessels in the ventromesial portion of the mesonephros (Figs. 117 and 139).

From the same primitive subcardinal plexus a minor longitudinal channel develops along the ventral border of each mesonephros. These vessels are called the ventral veins of the mesonephroi (Figs. 66 and 139). The fact that they are rather conspicuous in pig embryos at this stage sometimes leads to their confusion with the main subcardinal channels. Their characteristic superficial position on the ventral border of the mesonephros (Fig. 139, E) should preclude such a possibility. They are to be regarded as an incidental modifica-

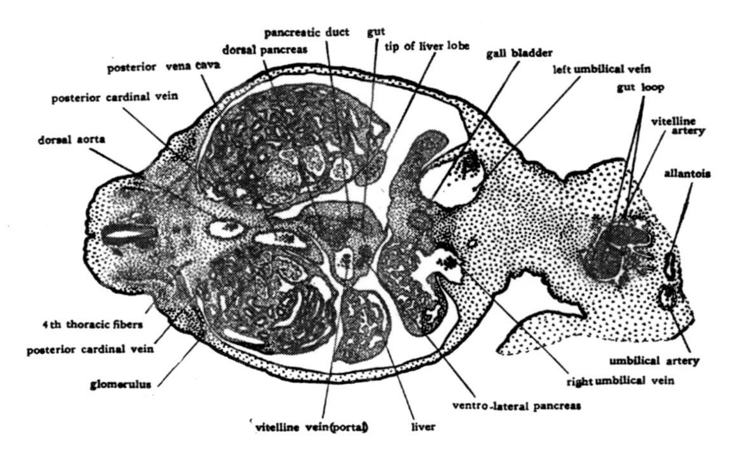

FIG. 76. Transverse section of 9.4 mm. pig embryo at level of pancreatic outgrowth from gut (× 15). Tail and part of belly-stalk omitted. (Serial No. 406.)

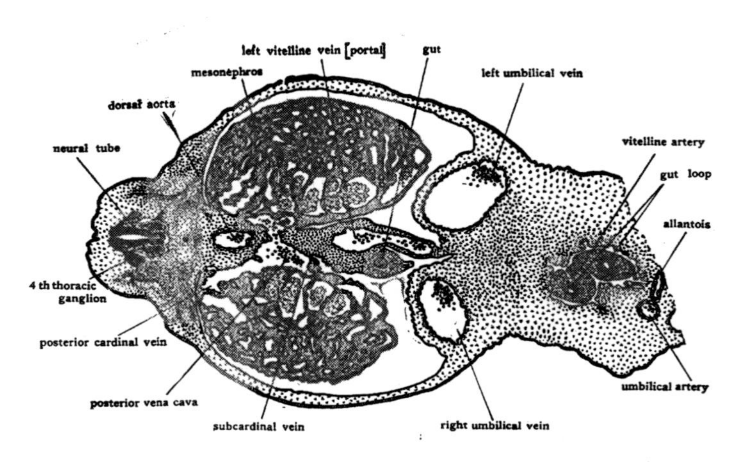

FIG. 77. Transverse section of 9.4 mm. pig embryo at level of inter-subcardinal anastomosis. (Cf. Fig. 139, D, F.) (Serial No. 426.)

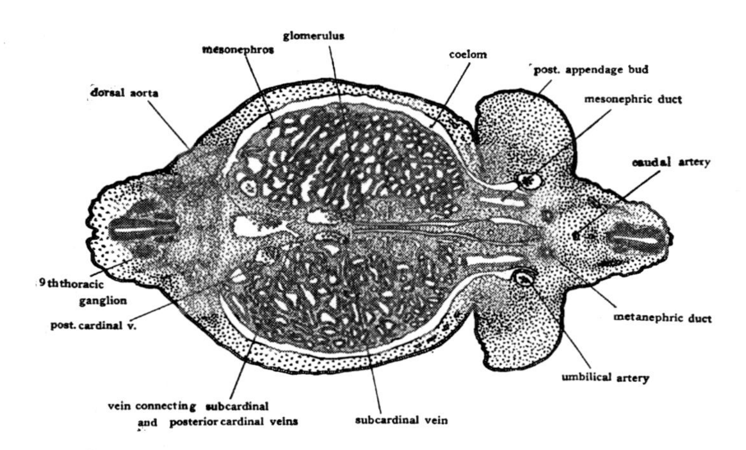

FIG. 78. Transverse section of 9.4 mm. pig embryo at level of metanephric primordia (× 15). (Serial No. 518.)

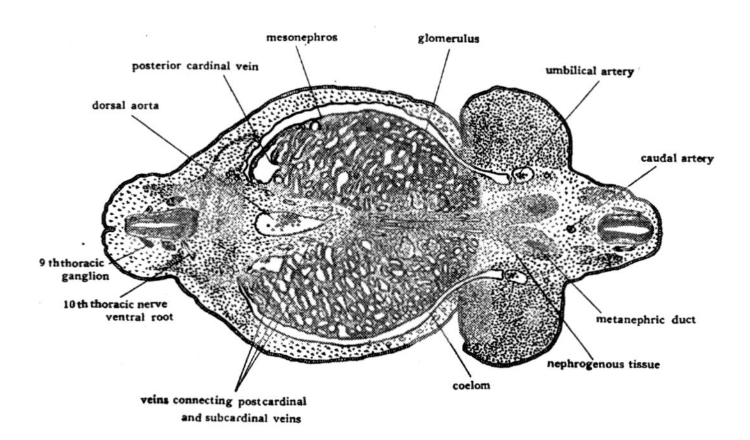

FIG. 79. Transverse section of 9.4 mm. pig embryo passing through caudal end of mesonephros (× 15). (Serial No. 529.)

tion of the plexus of small vessels connecting the sub- and postcardinals rather than as main channels of any special importance.

With the growth of the mesonephroi the rapidly enlarging subcardinal veins are brought very close to each other (cf. Fig. 139, B, E). Where they are approximated, cross-communication is established, first by small vessels and then by a broad anastomosis (Figs. 60, 68, 77, and 139, D–F). The large median venous sinus thus formed probably offers less resistance to the flow of blood than surrounding channels; in any case all the vessels connecting with it tend to drain toward it. The diversion of blood toward this sinus by way of the small vessels which connect the sub- and postcardinals is responsible for the breaking down of the postcardinal veins at this level (Figs. 66 and 139, D, F).

The Posterior Vena Cava. One might expect that the great volume of blood entering the subcardinal sinus would cause a corresponding enlargement of the cephalic portion of one or both subcardinal veins. Instead, a new and more direct channel toward the heart appears. In its growth the liver is crowded very close to the mesonephroi. The developing liver contains a maze of vascular channels, as does the mesonephros. Capillaries ramifying in the base of the mesentery between the liver and the right mesonephros form the connecting link between the vessels of these two organs (Fig. 140). Once the blood begins to find its way by this route, the small irregular channels are rapidly enlarged and straightened. The new and more direct channel thus established leads from the subcardinal sinus through the right subcardinal vein for a short distance and thence, by the newly excavated channels in the mesentery, through the liver to the heart (Fig. 139, D). This is the start of that embryologically composite[1] vessel which we know in the adult as the posterior, or inferior, vena cava (Figs. 66, 68, and 74–77).

The Omphalomesenteric Veins. Primarily the omphalomesenteric veins are the main channels into which the vitelline veins from the yolk-sac converge (Fig. 45). Two factors radically change their original relations. The early degeneration of the yolk-sac reduces their

[1] The developmental complexity of the posterior vena cava is recognized by designating the part of it which arises by the straightening of small channels in the liver as its *intra-hepatic portion;* the part which arises from the capillaries in the caval fold as its *mesenteric portion;* the part formed by the inter-subcardinal anastomoses as the *inter-renal portion;* and that part which at a later stage is added by the appropriation of the right supracardinal as the *post-renal portion.*

peripheral drainage area, with a resultant decrease in their relative size; and the growing liver envelops and breaks up their proximal portions (Fig. 141, A–C). Thus we find them in 10 mm. pig embryos reduced to small vessels which collect the blood delivered by the enteric arteries to the gut. Distally the omphalomesenteric (vitelline) veins are still paired but where they discharge into the liver they have been reduced to a single vessel, which we can now quite properly call by its adult name, the *portal vein* (Figs. 66, 76, and 77).

The Umbilical Veins. Except for a striking increase in size, the distal portions of the umbilical veins in 9 to 12 mm. pig embryos have undergone little change (cf. Figs. 45 and 66). Proximally, however, they have been rerouted through the liver. The underlying factor in this change is the extensive growth of the liver which brings it into contact with the lateral body-walls in which the umbilical veins are embedded in their course from the belly-stalk to the sinus venosus. Fusion follows the contact, and small vessels develop between the umbilicals and the network of channels in the liver (Fig. 141, B). As these new vessels develop the portions of the umbilical veins cephalic to them gradually drop out altogether and all the placental blood passes through the liver (Figs. 66, 68, 74, 75, and 76).

With the completion of this change in the umbilical circulation, the liver has become the common returning path for both of the original extra-embryonic circulatory arcs and most of the intra-embryonic circulation of the posterior half of the body. Only the dwindling current of the postcardinals and the unchanged anterior cardinal circulation now enter the sinus venosus without first passing through the liver. When we consider all this volume of blood passing through one organ, it leaves little room for surprise at the relatively enormous bulk attained by the liver in mammalian embryos.

CHAPTER 8

The Development of the Nervous System

I. The Functional Significance of the Various Parts of the Nervous System

Without some knowledge of the functional significance of the various parts of the nervous system to serve as a basis for correlation and interpretation, its study, from either the developmental or anatomical point of view, is barren and discouraging. Therefore, even though it involves reviewing some familiar facts and also introducing some material which is more neuroanatomical than embryological, it seems advisable to summarize here certain conceptions which are essential to an understanding of the nervous system.

Neurons. The nervous system is made up of cells which are highly specialized in two of the fundamental properties of protoplasm, irritability and conductivity. These cells develop long cytoplasmic processes which extend from one part of the body to another, acting in the manner of telephone lines keeping the various parts of the organism in touch with each other and making possible prompt and coördinated response to alterations in internal or external conditions. In the nervous system of animals as complex as the vertebrates most lines of communication involve chains of such cells arranged so the ends of the processes of one cell come into close relation with the processes or cell body of another. When a change in environmental conditions (a stimulus) starts a wave of electro-chemical change (a nerve impulse) in the protoplasm of one cell in the chain, the wave traverses the processes of the cell in which it was initiated and passes on to the next cell in the chain, and so on. Each link in the chain, that is, each nerve cell with its processes, is called a neuron.

Synapses. The point at which the nerve impulse passes from the processes of one cell to the processes of another is known as a synapse. Synapses between neurons appear to be in the nature of "contacts" sufficiently intimate to permit the passage of a nerve impulse, but not ordinarily involving structural continuity of the cell processes. This contact type of relation at a synapse, which apparently is "made" or "broken" under varying physiological conditions, underlies such phenomena as alternative responses to a given stimulus. It implies the possibility of selective routing of the impulse over one of several neuron chains by the occurrence of physiological contact at certain synapses and physiological disjunction at others.

The arrangement of neuron chains or arcs, as they are frequently called, is exceedingly complex in the vertebrate nervous system. Consideration of its details would carry us far afield, but it is quite possible to take up enough of the basic scheme of neuron arrangement so that the various parts of the nervous system assume some meaning in terms of function rather than remaining as mere names, without significance and readily confused.

Functional Classes of Neurons. All the myriad neurons which go to make up the central and peripheral nervous system are alike in that they are cells with attenuated processes specialized in conductivity. Among themselves they differ greatly as to location, relations, length, number and distribution of processes, and type and direction of impulses transmitted. Functionally neurons can be divided into three main groups. Some neurons carry impulses from sensory nerve endings and sense organs (receptors) toward the cord and brain. These are said to be *afferent* neurons. Others conduct motor impulses away from the cord and brain to muscles or glands (effectors) which respond by appropriate activity. Such neurons are said to be *efferent.* In the cord, and especially in the brain, are countless neurons having many relatively short processes which can transfer an incoming sensory impulse to any one of a number of efferent neurons with which their various processes connect. These are *association* neurons. These three functional categories of neurons, afferent, association, and efferent, together with the receptors attuned to pick up various changes in internal or external conditions, and the effectors capable of carrying out the appropriate responses, constitute what we may call the action system of the organism.

Nerves. The various anatomical parts of the nervous system can be translated into terms of sensory, motor, and association neurons, and their characteristic activities. Thus, what in the dissecting room we call "nerves," are bundles of delicate neuron processes projected to various peripherally located structures. The nuclei and the bulk of the cytoplasm (cell bodies, cytons) of the neurons are either massed at some point on the nerve to form a ganglion, or buried in the central nervous system where they are spoken of as nuclear masses. The nerve itself consists only of the long slender neuron processes (nerve fibers) and the sheaths which protect them.

The Spinal Cord and Reflexes. The spinal cord is at once a center for automatic local responses to local stimulation, and a conduction pathway. In its capacity as a local center it receives sensory impulses coming to it by way of afferent neurons and sends out motor impulses over efferent neurons which activate effectors in the region stimulated. Such an automatic local response is known as a simple *intrasegmental reflex* and represents the most primitive type of mechanism in the vertebrate action system (Fig. 80, arc 1).

By transmission of a sensory impulse longitudinally along the cord through the agency of association neurons, the outgoing impulse may be imparted to efferent neurons in several adjacent metameres. Such a mechanism is termed an *intersegmental reflex*. It is a distinct step in advance over the strictly local reflex in that it affords a concerted response on the part of a group of effectors (Fig. 80, arc 2). The part of the cord primarily involved in segmental and intersegmental responses is the centrally located "gray matter." This gray matter is composed chiefly of association neurons and of the cell bodies of motor neurons which send their processes into the spinal nerves.

Cerebrospinal Conduction Paths. The peripheral "white matter" of the cord is composed of nerve fibers which run longitudinally, constituting pathways of intercommunication between the spinal cord and nerves, and the brain. The color from which this part of the cord takes its name and which sets it off so sharply from the richly cellular gray matter, is due to the sheaths enclosing the fibers. These sheaths are rich in a fatty substance (myelin) which imparts the characteristic whitish and glistening appearance.

Phylogenetically these conduction paths in the peripheral part of the cord increase in conspicuousness concomitantly with the increasing extent to which the brain assumes a coördinating control over the basic reflexes which constitute the primary function of the cord. The sensory (afferent) conduction paths are for the most part grouped in the dorsal portion, whereas the motor (efferent) paths are found in the lateral and to a less extent in the ventral portion of the cord. (Fig. 86, E.)

Reflexes Involving Organs of Special Sense. Much of the control exercised by the brain on the activities of the body as a whole depends on afferent impulses entering from sense organs of a much more highly

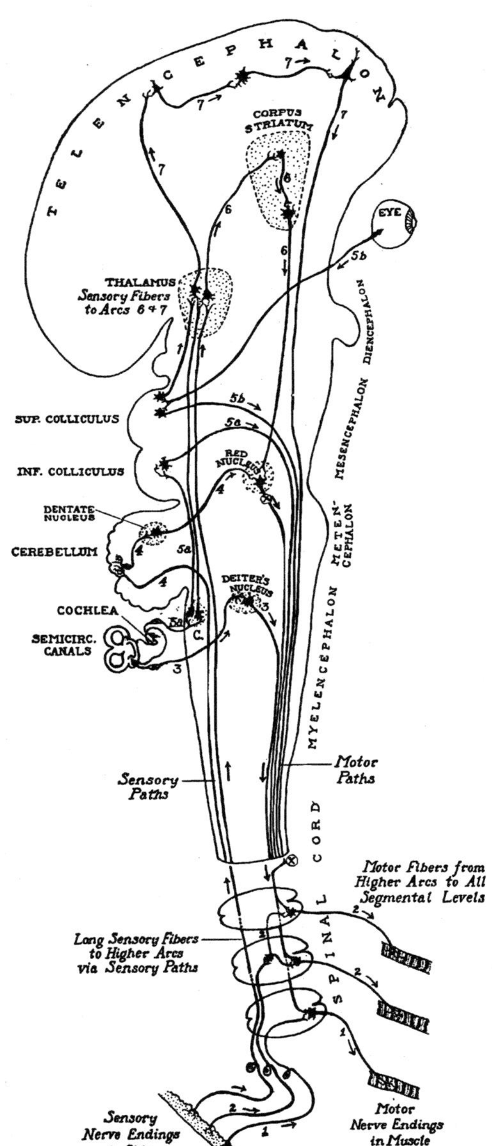

Arc 7

VOLUNTARY AND INHIBITORY CONTROL

Choice of response based on memory of past experiences. (Via pyramidal ract.)

Arc 6

AUTOMATIC ASSOCIATED CONTROL

of complex muscular actions. (Via striato-rubro-spinal tract.)

Arc 5

a. AUDITORY REFLEXES

e.g. Automatic response to sudden noise.

b. VISUAL REFLEXES

e.g. Automatic response to blinding flash of light. (Both via tecto-spinal tract.)

Arc 4

SYNERGIC CONTROL

Automatic coördinating control of muscular actions. (Via rubro-spinal tract.)

Arc 3

EQUILIBRATORY CONTROL

Automatic balancing reactions. (Via vestibulo-spinal tract.)

Arc 2

INTERSEGMENTAL REFLEX

Impulse carried by association neurones to neighboring segments causing coördinated response of muscles in several segments.

Arc 1

INTRASEGMENTAL REFLEX

Response limited to segment stimulated.

FIG. 80. Schematic diagram showing the nature of the activities carried out in various parts of the central nervous system. (Collaboration of Dr. L. J. Karnosh.)

specialized type than the endings in the general skin surfaces. Such organs as those of equilibrium, hearing, and sight play an exceedingly important part in determining the appropriate reactions of the various segmental effectors and thereby regulating the reactions of the animal as a whole.

Positional changes affect special sensory end organs in the semicircular canals. Through sensory neurons the impulse passes to a synaptic center (Deiters' nucleus) at the boundary between myelencephalon and metencephalon (Fig. 80, arc 3). Thence conduction fibers carry the impulse along an efferent path from which it may enter the appropriate local motor neurons at any segmental level. This arc would be involved in the automatic balancing reaction to a sudden upset in equilibrium such as slipping unexpectedly on a patch of ice.

Auditory and visual stimuli are conducted to centers in the mid-brain. The auditory stimuli received in the cochlea are transmitted through a series of sensory neurons to synaptic centers in the inferior colliculi on both sides of the mesencephalon and thence by motor paths to the appropriate local motor fibers at any or all segmental levels (Fig. 80, arc 5a). This arc is the one responsible for the involuntary reaction to a sudden noise.

Responses to sudden visual stimulation involve an arc much similar to that described for reactions initiated by auditory stimuli. The visual arc (Fig. 80, arc 5b) starts with sensory fibers arising in the retina of the eye. In the superior colliculi of the mesencephalon these afferent fibers have synapses with association and motor fibers. The motor fibers communicate with the effector mechanism of the body over paths paralleling those involved in the responses to auditory stimuli. Through this arc would be brought about the automatic recoil from a sudden blinding flash of light.

Coördinating Centers. So far the action mechanisms described are purely reflex in character and relatively simple. In the complex and more deliberate actions involving large groups of muscles in several segments, a coördinating effect is exercised from a center in the cerebellum. This center receives sensory impulses (position sense) from all segmental levels in the body. These impulses pass by way of a series of neurons in the cerebellum (metencephalon), to a synaptic center (red nucleus) in the mid-brain. Thence (Fig. 80, arc 4) the impulse is conducted by motor fibers through the common motor pathways to the general body musculature. Such an arc provides for a smooth and precise, but unconscious, coördination of muscular action, such as that involved in slowly bringing together the tips of the forefingers of the two hands. This is the so-called synergic type of muscular control.

Certain fairly complex responses are, nevertheless, largely automatic or unconscious, such as the rhythmic swaying of the trunk and swinging of the arms in walking. These are regulated by association neurons passing through the thalamic region of the diencephalon and the corpus striatum of the telencephalon. Sensory fibers enter the thalamus from all afferent pathways. Thence the impulses (Fig. 80, arc 6) are transmitted to synaptic centers in the corpus striatum. From the corpus striatum they pass over motor fibers to the red nucleus whence they enter the common motor paths along with the motor fibers from the cerebellar centers.

Voluntary and Inhibitory Control. Superimposed on the primitive types of response is a mechanism affording a wide choice of reactions in response to the stimuli entering from the various afferent pathways. The centers for this, the highest and most plastic system, are in the cortical areas of the telencephalon. Fibers from practically all receptors enter this system by way of the thalamus (Fig. 80, arc 7). From synapses in the thalamus the afferent fibers are dispersed according to their special functions into localized areas in the cerebral cortex. By myriads of association neurons these various centers are in free intercommunication. These centers are responsible for memory and for all choices of action conditioned by previous experience. In short, they are the centers of intelligent response, in distinction to reflex reactions to existing conditions. Through the intercommunicating association neurons of this system motor impulses may be transmitted by way of the common motor paths of the spinal cord to any parts of the action system.

In becoming acquainted with the various regions of the central nervous system, therefore, we should think of the spinal cord and the myelencephalon as carrying out the dual rôle of reflex centers and conduction pathways to and from the higher brain centers. The cerebellum acts as a coördinating center for complex muscular actions such as those concerned with the maintenance of normal posture. The floor of the mesencephalon is made up primarily of great fiber tracts passing to and from the higher brain centers. Specialized regions in the dorsal walls of the mesencephalon are concerned with visual and auditory reflexes. Of the two pairs of conspicuous elevations (corpora quadrigemina) the more cephalic pair (superior colliculi) are visual reflex centers and the more caudal pair (inferior colliculi) are auditory reflex centers. The thalamic region of the diencephalon serves as the gateway for fibers having cerebral connections. In the deeper part of each cerebral hemisphere is the corpus striatum which is concerned with muscle tonus and automatic associated movements. The more superficial portions of the cerebral hemispheres become specialized as the cerebral cortex. Certain cortical areas are the highest terminal centers for the reception of incoming impulses which keep us in touch with our environment, such as those resulting from auditory, visual, and tactile stimulation. Other cortical areas contain the cell bodies of neurons which are the first units in efferent chains. The connection of afferent and efferent areas by association neurons affords the connecting link placing the effector mechanisms of the body under voluntary control. Thus the cerebral hemispheres act as association centers superimposed on the lower reflex mechanisms and affording the possibility of intelligent choice of response based on experience.

II. Review of the Early Stages in the Establishment of Nervous System

The initial steps in the formation of the nervous system take place very early in development. Directly or indirectly, many points of importance in connection with its establishment and early differentiation have already been dealt with. We have seen the origin of the neural groove by the infolding of a thickened plate of ectoderm in the mid-dorsal line of the embryo; the closure of the neural groove to form the neural tube, and the coincident separation of the tube from the parent ectoderm (Figs. 24, 25, 28, 29, and 35).

In the closure of the neural groove, certain cells lying near its margins remain independent, being included neither in the walls of the neural canal nor in the superficial ectoderm as it closes above the newly established neural tube. These ribands of cells come to lie on either side in the angles between the superficial ectoderm and the neural tube and constitute the neural crests (Fig. 35). They are the primordia of the sensory root ganglia of the spinal and cerebral nerves and indirectly of the sympathetic ganglia.

Almost as soon as it is independently established the neural tube becomes markedly enlarged cephalically. This dilated portion is the primordium of the brain. Caudally the neural tube remains of relatively uniform diameter as the forerunner of the spinal cord.

In its enlargement the brain at first exhibits three regional divisions—the primary fore-brain, mid-brain, and hind-brain; or, to use their more technical synonyms, the prosencephalon, mesencephalon, and rhombencephalon (Fig. 36). This three-vesicle stage of the brain is short-lived. The prosencephalon is subdivided into two regions, telencephalon and diencephalon; the mesencephalon remains undivided; and the rhombencephalic region becomes differentiated into metencephalon and myelencephalon. Thus in place of three vesicles, five are established. This stage in the development of the brain is well shown in embryos between 9 and 12 mm. in length (Figs. 59, 60, and 65). Starting with these familiar conditions as a basis we are ready to trace the later differentiation of some of the more important parts of the nervous system.

III. The Histogenesis of the Spinal Cord and the Formation of the Spinal Nerves

The Establishment of Ependymal, Mantle, and Marginal Layers. The ectoderm of the open neural groove is at first but a single layer of cells in thickness (Fig. 81, A). These original cells proliferate very rapidly and by the time the neural tube has become closed, its wall consists of many cell layers (Fig. 81, B). The individual cells, meanwhile, tend to lose their originally clear-cut outlines. In the older studies it was generally stated that at this stage the cells merged into a syncytium as suggested by the classical illustration of Hardesty reproduced here as figure 81. More recently Sauer has restudied such material and maintains that although the membranes become delicate and inconspicuous, if sufficiently well-preserved material is carefully studied the cells can be seen to retain their membranes intact. Toward the lumen the neural tube is bounded by an internal limiting membrane, and peripherally its extent is sharply marked by an external limiting membrane (Fig. 81, C).

Certain of the cells lying near the lumen of the neural tube can, at this stage, be seen undergoing mitosis. They are called germinal cells (Fig. 81, C) although they are probably merely the cells in this region that happened to be dividing at the time the material was fixed. Most of the new cells formed by these cell divisions are crowded somewhat away from the internal limiting membrane into a zone in the cord which becomes densely packed with nuclei. This zone is called the *mantle layer* (Fig. 81, D). The cells which remain nearest to the internal limiting membrane become more or less elongated and radially arranged about the lumen of the neural tube. They constitute

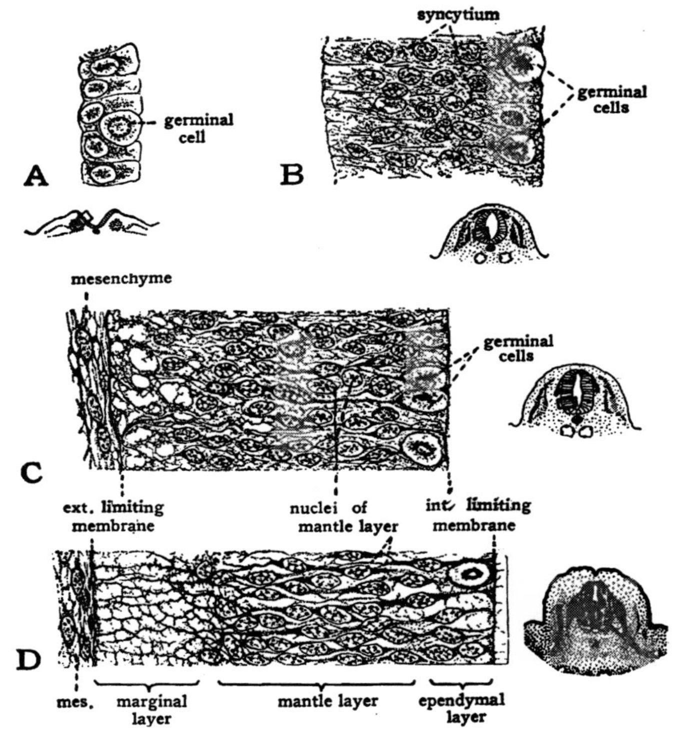

FIG. 81. Stages in the histogenesis of the spinal cord. (After Hardesty.)
A, From open neural plate of rabbit embryo.
B, From wall of recently closed neural tube, 5 mm. pig embryo.
C, From neural tube of 7 mm. pig.
D, From neural tube of 10 mm. pig. (All drawings × 550.)

a zone within the mantle layer known as the *ependymal layer* of the cord. Outside the mantle layer is a peripheral region into which practically no nuclei enter. This is the *marginal layer* (Figs. 81, D, and 83).

Spongioblasts and Neuroblasts. Of these three primary zones in the developing spinal cord the mantle layer is the first to show striking differentiation. Its cells continue to divide rapidly and undergo divergent specialization. Some of them become spongioblasts which are destined to form merely supporting tissue, and some of them neuroblasts, which will become the functionally active nerve cells.

These two types of cells can first be differentiated from each other by the fact that the neuroblasts develop large nuclei while the nuclei of the spongioblasts remain small.

Neuroglia. The formation of supporting tissue from the spongioblasts takes place by the development of exceedingly slender and irregular cytoplasmic processes. Some of these processes may eventually lose their association with the parent cells and appear as separate fibers. The majority of them, however, retain some connection with the cells from which they were derived. In this respect as well as in its ectodermal origin, neuroglia, as this peculiar connective tissue of the central nervous system is called, differs from the other connective tissues of the body. The fibers and processes formed from the spongioblasts are so delicate that they are exceedingly difficult to demonstrate in material stained by routine histological methods. But when they are subjected to metallic impregnation (e.g., the Golgi silver nitrate method) the 'glia cell processes and fibers appear as blackened strands forming an elaborate tracery of supporting elements throughout the substance of the cord (Fig. 82).

All neuroglia cells exhibit processes of one sort or another and all of them are supporting in function, but the cells differ much among themselves as to shape and arrangement of processes. For convenience in description they are commonly designated as belonging

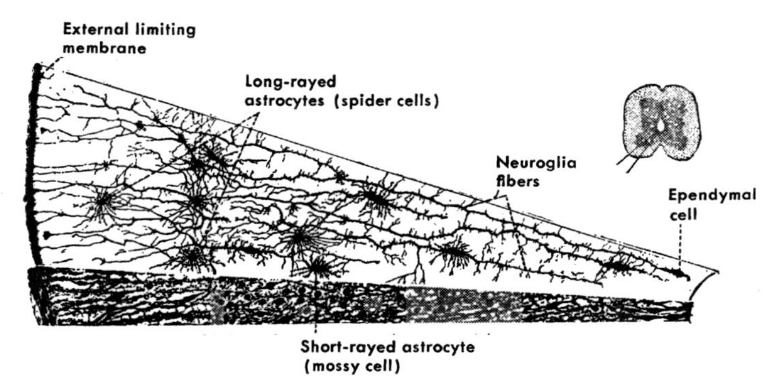

Fig. 82. Segment of the spinal cord of a 70 mm. pig showing the differentiation of the neuroglial elements. (After Hardesty.) The upper part of the segment is drawn to show the 'glia cells and fibers as demonstrated by silver impregnation. The lower part of the segment indicates their appearance after routine staining with hematoxylin and eosin.

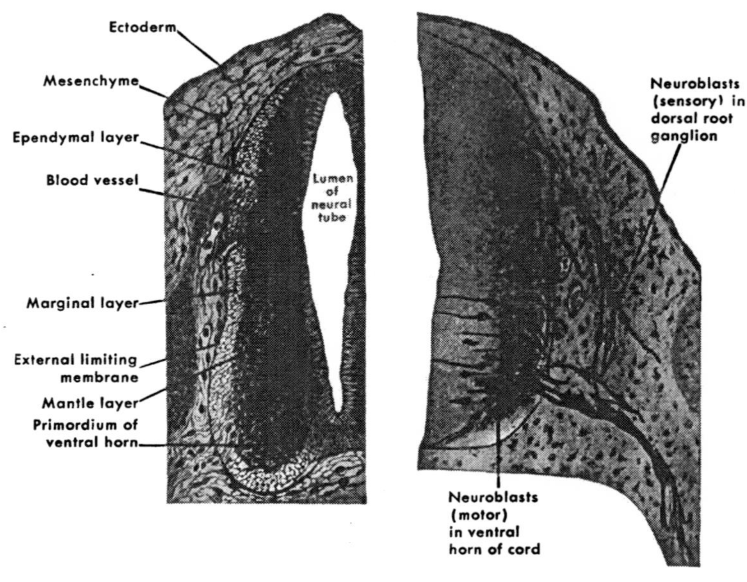

FIG. 83 FIG. 84

FIG. 83. Wall of the neural tube of a 9 mm. pig embryo drawn (× 100) to show the ependymal, mantle, and marginal layers at an early stage in their differentiation. (After Hardesty.)

FIG. 84. Wall of the neural tube of a 10 mm. pig embryo treated by a special technique which brings out the developing neuroblasts and their processes. (After Held.) Compare with figure 85 and note the manner in which the dorsal root of the spinal nerve is formed from processes arising from neuroblasts in the dorsal ganglion, while the ventral root is composed of fibers arising from neuroblasts in the cord.

to one of four types, *ependymal cells*, *fibrous astrocytes*, *protoplasmic astrocytes*, or *oligodendroglial cells*. The ependymal cells arise from spongioblasts which have themselves remained close to the internal limiting membrane, but have sent out long processes all the way to the external limiting membrane of the cord (Fig. 82). Both types of astrocytes arise from spongioblasts in the mantle layer and gradually assume the characteristic shapes suggested by their names. Oligodendroglial cells become recognizable later in development than astrocytes. They appear as satellites around the cell bodies of neurons and are scattered along the developing myelinated nerve tracts of the white matter of the central nervous system.

Neuroblasts. Preparations of the spinal cord made by the usual methods do not show the neuroblasts to advantage (Fig. 83). If, however, a special technique such as intra-vitam staining by methylene blue or one of the metallic impregnation methods is used, the slender processes of the neuroblasts can readily be seen (Fig. 84). With the development of these characteristic nerve fibers we can think of the neuroblasts as having become young neurons.

The Formation of the Spinal Nerves. The outgrowth of processes from neuroblasts lying in the ventral and lateral portion of the mantle layer of the cord establishes the motor fibers which compose the ventral roots of the spinal nerves (Fig. 85). Neuroblasts in the dorsal root ganglia send to the cord afferent processes which constitute the dorsal roots of the spinal nerves, and send other processes peripherally which end in connection with various types of receptors (Fig. 85). Neuroblasts which have migrated from the cord and from the neural crest to form the sympathetic ganglia develop processes which relay efferent impulses thence to their destination (Fig. 85). Reference to figure 80 will show the relations of the neurons of the spinal nerves to the central nervous mechanism as a whole. Those interested in working out in more detail the functional significance of the various types of neurons encountered in the spinal nerves will find a brief analysis appended to the legend of figure 85. For more comprehensive information along these lines reference should be made to the discussion of "spinal nerve components" in a textbook of neurology.

The Development of the White and Gray Matter of the Cord. During the period of development when the neurons are being differentiated, the appearance of the spinal cord as seen in sections undergoes very marked changes. Some of the neuroblasts in the mantle layer of the cord, as we have seen, send out processes very early in development. Others remain undifferentiated and continue to proliferate for a time, causing continued growth in the mantle layer. As it grows in mass the mantle layer takes on a very characteristic configuration, becoming butterfly-shaped in cross-section. With this change in shape, and with the transformation of its spongioblasts into neuroglia and its neuroblasts into characteristic nerve cells, the mantle layer becomes the so-called "gray matter" of the spinal cord (Fig. 86).

During the growth of the mantle layer the originally extensive

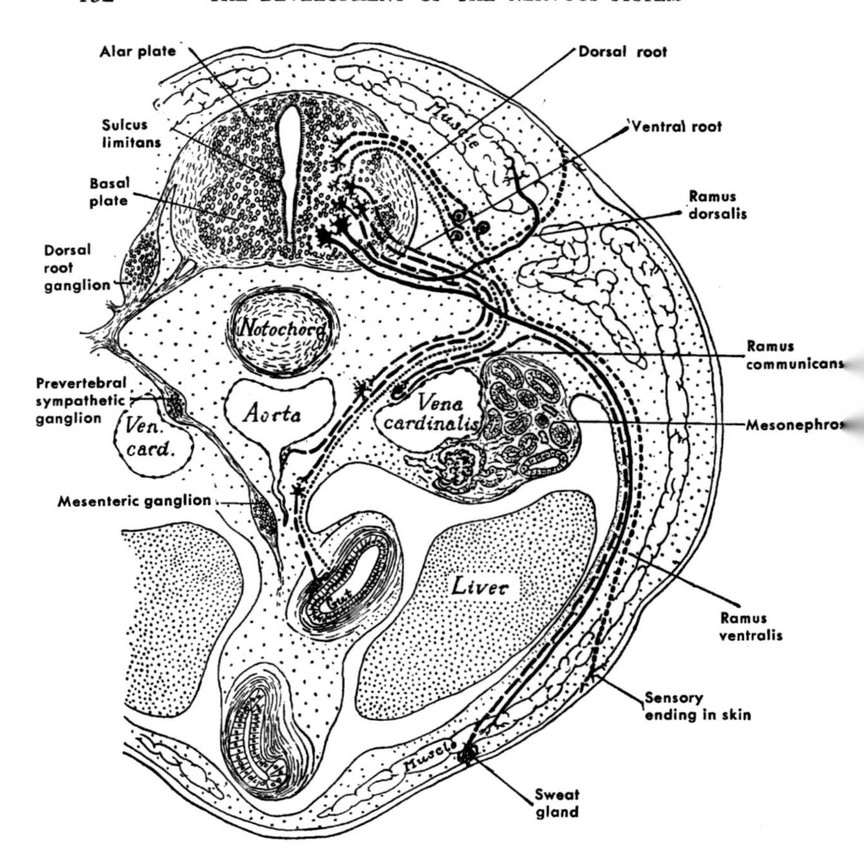

FIG. 85. Schematic diagram indicating the various connections made by the neurons which develop in a typical spinal nerve. (Modified from Froriep.)

The neurologist classifies the fibers in a spinal nerve according to their relations and functions. The components of a typical spinal nerve on this basis are:

I. AFFERENT

A. GENERAL SOMATIC AFFERENT

(1) *Exteroceptive*, i.e., fibers conducting impulses from the external surface of the body such as touch, pain, temperature. (Represented in this figure by short broken lines.)

(2) *Proprioceptive*, i.e., fibers carrying impulses of position sense from joints, tendons, and muscles. (Not represented in this diagram.)

lumen of the neural tube is reduced, by obliteration of its dorsal portion, to the small central canal characteristic of the adult cord (Fig. 86). The cells of the ependymal layer now constitute a sort of epithelioid lining of the central canal.

Meanwhile the outer or marginal layer of the cord has been increasing extensively in mass. Its growth is due to the secondary ingrowth of longitudinally disposed neuron processes which constitute the conduction paths between the various levels of the spinal cord and the brain (Fig. 80). Each of these fibers being enveloped in a

---

FIG. 85—(*Continued*)

B. GENERAL VISCERAL AFFERENT
Fibers from viscera (interoceptive) by way of sympathetic chain ganglion, white ramus communicans, and dorsal root; cell bodies in dorsal root ganglion; no synapse before reaching cord. (Illustrated in this figure by dotted line.)

II. EFFERENT

A. GENERAL SOMATIC EFFERENT
Motor neurons to skeletal muscle; cell bodies in ventral columns of gray matter; fibers emerge by ventral roots. (Illustrated in this figure by solid lines.)

B. GENERAL VISCERAL EFFERENT
Two-neuron chains from cord to glands and to smooth muscle of viscera and blood vessels. The first neurons (preganglionic) have their cells of origin in lateral column of gray matter of cord from first thoracic to third lumbar level. Fibers leave cord by ventral root, turn off in white ramus communicans to end in synapse with the second neurons (postganglionic) of the two-neuron chain.

Note the various destinations of the visceral efferent paths, e.g.:

(1) fibers to smooth muscle of gut wall; impulse relayed by second order motor neurons from synapses in mesenteric ganglion.

(2) fibers (vaso-motor) to smooth muscle of blood vessel wall; impulses relayed by second order motor neurons from synapses in prevertebral ganglia.

(3) fibers (pilo-motor) to muscles about hair follicles, and fibers (sudo-motor) to sweat glands in skin. Impulses in both these cases relayed from synapses in prevertebral ganglia by second order motor fibers passing back over the ramus communicans and thence to periphery via branches of spinal nerve.

Most of the fibers in the spinal nerve are medullated and therefore whitish in appearance. The visceral second order motor ("post-ganglionic") fibers, however, lack a myelin sheath and are grayish in appearance. Certain of such second order visceral motor fibers (e.g. pilo-motor and sudo-motor) which run back along the ramus communicans after synapse in a prevertebral ganglion account for the so-called "gray bundle of the ramus communicans."

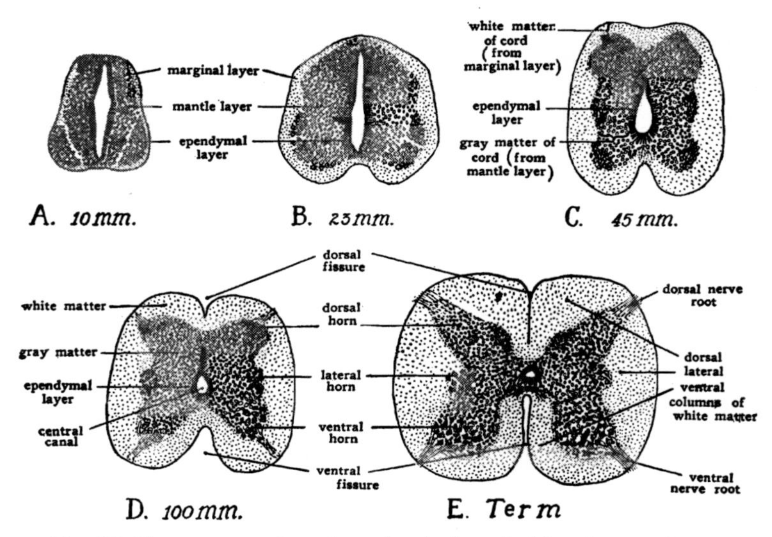

FIG. 86. Transverse sections through spinal cord of the pig at various ages. Note especially the parts of the adult cord derived from the ependymal, mantle, and marginal layers of the embryonic neural tube.

sheath rich in myelin, the region of the cord in which they lie has a characteristic whitish appearance which contrasts strongly with the gray color of the richly cellular portion of the cord derived from the mantle layer. For this reason the fibers which develop in the marginal layer of the cord are said to constitute its white matter. The main groups of these fibers are more or less marked off from each other by the dorsal and ventral horns of the gray matter. They are known as the dorsal, lateral, and ventral columns of the white matter of the cord (Fig. 86). The dorsal columns contain the main tactile and proprioceptive paths to the brain; the ventral columns are primarily motor; and the lateral columns contain important ascending fiber tracts to the brain and also some of the main motor paths from brain to cord and thence to spinal nerves (Fig. 80).

IV. The Regional Differentiation of the Brain

It will be recalled that in embryos of about 5 mm. the brain was just beginning to progress from the three- to the five-vesicle stage. In embryos of the 9 to 12 mm. range we saw the five-vesicle condition of the brain well established. These same five basic regions of the brain will continue to be recognized as the major divisions of the adult brain. During their later development they become greatly altered in appearance and certain specialized parts of them receive new names, but their fundamental relations remain the same. The details of all the structural features which appear in the various parts of the brain

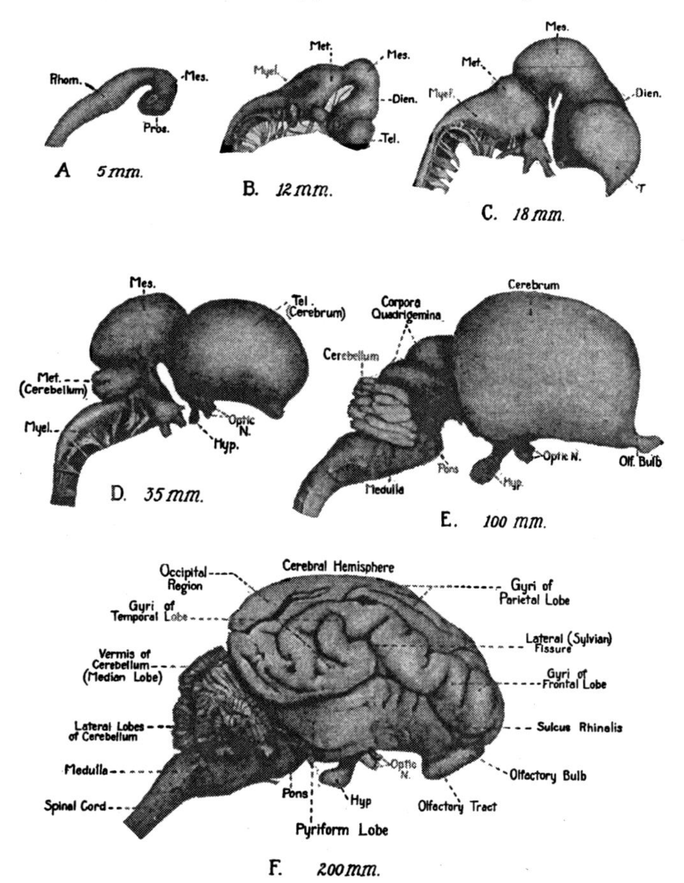

FIG. 87. Lateral views of brains of pig embryos at various stages of development. The younger stages are more highly magnified than the older. (B, after Lewis; C and D, after Prentiss.)

constitute too complex a story to be satisfactorily handled in brief compass. Accordingly we shall confine ourselves to becoming acquainted with the main morphological landmarks and the locations of the principal functional centers of the brain—the bare foundation on which subsequent work may build a fuller knowledge of this interesting system.

The Myelencephalon. The myelencephalon of the embryo becomes the medulla of the adult brain (Fig. 87). Very early in development the lumen of this part of the neural tube becomes dilated, foreshadowing its ultimate fate as the large cavity in the medulla known as the fourth ventricle (Fig. 88). At the same time its roof becomes very thin (Fig. 88, A). Small blood vessels develop against this membranous roof and push it ahead of them into the lumen of the fourth ventricle. The freely branching group of vessels thus formed is known as the choroid plexus of the fourth ventricle (Figs. 65, 99, and 106).

The walls of the neural tube in the brain region show the same early histological changes which occur in the walls of the spinal cord, with the resulting establishment of ependymal, mantle, and marginal layers. The ependymal layer of the myelencephalon becomes the epithelioid lining of the fourth ventricle. Its mantle layer gives rise in part to continuous columns of gray matter as in the cord, and also forms more or less distinct cell masses (nuclei) associated with the roots of the more posterior cranial nerves (Fig. 91). The marginal layer receives an ingrowth of longitudinally disposed medullated fibers constituting the conduction pathways between the spinal cord and nerves and the more rostral parts of the brain (Fig. 80).

In dealing with the topography of the neural tube as it appears in cross-sections it is customary to designate its thickened side-walls as the lateral plates, its thin dorsal wall as the roof plate, and its thin ventral wall as the floor plate. On this basis the membranous covering of the fourth ventricle represents a roof plate greatly stretched out by the divergence of the lateral plates dorsally. (Compare the configuration of the cord, as shown in figure 85, with that of the medulla as diagramed in figure 91.) The deep ventral groove in the floor of the ventricle overlies a very thin floor plate. The great bulk of the myelencephalic wall consists of thickened lateral plates. Extending along the inner surface of each lateral plate is a longitudinal sulcus (*sulcus limitans*) which suggests a division of the lateral plate into a dorsal part (*alar plate*) and a ventral part (*basal plate*, Fig. 91). The sulcus limitans is especially strongly marked during the early stages of the development of the myelencephalic region. Later it becomes masked in certain regions by the growth of underlying nuclei, but wherever it persists it is a valuable landmark in dealing with the location of nuclei and fiber tracts. In the brain, as in the cord, afferent centers develop dorsal, and efferent centers ventral to the sulcus limitans (Fig. 91).

The Metencephalon. The dorso-lateral walls of the neural tube in the metencephalic region undergo very extensive growth and give rise to the cerebellum of the adult brain. The early smooth outline of this region (Fig. 87, C) is broken up by the development of a complex series of folds (Fig. 87, D, E). Eventually three main lobes are formed, each of which is subdivided into a great number of minor folds which impart a very characteristic appearance to the cerebellum (Fig. 87, F). In this part of the brain are developed the synaptic centers concerned with the coördination of complex muscular movements (Fig. 80, arc 4).

Relatively late in development great groups of fibers which form the paths of intercommunication between the cerebellum and other parts of the nervous system appear superficially in the walls of the metencephalon. These form the ventral prominence known as the *pons* (Fig. 87, E, F), and the cerebellar peduncles which extend over the lateral walls of the metencephalon. Deep to, and partly intermingled with, these superficial groups of fibers lie continuations of the same longitudinal fiber tracts which, on their way to and from the brain, traverse the marginal layer of the medulla. Still deeper are the masses of cells which originate from the mantle layer of this part of the neural tube. These cells are clustered in definite centers (nuclei) associated with the cranial nerves of the metencephalic level (see Fig. 91).

The original lumen of the neural tube in the metencephalic region remains of considerable size. Since there is no line of demarcation between it and the lumen of the medulla, it is regarded as the anterior part of the fourth ventricle (Fig. 88).

The Mesencephalon. The dorso-lateral walls of the mesencephalon give rise to two pairs of rounded elevations known as the *corpora quadrigemina* (Fig. 87, E). The two more rostral prominences, called the *superior colliculi*, are the synaptic centers for visual reflexes

(Fig. 80, arc 5b); and the two more caudal prominences, called the *inferior colliculi*, are the synaptic centers for auditory reflexes (Fig. 80, arc 5a).

The ventro-lateral parts of the mesencephalic walls constitute the main pathway over which fibers pass to and from the more anterior parts of the brain (Fig. 80, arcs 6 and 7). The fact that these fiber tracts in the mesencephalon are continuations of the longitudinal tracts encountered in the myelencephalon and in the floor of the metencephalon should be especially emphasized. These tracts in the mesencephalic floor are designated as the *cerebral peduncles*.

With the great thickening of its walls, the lumen of the mesencephalon becomes relatively reduced to form a narrow canal joining the lumen of the metencephalon and myelencephalon (fourth ventricle) with the lumen of the diencephalon (third ventricle). This canal is known as the *cerebral aqueduct* or *aqueduct of Sylvius*.

The Diencephalon. Although the diencephalon undergoes striking local modifications, its original name is still retained in the terminology of adult anatomy. Its roof becomes thin and vessels developing on its outer surface force it ahead of them in finger-like processes which project into the third ventricle as the *anterior choroid plexus* (Fig. 106).

In the median part of the diencephalic roof, caudal to the point of origin of the choroid plexus, the epiphysis appears as a small local evagination (Fig. 100). Later in development the walls of the epiphysis become thickened and its lumen is practically obliterated.

In the floor of the diencephalon is formed a median diverticulum called the *infundibulum*. The distal portion of Rathke's pocket loses its original connection with the stomodaeal ectoderm and becomes closely applied to the infundibulum (cf. Figs. 65 and 138). Somewhat later these two structures become intimately fused to form an endocrine gland known as the *hypophysis* (Fig. 100).

Very early in development the optic vesicles arise as outgrowths from the ventro-lateral walls of the prosencephalon (Figs. 36, E, and 41, D). When the prosencephalon is divided into telencephalon and diencephalon the optic stalks open into the brain very near the new boundary (Figs. 60 and 67). In fact the median depression in the floor of the brain opposite their point of entrance is regarded as the ventral landmark which establishes the demarcation between telencephalon and diencephalon (*recessus opticus*, Figs. 65 and 88, A). Immediately caudal to the optic recess there is a marked thickening

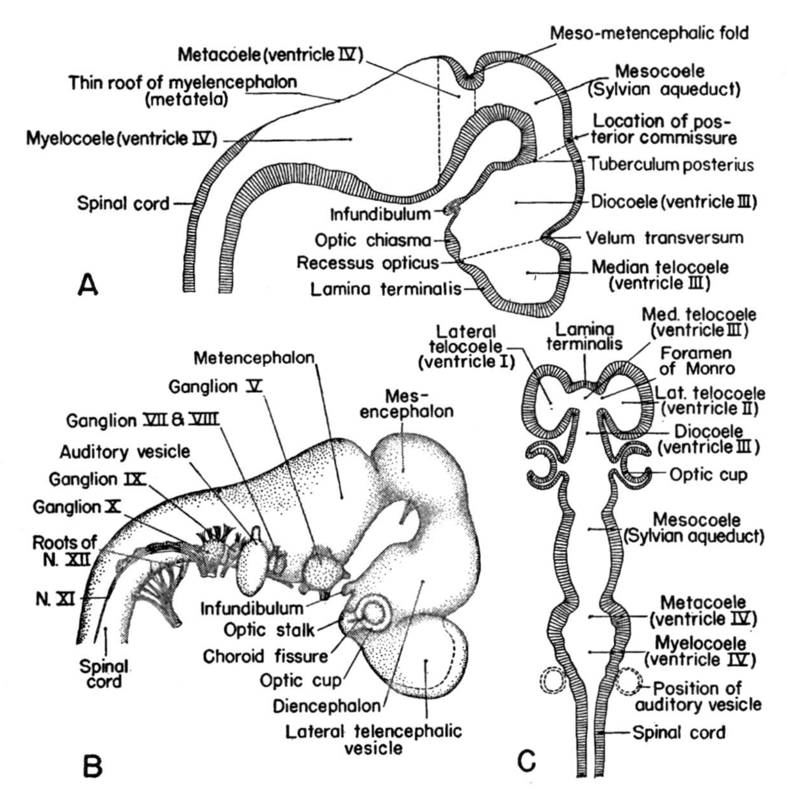

Fig. 88. Diagrams to show the topography of the brain shortly after the transition from the three- to the five-vesicle stage. A, Sagittal section. The conventional lines of demarcation between adjacent brain vesicles are indicated by broken lines. B, Surface view of brain with position of cranial ganglia and nerve roots indicated. C, Schematic frontal plan of brain as it would appear if the flexures had all been straightened out before cutting.

in the floor of the diencephalon where part of the fibers of each optic nerve cross to the other side of the mid-line. This point of crossing of optic nerve fibers is known as the *optic chiasma* (Figs. 65 and 88, A). Beyond the chiasma the crossed and uncrossed fibers on either side run together as the *optic tracts*. The optic tracts pass along the lateral walls of the diencephalon where some of their fibers end. Others pass to the visual reflex centers in the superior colliculi (Fig. 80, arc 5b). Still others concerned with the interpretation and memory of visual impulses pass by way of the geniculate nuclei to the visual areas of the cerebral cortex.

The dorsal parts of the lateral walls of the diencephalon become greatly thickened by multiplication of neuroblasts in the mantle layer. These thickened regions are known as the *thalami* (Fig. 100). The dorsal portion of the thalamus is the gateway of fibers passing from the cord and the brain-stem[1] to the cerebral hemispheres (Fig. 80, arcs 6 and 7). In it are large synaptic centers acting as relay stations. Superficial to these nuclear masses are fiber tracts radiating through the lateral diencephalic walls.

The thickening of the lateral walls of the diencephalon greatly reduces the width of its lumen. In its central portion the two walls come in contact and fuse, forming across the third ventricle a conspicuous connection known as the *massa intermedia.*

The Telencephalon. The telencephalon consists of the most rostral part of the neural tube together with paired dorso-lateral outgrowths from the primary median portion. These outgrowths first appear as roughly hemispherical evaginations called the *lateral telencephalic vesicles* (Figs. 60 and 88, C). Although the division of the lateral walls of the neural tube into alar and basal plates is not clearly marked this far forward in the brain, the telencephalic evaginations, because of their general relations, are regarded as involving the alar plates.

At first the cavities within the two telencephalic vesicles are broadly continuous with the primary lumen of the neural tube (Fig. 67). Later in development these openings into the lateral vesicles appear relatively much smaller; nevertheless they persist, even in the adult, as the so-called *foramina of Monro.* Thus in spite of extensive local modifications the original neural canal remains open throughout the entire length of the central nervous system. Its most rostral parts, the cavities in the telencephalic vesicles (*first and second ventricles* of the adult brain), communicate with the median telencephalic lumen by way of the foramina of Monro. Since there is no line of demarcation between this small median telocoele and the diocoele, both are included in the cavity of the adult brain known as the *third ventricle*

---

[1] Brain-stem is a commonly used term for designating those portions of the brain other than the telencephalon, diencephalon, and cerebellum.

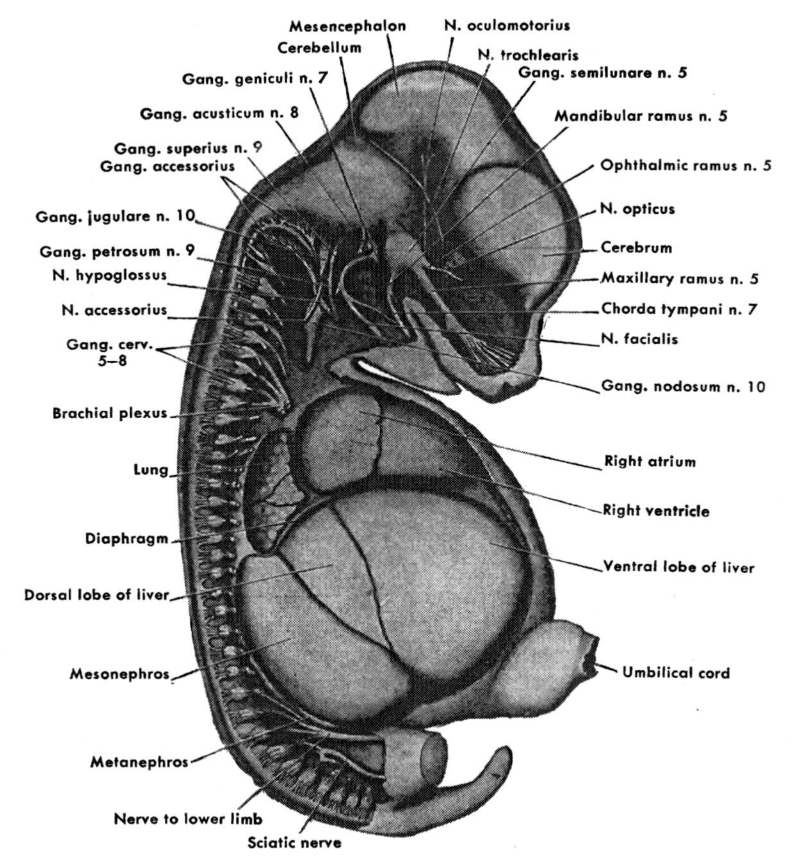

FIG. 89. Model made from a dissection exposing the nervous system of an 18 mm. pig embryo. (After Prentiss.)

(Fig. 88, C). From the third ventricle the cerebral aqueduct leads through the mesencephalon to the *fourth ventricle*, which is the adult term for the confluent lumina of the metencephalon and myelencephalon. The fourth ventricle becomes narrowed caudally and is directly continuous with the central canal of the spinal cord (Fig. 88, C).

The lymph-like fluid which fills these cavities in the nervous system appears to be derived primarily from plexuses of small blood vessels which invade thin places in the dorsal wall of the brain. One such mass of vessels we have already encountered under the name of the *posterior choroid plexus* or choroid plexus of the fourth ventricle. Another appears in the rostral part of the diencephalic roof (Fig. 106). This *anterior choroid plexus* grows into the third ventricle. Closely associated plexuses push through each foramen of Monro into the lateral ventricles of the telencephalic lobes, constituting the *lateral choroid plexuses* (Fig. 100). It should perhaps be emphasized that the blood vessels of the choroid plexuses do not break through the brain roof but push it ahead of them so that although they appear to lie in the ventricles they are always separated from the lumen by a thin enveloping layer derived from the ependymal layer of the dorsal wall of the neural tube.

Once established, the lateral lobes of the telencephalon undergo exceedingly rapid growth. Their extension rostrally conceals the median portion of the telencephalon, and their even greater expansion dorsally and caudally eventually covers the entire diencephalon and mesencephalon (Fig. 87). At first the telencephalic lobes are smooth in contour and without striking local differentiations (Fig. 87, B–E). Relatively late in development they become much convoluted and certain regional divisions become clearly marked.

A conspicuous fissure called the sulcus rhinalis divides the ventral part of the telencephalic lobes from the dorsal (Fig. 87, F). The region ventral to the sulcus rhinalis is chiefly concerned with the olfactory sense and is, therefore, often called the *rhinencephalon*. It includes the olfactory bulb, the olfactory tract, and the pyriform lobe (Fig. 87, F). The rhinencephalon reaches its maximum development in lower forms. In higher mammals it becomes largely overshadowed by the tremendous growth of the more dorsal portions of the cerebral cortex.

Dorsal to the sulcus rhinalis the outer walls (pallium) of the telencephalic vesicles constitute the non-olfactory portions of the cerebral cortex. These cortical areas are phylogenetically the newest portions of the brain. In them are located the suprasegmental centers concerned with memory, voluntary action, and inhibitory control (Fig. 80, arc 7).

In forms such as the mammals, where the cerebral hemispheres are especially highly developed, the cortex is extensively folded. All the principal folds (gyri) and grooves (sulci) are named and the association centers for various special functions have, with considerable

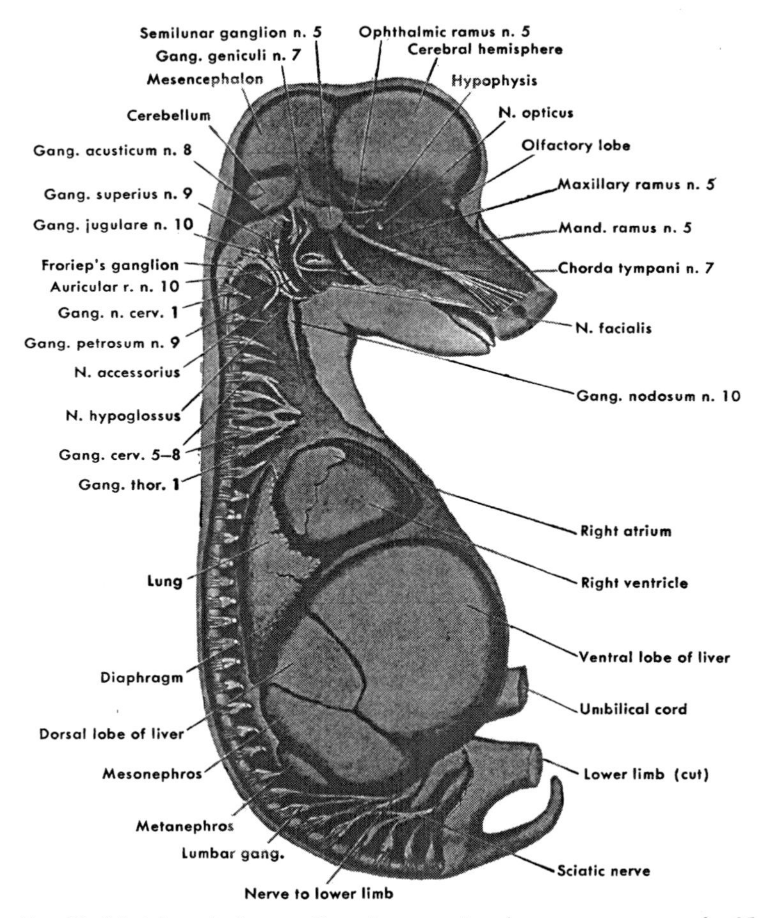

FIG. 90. Model made from a dissection exposing the nervous system of a 35 mm. pig embryo. (After Prentiss.)

accuracy, been located in specific areas. It would carry us beyond the scope of this book, however, were we to attempt to do more than become familiar with the major regional divisions of the cerebral hemispheres known as the frontal, parietal, temporal, and occipital lobes (Fig. 87).

The *corpus striatum* is developed in the mantle layer of the ventro-lateral walls of the telencephalon. Its location in the telencephalic floor makes it more closely associated positionally with the rhinencephalic region than with the pallial region which gives rise to the cerebral cortex. In its development it tends to bulge from either side in toward the lumen of the lateral ventricles (Figs. 99 and 100), becoming eventually one of the conspicuous internal landmarks in the telencephalic region. It derives its name from the fact that when sectioned it shows alternate layers of fibers and cellular substance arranged in more or less regular bands. Two large nuclear masses, the *caudate nucleus* and the *lentiform nucleus*, make up the major part of the corpus striatum. Its connections are very complex and many of them are not as yet entirely worked out. Whatever its other activities may be, it is clearly involved in the coördination of certain complex muscular activities (Fig. 80, arc 6). It also appears to exert a steadying influence on voluntary muscular actions generally, because interference with it is followed by the appearance of tremors during movement.

V. The Cranial Nerves

In dealing with the spinal nerves we recognized four functional types of neurons, somatic afferent, somatic efferent, visceral afferent, and visceral efferent (Fig. 85). In the cranial nerves we find these same types of neurons and in addition other subtypes with more restricted distribution and more specialized function. The eye and the ear, for example, are very highly differentiated and sharply localized somatic sense organs. Therefore their fibers are set apart from the general somatic afferent category as SPECIAL somatic afferent fibers. The musculature in the pharyngeal region differs from other visceral musculature in that it is striated. So the motor fibers to it are distinguished from other visceral efferent fibers by calling them SPECIAL visceral efferent neurons.

The spinal nerves are segmentally arranged and all of them are built on the same general plan. The cranial nerves have lost their segmental arrangement and become very highly specialized. Some of them contain both sensory and motor fibers as is the case with the spinal nerves. These are called mixed nerves. Some contain only motor fibers and others only sensory fibers. No single nerve contains all the types of fibers which occur in the cranial nerves as a group as diagramed in figure 91.

In both spinal and cranial nerves, afferent fibers arise from cell bodies outside the neural tube (cf. Figs. 85 and 91). Thus the cranial nerves which carry afferent fibers have ganglia composed of clusters of their cell bodies situated just outside the brain wall (Fig. 92). Likewise the efferent fibers in the spinal and the cranial nerves are similar in that they arise from cell bodies inside the wall of the neural tube. In the spinal nerves these cell bodies lie in the ventral and lateral horns of the gray matter of the cord. In the cranial nerves their position is homologous for they lie in clusters (nuclei) in the basal plate of the brain wall (cf. Figs. 85 and 91).

If one considers these general facts it becomes apparent that, although the cranial nerves differ from the spinal nerves in many respects, we find the same types of neurons involved and the same characteristic difference in the position of the cell bodies which so clearly sets apart afferent and efferent fibers. Put in another way, the cranial and spinal nerves contain the same types of components differently grouped. It will be helpful to keep this in mind in considering the various cranial nerves.

The Olfactory Nerve (I). Unlike other sensory nerves the olfactory nerve lacks a ganglion. It is peculiar, also, in that all its fibers are non-medullated. These fibers arise from cells in the epithelial layer lining the olfactory pits (Fig. 93). Thence they grow centripetally into the olfactory bulbs. In the olfactory bulbs the fibers of the nerves terminate in synapses with other neurons which relay the impulses along the olfactory tracts (Fig. 87) to centers in the rhinencephalon.

The Optic Nerve (II). As is the case with the olfactory nerve, the optic nerve fibers arise from peripherally located cells and grow centripetally. Neuroblasts situated in the sensory layer of the retina in close association with its photosensitive cells send out processes which leave the optic cup through the choroid fissure. These fibers then traverse the grooved ventral surface of the optic stalk and enter the brain in the diencephalic floor. At their point of entrance the two optic nerves[2] intersect. At the intersection, part of the fibers from each nerve cross over to the opposite side so that each eye has central connections with both sides of the brain. It will be recalled that the point of the nerve intersection and fiber crossing is known as the optic

[2] Since both the photosensitive cells themselves and the retinal ganglion cells sending out their associated nerve fibers arise in the walls of the optic cup, which is in turn an evagination of the embryonic fore-brain, what we commonly call the optic nerve is, strictly speaking, not a nerve but a fiber tract arising within a modified portion of the brain wall.

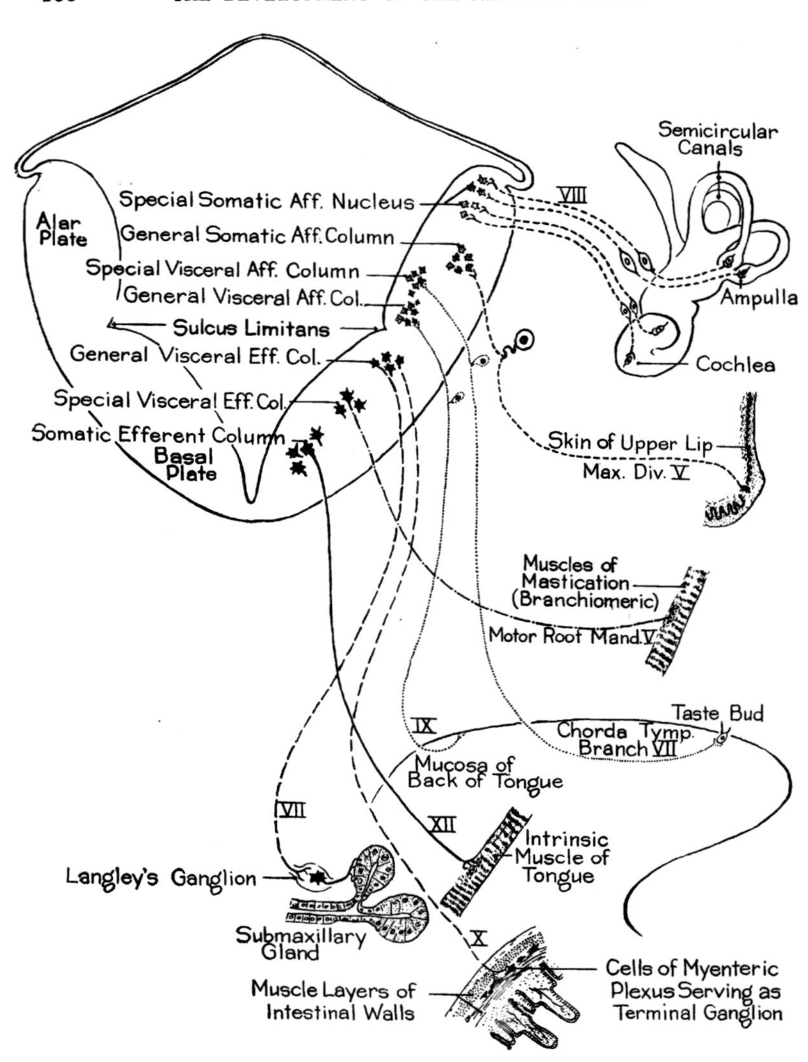

FIG. 91. Diagram showing the central relations of the various types of fibers in cranial nerves. (Patten: "Human Embryology," The Blakiston Company.) It should be emphasized that the diagram is a schematic composite and that no one cranial nerve contains all the types of fibers shown.

(*Continued on facing page.*)

chiasma (Fig. 65). From the chiasma the fibers pass along the lateral walls of the diencephalon to the visual reflex centers in the superior colliculi (Fig. 80, arc 5b), and by way of relay centers (geniculate nuclei) impulses are sent also to visual correlation centers in the cerebral cortex.

The Oculomotor Nerve (III). As its name implies, the oculomotor nerve contains efferent fibers to muscles moving the eye. The cluster of neuroblasts from which it arises is located in the basal plate of the mesencephalon. Its fibers have internal relations in general comparable to those indicated in figure 91 for the fibers of nerve XII leading from the somatic efferent nucleus to the intrinsic muscles of the tongue. Emerging from the floor of the mesencephalon (Figs. 92 and 93) they pass directly to the orbital region and innervate the inferior oblique, and the superior, inferior, and internal rectus muscles of the eyeball.

The Trochlear[3] Nerve (IV). The trochlear nerve is a motor nerve to the superior oblique muscle of the eye. Its nucleus of origin, like that of the third nerve, is located in the basal plate of the mesencephalon. It is peculiar in that its fibers do not leave directly from the ventro-lateral walls of the brain as usually happens in a motor nerve. Instead they pass to the dorsal wall of the mesencephalon (Figs. 92 and 93) and cross before emerging.

The Trigeminal Nerve (V). The trigeminal nerve takes its name from the fact that it has three main divisions, the ophthalmic, the maxillary, and the mandibular (Figs. 89, 90, and 92). As is indi-

[3] Not even sufficient knowledge of Greek to recognize in the word *trochlear* the root meaning pulley is of much immediate help in understanding its significance as used in anatomy. One has to find that the nerve takes its name from an old appellation of the superior oblique muscle to which it runs. This muscle was formerly known as the trochlear muscle because its tendon passes through a pulley-like fibrous loop attached to the eye-socket.

FIG. 91—(*Continued*)

As was the case with the spinal nerves (Fig. 85), the efferent nuclei or clusters of efferent nerve cells are located ventro-laterally, in the basal plate of the neural tube wall. The afferent fibers of the cranial nerves have their cell bodies located outside the neural tube in ganglia (cf. Fig. 92). The afferent nuclei or columns in the alar plate of the walls of the neural tube are clusters of cell bodies belonging to neurons which relay the incoming impulses to other parts of the brain. (Cf. Fig. 80, arc 5a.) For further explanation see text.

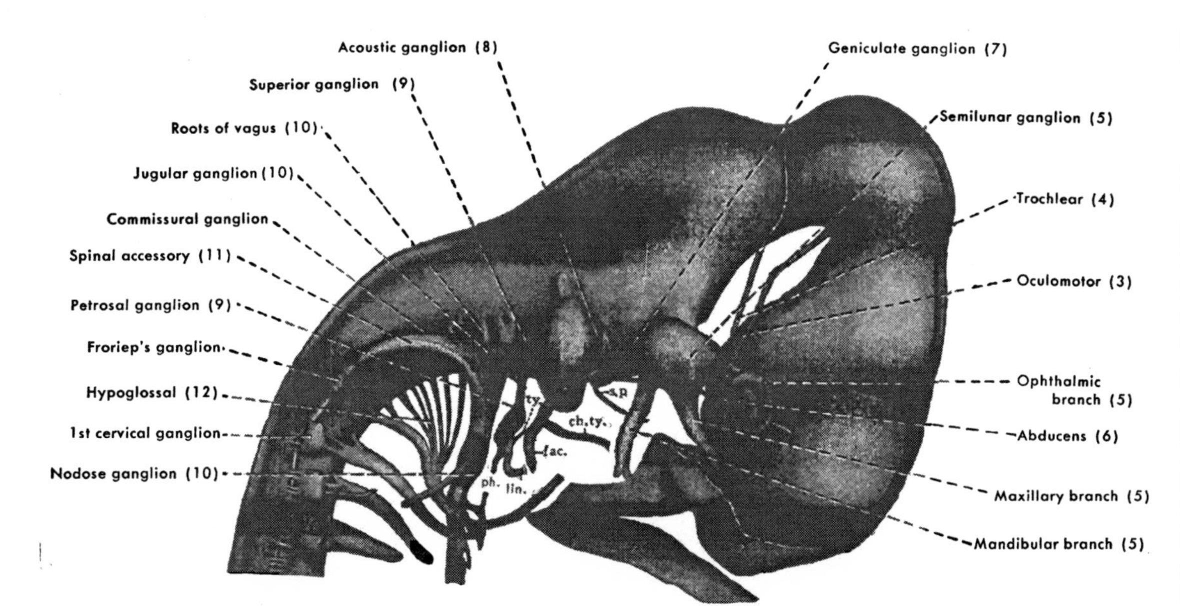

Fig. 92. Reconstruction of the brain and cranial nerves of a 12 mm. pig embryo. (After F. T. Lewis, *Am. Jour. Anat.*, Vol. 2.)

Abbreviations: ch.ty., chorda tympani branch of the 7th (facial) nerve; fac., facial nerve; lin., lingual branch of 9th nerve; ph., pharyngeal branch of 9th nerve; s.p., greater superficial petrosal nerve; ty., tympanic branch of 9th nerve.

cated by its large semilunar ganglion, the fifth nerve has great numbers of sensory fibers. There are, nevertheless, enough motor fibers associated with its mandibular branch so it must be regarded as a mixed nerve (Figs. 91 and 94). The names of its branches clearly indicate its distribution to the facial region. Its sensory fibers have relations of the nature indicated in figure 91 by the fiber leading from the skin of the lip to the general somatic afferent column.

The Abducens Nerve (VI). The abducens nerve takes its name from the fact that it controls the external rectus muscle, contraction of which makes the eyeball rotate outwards. Its nucleus lies in the basal plate of the myelencephalon from which its fibers emerge ventrally just caudal to the pons, and pass toward the orbit (Figs. 92 and 94).

The Facial Nerve (VII). The facial nerve is primarily motor but the presence of the geniculate ganglion on its root shows that it carries also some sensory fibers. A large part of its sensory fibers pass by way of the chorda tympani branch (Figs. 91 and 92) to join the mandibular branch of the fifth nerve. These fibers are concerned with the sense of taste. Its motor fibers arise from a nucleus situated in the basal plate of the myelencephalon and innervate the muscles of facial expression.

The Auditory Nerve (VIII). At first the ganglionic mass from which the fibers of the eighth nerve arise is closely associated with the geniculate ganglion of the seventh nerve (Fig. 92). Gradually these ganglia become entirely distinct. Still later the ganglion of the eighth nerve divides into two parts, a vestibular ganglion and a spiral ganglion. With the division of the ganglion, the nerve fibers arising from its cells become grouped into two main bundles, one associated with each ganglion. Meanwhile the otic vesicle has differentiated into two distinct parts, the cochlea, which is the organ of hearing, and the group of semicircular canals which, together with the utriculus and sacculus, constitute an organ of equilibration. The spiral ganglion and the cochlear branch of the eighth nerve become associated with the auditory part of the mechanism. The vestibular ganglion and its branch of the eighth nerve become associated with the semicircular canals. The auditory and equilibratory fibers have the general relations indicated in figure 91 by the sensory fibers connecting, respectively, with the cochlea and with the ampullae of the semicircular canals.

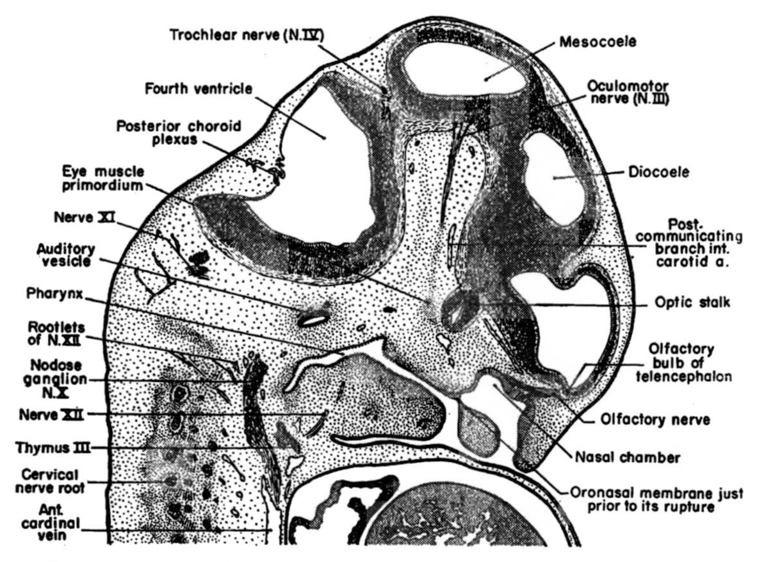

FIG. 93. Drawing (× 14) of parasagittal section of head of 15 mm. pig embryo. The section is to the right of the mid-line, in a plane especially favorable for showing the relations of the nasal pits and of the olfactory (I) and oculomotor (III) nerves.

The Glossopharyngeal Nerve (IX). The glossopharyngeal is a mixed nerve but by far the greater number of its fibers are sensory. The ganglion cells from which these sensory fibers arise are grouped in two clusters, one near the root of the nerve (superior ganglion), and one farther peripherally on its course (petrosal ganglion) (Figs. 92 and 94). The cell bodies in the superior ganglion give rise to fibers which innervate a small cutaneous area of the external ear. These neurons are, therefore, general somatic afferent. The petrosal ganglion contains cell bodies which give rise to visceral afferent fibers. Some of these, being concerned with general sensibility in the region about the root of the tongue, are general visceral afferent (Fig. 91). Other fibers innervate taste buds in the back parts of the tongue and so are classified as special visceral afferent. The efferent fibers arise from nuclei in the basal plate of the myelencephalon. Some of these fibers innervate the stylopharyngeus muscles. These are special visceral efferent neurons. Others are secretory fibers to the parotid gland by way of the otic ganglion. These are general visceral efferent fibers.

The Vagus Nerve (X). The vagus is a mixed nerve carrying five different types of fibers. General somatic afferent fibers arise from cells in its jugular ganglion and extend peripherally to the skin in the region of the external ear. General visceral afferent fibers arise from cells in the nodose ganglion and extend peripherally to the pharynx, larynx, trachea, esophagus, and the thoracic and abdominal viscera. Special visceral afferent fibers having cells of origin in the nodose ganglion carry gustatory impulses from scattered taste buds in the region of the epiglottis. From nuclei of origin in the myelencephalon special visceral efferent fibers extend to striated muscles of the pharynx and larynx. General visceral efferent fibers run to various terminal ganglia whence the impulses are relayed to the visceral musculature over second-order motor neurons (Fig. 91).

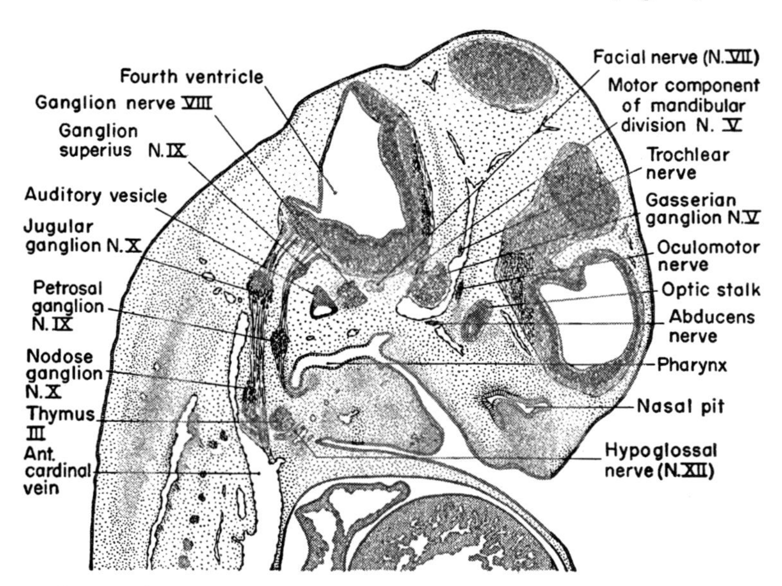

FIG. 94. Drawing (× 14) of parasagittal section of head of 15 mm. pig embryo. The plane of section is slightly farther to the right than that shown in the preceding figure. It is particularly favorable for showing the position of origin, and the ganglia, of the trigeminal (V), glossopharyngeal (IX), and vagus (X) nerves.

The Accessory Nerve (XI). The commissural ganglion (Fig. 92) which appears so closely associated with the accessory nerve is really a continuation of the jugular ganglion of the vagus. Froriep's ganglion usually disappears in the adult so the accessory nerve is left without ganglia. Its fibers, practically all efferent, originate not only from the posterior part of the myelencephalon but also from the first five or six segments of the spinal cord (Fig. 92). A large number of the fibers of the accessory nerve run with the general visceral efferent fibers of the vagus nerve to sympathetic ganglia from which the motor impulses are relayed to the smooth muscle of the viscera. Other fibers (special visceral efferent) from the accessory join similar vagus fibers to striated muscles in the pharynx and larynx. Most of the fibers arising from the cervical part of the cord turn off in the external ramus to end in the trapezius and sterno-cleido-mastoid muscles.

The Hypoglossal Nerve (XII). The hypoglossal nerve is composed practically entirely of somatic motor fibers. They arise from an elongated nucleus in the posterior part of the myelencephalon (Fig. 91) and emerge in several separate roots which join to form a single main trunk (Fig. 92). Peripherally they are distributed to the muscles of the tongue.

CHAPTER 9

The Development of the Digestive and Respiratory Systems and the Body Cavities

I. The Digestive System

In considering the structure of young embryos we traced the walling in of the primitive gut tract by entoderm, its regional division into fore-gut, mid-gut, and hind-gut, and the establishment of the oral and anal openings by the breaking through of the stomodaeal depression cephalically and the proctodaeal depression caudally (Figs. 16, 37, and 65). In embryos of from 9 to 12 mm., local differentiations in the gut tract clearly foreshadowed the development of certain organs and gave indications of the impending establishment of others. Starting with these now familiar conditions as a basis we shall trace briefly the more important steps by which the adult structure and relations of the various organs are established.

Oral Cavity. The oral cavity of the adult and its various special structures are derived from the stomodaeal region of the embryo. The entire face and jaw complex is formed from processes which arise about the margins of the stomodaeum. The progressive growth of these abutting structures results in a deepening of the originally shallow stomodaeal depression to form the oral cavity. An idea of the extent to which this growth progresses can be gained from the fact that the point of rupture of the stomodaeal membrane (oral plate) comes to lie, in the adult, at about the level of the tonsils. So many processes of special interest are involved in the changes which go on in this region that it has seemed wise to dismiss them for the present with this general statement and return to them later for special consideration (Chap. 13).

The Pharyngeal Region. In embryos of about 4 to 6 mm. the cephalic part of the fore-gut has become differentiated as the pharynx. Greatly compressed dorso-ventrally, the pharynx has a wide lateral extent with a series of pouch-like diverticula pushing out on either side between the branchial arches (Figs. 41, B, and 95). This stage of

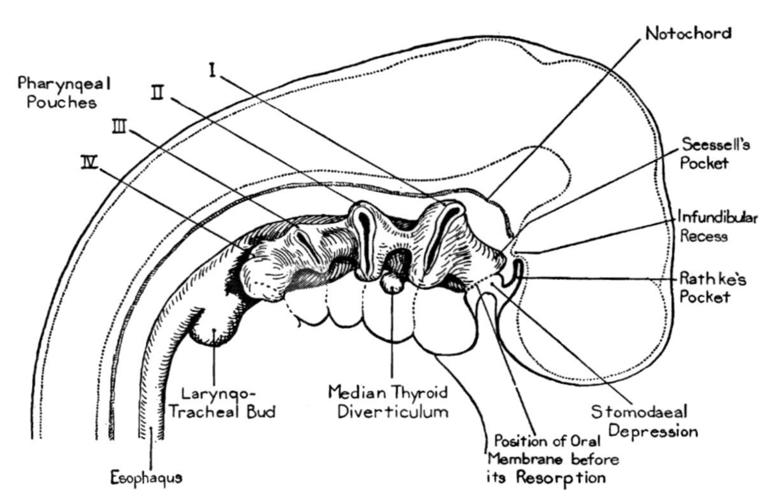

FIG. 95. Lateral view of the pharynx of a young mammalian embryo with its relations to important adjacent structures indicated. (Patten: "Human Embryology," The Blakiston Company.) The contours of the visceral arches are suggested and the broken lines between them indicate the location of the external gill furrows. The drawing is semischematic and equally applicable to conditions in a 4-week human embryo or a 5 mm. pig embryo.

the pharynx is a recapitulation of conditions which had an obvious functional significance in water-living ancestral forms. For the pharyngeal pouches of the mammalian embryo are homologous with the inner portion of the gill slits. The repetition of race history is here, as so frequently happens, slurred over. Although in the mammalian embryo the tissue closing the gill clefts becomes reduced to a thin membrane consisting of nothing but a layer of entoderm and ectoderm with no intervening mesoderm whatever (Fig. 62), this membrane rarely disappears altogether. Occasionally the more cephalic of the pharyngeal pouches break through to the outside, establishing open gill slits, but in such cases the opening is very short-lived and the clefts promptly close again.

Like many other vestigial structures which appear in the development of higher forms, the pharyngeal pouches give rise to organs having a totally different functional significance from the ancestral structures they represent. It is as if, to speak figuratively, nature was too economical to discard entirely structures rendered functionally obsolete by the progress of evolution, but rather conserved them in part at least and modified them to carry on new activities.

Discussion of the processes whereby various parts of the original pharyngeal apparatus become converted into other structures would involve too many details to permit of inclusion here. A bare statement of what these pharyngeal derivatives are and where they arise must suffice.

The main pharyngeal chamber of the embryo, that is, the central portion in distinction to its various diverticula, becomes converted directly into the pharynx of the adult. In this process its lumen is simplified in configuration and relatively reduced in extent. An important factor in these changes is the separation of various diverticula from the main part of the pharynx. The cell masses thus originating migrate into the surrounding tissues and there undergo divergent differentiation.

The first pair of pharyngeal pouches, extending between the mandibular and hyoid arches, come into close relation at their distal ends with the auditory vesicles (Fig. 60). They give rise, on either side, to the *tympanic cavity* of the middle ear and to the *Eustachian tube.*

The second pair of pouches become progressively shallower and

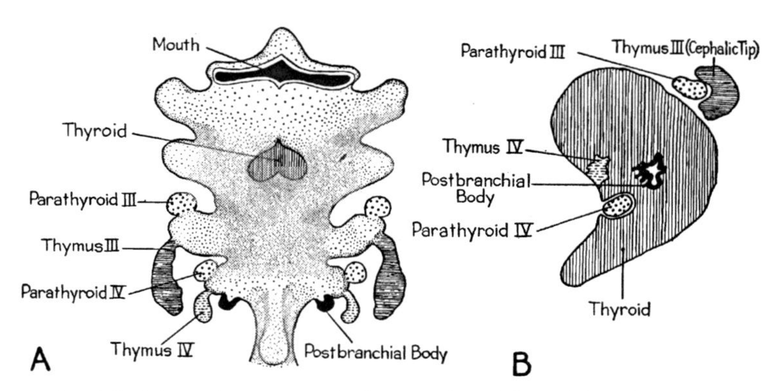

FIG. 96. Schematic diagrams indicating the origin and later interrelations of some of the derivatives of the embryonic pharynx. (Modified from Swale-Vincent.) In the schematic cross section of one lobe of the adult thyroid (B) the numbers attached to parathyroids and thymus refer to their pouches of origin as indicated in A. Note that the thymic tissue which arises from the fourth pouch is drawn in lightly to indicate that it is not well developed in all mammals.

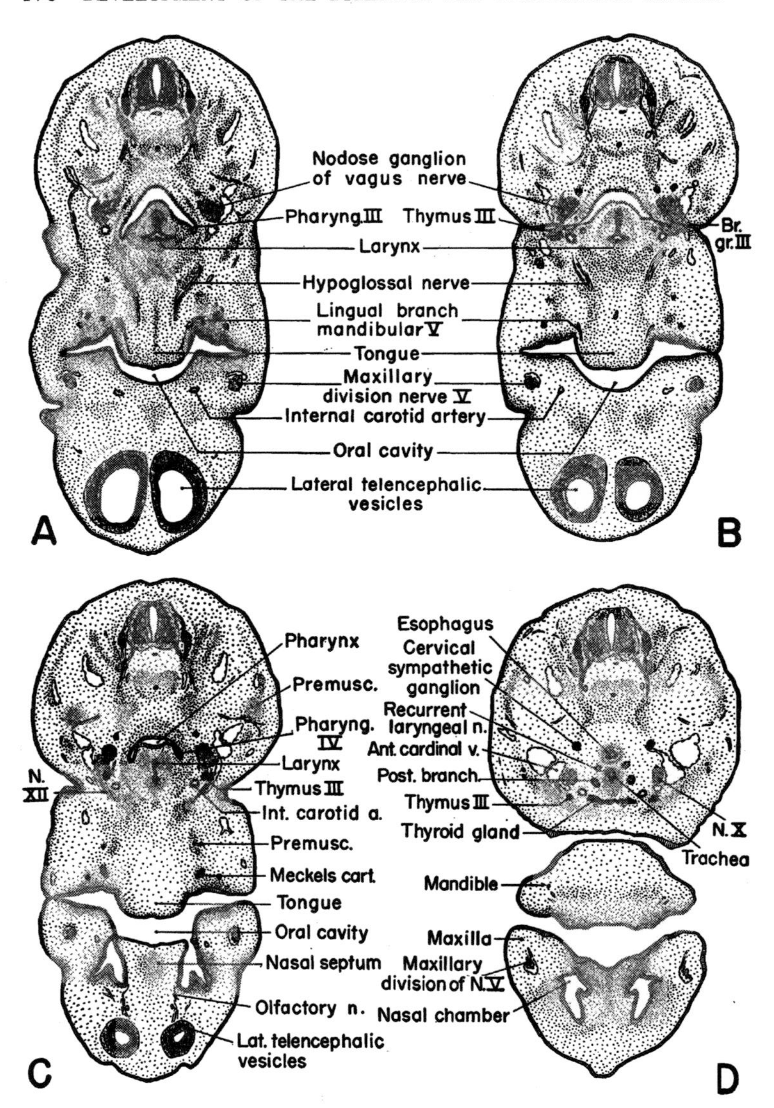

FIG. 97. Drawings (× 11) of transverse section through the pharyngeal region of a 15 mm. pig embryo. A, Upper laryngeal level. B, Level of third

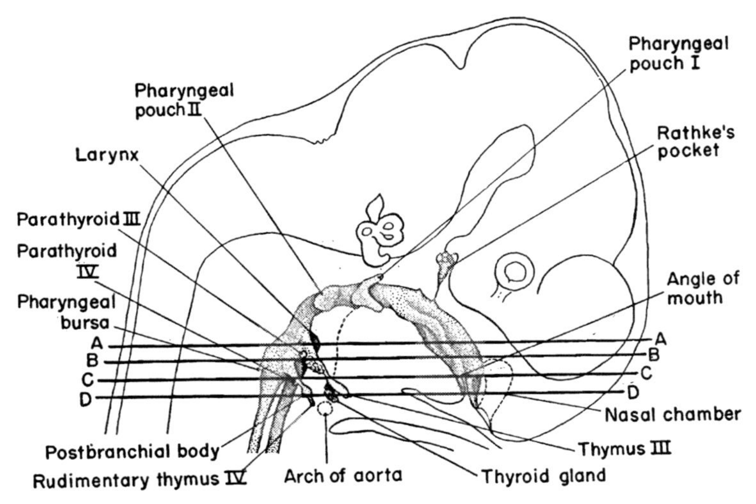

FIG. 98. Pharynx of 15 mm. pig embryo schematically represented in relation to the outlines of other cephalic structures. (Adapted from several sources.) The heavy horizontal lines indicate the levels of the correspondingly lettered sections in the preceding figure.

less conspicuous. Late in fetal life the faucial tonsils are formed by the aggregation of lymphoid tissue in their walls, and vestiges of the pouches themselves persist as the *supratonsillar fossae.*

From the floor of the pharynx, in the mid-line at about the level of the constriction between the first and the second pair of pharyngeal pouches, a diverticulum is formed which is destined to give rise to the thyroid gland (Figs. 95 and 96). The mass of epithelial cells making up the walls of this evagination push into the underlying mesenchyme, break away from the parent pharyngeal epithelium, and migrate down into the neck (Figs. 97, D, 98, and 138). Only after arriving in its

---

FIG. 97—(*Continued*)

pharyngeal pouch. C, Level of fourth pharyngeal pouch. D, Through the neck caudal to the level of the pharynx and larynx.

The level of each of the sections represented in this figure is indicated by the correspondingly lettered line in the next figure.

Abbreviations: Br. gr. III, third branchial groove; N. XII, hypoglossal nerve; Pharyng. III, third pharyngeal pouch; Pharyng. IV, fourth pharyngeal pouch; Premusc., premuscular concentration of mesenchyme.

definitive location, relatively late in development, does the thyroid primordium undergo its final characteristic histogenetic changes.

The third and fourth pairs of pharyngeal pouches give rise to outgrowths which are involved in the formation of the parathyroid glands, the thymus, and the post-branchial bodies. There are two pairs of parathyroid glands, usually spoken of as parathyroids III and parathyroids IV because they arise from the third and the fourth pharyngeal pouches (Figs. 96, A, and 98). As was the case with the thyroid, the parathyroid primordia soon break away from their points of origin and migrate into the neck. Here, as their name implies, they are positionally more or less closely associated with the thyroid. Parathyroids IV are particularly likely to become adherent to the thyroid capsule or even to become partially embedded in the substance of the gland (Fig. 96, B).

The thymus in the mammalian group is derived from outgrowths from the more ventral portions of the third and fourth pharyngeal pouches (Figs. 96, A, and 98). In different species there is considerable difference in the relative conspicuousness of the two pairs of primordia. In most of the higher mammals the primordia arising from the third pouches are much the more important thymic contributors. This is the situation for the pig as well as for man. In pig embryos of the 15–17 mm. range, however, it is usually possible to make out a rudimentary thymus IV (Fig. 98). The characteristic histogenetic changes in the thymus occur relatively late in development, and even in 15–17 mm. embryos thymus III is but a slender pair of cell cords growing into the tissue at the base of the neck (Fig. 97, C, D).

The post-branchial bodies are structures of problematical significance. Arising as they do on the caudal face of the fourth pharyngeal pouches, many observers regard them as rudimentary fifth pharyngeal pouches. When the post-branchial bodies detach themselves from their site of origin they lie in the loose mesenchymal tissue (Fig. 97, D) close to the route followed by the thyroid gland in its descent. As the thyroid expands laterally the tissue of the post-branchial bodies becomes embedded in it on either side (Fig. 96, D). There is still difference of opinion as to whether this post-branchial tissue contributes to the formation of true thyroid glandular tissue or remains merely as an inconspicuous vestigial cell mass in the substance of the thyroid gland. Those who are convinced that these buds from the caudal face of the fourth pouches form true thyroid tissue generally designate them as lateral thyroid primordia. The non-committal term post-branchial bodies seems preferable until the real significance of these structures is more satisfactorily known.

Esophagus. The pharynx becomes abruptly narrowed just caudal to the most posterior pouches. It is at this point that the primitive gut gives rise ventrally to the tracheal outgrowth (Figs. 40 and 65).

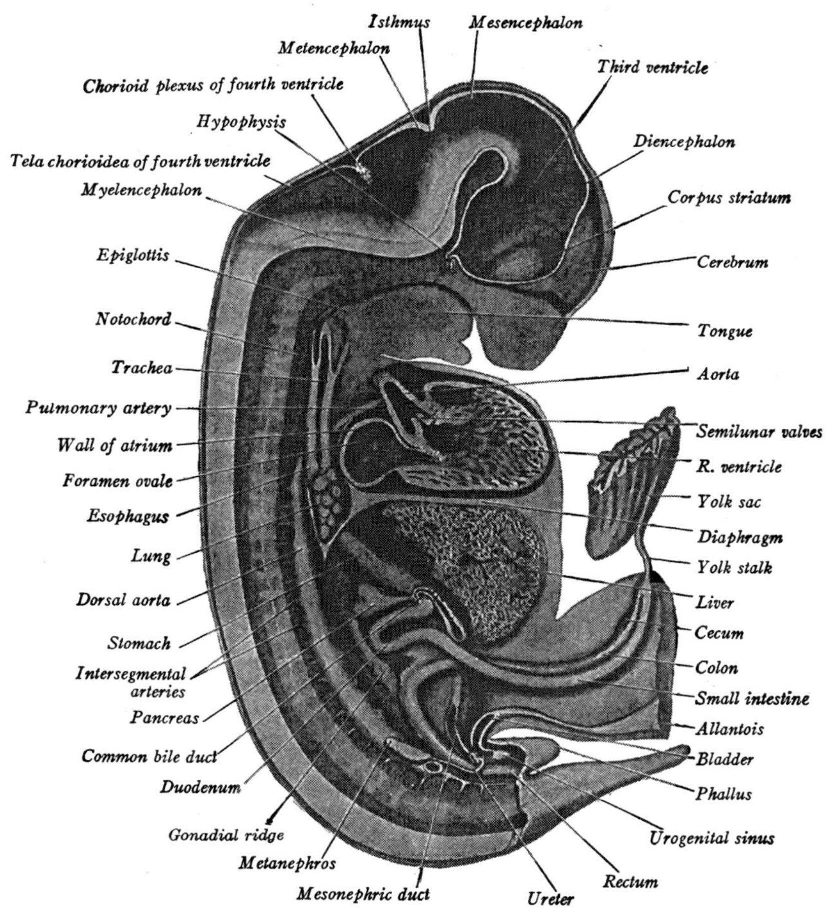

FIG. 99. Pig embryo of 18 mm. dissected to the mid-line to show the relations of the alimentary tract. (After Prentiss.)

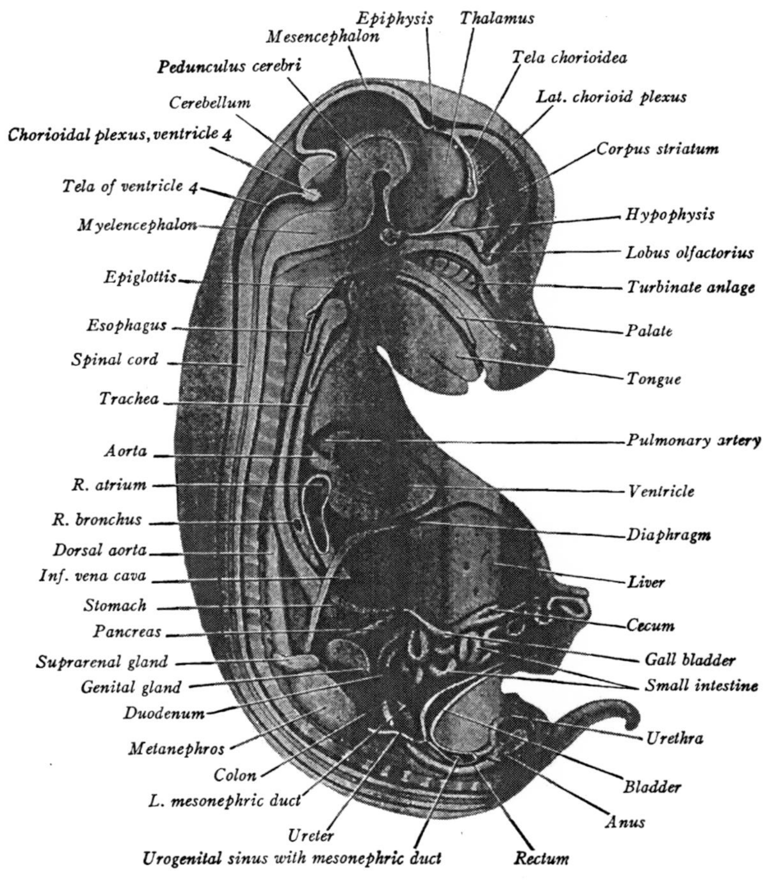

FIG. 100. Pig embryo of 35 mm. dissected to the mid-line to show the relations of the alimentary tract. (After Prentiss.)

The region of narrowing where the trachea becomes confluent with the gut tract may be regarded as the posterior limit of the pharynx. From this point to the dilation which marks the beginning of the stomach the gut remains of relatively small and uniform diameter and becomes the esophagus (Figs. 99 and 100). The original entodermal lining of the primitive gut gives rise only to the epithelial lining of the esophagus and to its glands. The connective tissue and muscle coats of the esophagus are derived from mesenchymal cells which gradually become concentrated about the original epithelial tube (Fig. 97, D).

Stomach. The region of the primitive gut which is destined to become the stomach is, in embryos of 10 mm., more or less clearly marked by a dilation (Figs. 60 and 64). Its shape, even at this early stage, is strikingly suggestive of that of the adult stomach. Its position is, however, quite different.

In young embryos the stomach is mesially placed with its cardiac (esophageal) end somewhat more dorsal in position than its pyloric (intestinal) end. It is slightly curved in shape, with the convexity facing dorsally and somewhat caudally and the concavity facing ventrally and somewhat cephalically (Fig. 99). The positional changes by which it reaches its adult relations involve two principal phases: (1) the stomach is bodily shifted in position so its long axis no longer lies in the sagittal plane of the embryo but diagonally across it, and (2) there is a concomitant rotation of the stomach about its own long axis so that its original dorso-ventral relations are altered as well. These changes in position are schematically indicated in figure 101. The shift in axis takes place in such a manner that the cardiac end of the stomach comes to lie to the left of the mid-line and the pyloric end to the right. Meanwhile rotation has been going on. In following the progress of rotation the best point of orientation is the line of attachment of the primary dorsal mesentery (Fig. 101). While the stomach occupies its original mesial position the mesentery is attached to it mid-dorsally, along its convex curvature (Fig. 111). As the stomach continues to grow in size and depart from the sagittal plane of the body it rotates about its own long axis. The convex surface to which the mesentery is attached and which was at first directed dorsally, now swings to the left. Since the long axis itself has in the meantime been acquiring an inclination, the greater curvature of the stomach comes to be directed somewhat caudally as well as to the left (Fig. 101, D).

The Omental Bursa. The change in position of the stomach necessarily involves changes in that part (*dorsal mesogastrium*) of the primary dorsal mesentery which suspends it in the body cavity (Figs. 101 and 111). The dorsal mesogastrium is pulled after the stomach and forms a pouch, known as the omental bursa. The opening from the general peritoneal cavity into the bursa is known as the *epiploic foramen* (*foramen of Winslow*). (See arrow in Fig. 101, D.)

The Intestines. The primitive gut is at first a fairly straight tube extending throughout the length of the body. Near its midpoint it

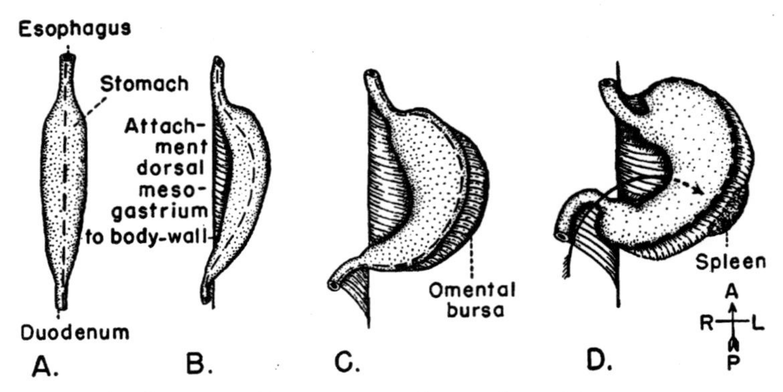

Fig. 101. Diagrams illustrating the changes in position of the stomach, and the formation of the omental bursa. The broken line indicates the attachment of the mesogastrium along that surface of the stomach which is primarily mid-dorsal. The arrow passes through the epiploic foramen into the omental bursa.

opens ventrally into the yolk-sac (Figs. 37 and 40). The first conspicuous departure from this condition is the formation of a hairpin-shaped loop in the future intestinal region. The closed end of this loop extends into the belly-stalk (Figs. 60 and 64). The yolk-stalk connects with the gut at the bend of the loop and forms an excellent point of orientation in following the series of foldings and kinkings by which the definitive configuration of the intestinal tract is established. The attachment of the yolk-stalk is just cephalic to what will be the point of transition (*ileo-cecal valve*) from small to large intestine. Thus all the gut between the yolk-stalk and the stomach becomes small intestine, and, except for about 2 feet of the terminal part of the small intestine, the gut caudal to the yolk-stalk goes to form the large intestine.

The characteristic coiling of the small intestine is the first to become evident (Figs. 100 and 102, B). The only change of significance which has taken place meanwhile in the large intestine is the establishment of the cecum as a definite pouch-like diverticulum (Fig. 102, B). But the large intestine does not remain long uncoiled. In the pig it attains a greater length and a more complicated configuration than in man, its final condition being that of a loop closely spiraled on itself (Fig. 102, E, F) and occupying a very conspicuous position among the abdominal viscera (Fig. 1).

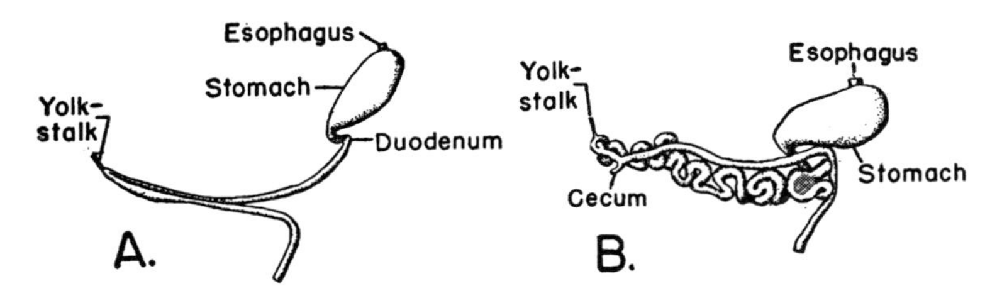

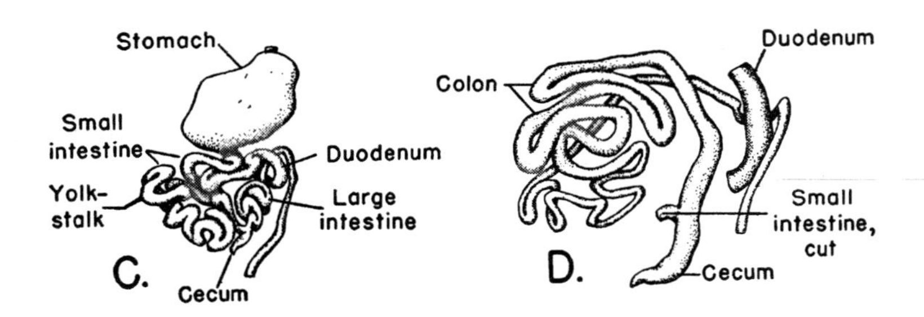

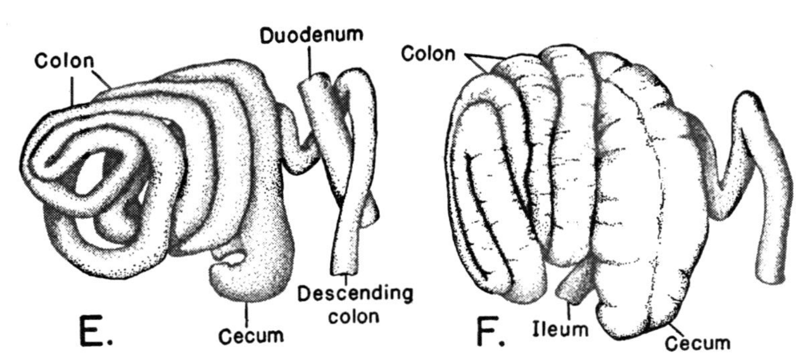

Fig. 102. Diagrams illustrating the development of the intestinal tract in pig embryos. (After Lineback.) A, 12 mm.; B, 24 mm.; C, 35 mm.; D, 75 mm.; E, 110 mm.; F, 1 month after birth.

A to C show the entire gastro-intestinal tract viewed from the left side. D to F show the large intestine only. The relations of the last three figures will be made apparent by comparing C and D, taking for orientation the cecum and that part of the duodenum which loops across the large intestine.

Rectum and Anus. The attainment of adult conditions at the extreme caudal end of the digestive tract is so intimately associated with the development of the urogenital openings that changes in the cloacal region as a whole can more profitably be taken up later in connection with the reproductive organs.

The Liver. Very early in development the diverticulum which gives rise to the liver is budded off from the entoderm of the primitive gut tract. In embryos as small as 4 mm. the hepatic diverticulum can be identified extending ventrad from the duodenal portion of the gut

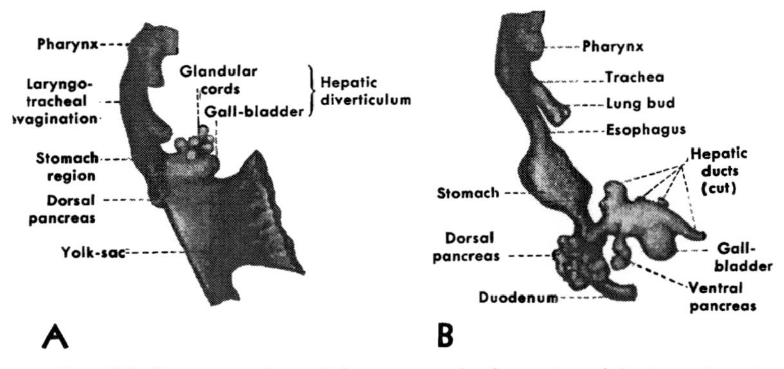

FIG. 103. Reconstructions of the gut tract in the region of the hepatic and pancreatic diverticula.
A, From 4 mm. pig embryo in the Carnegie Collection.
B, From 5.5. mm. pig embryo. (After Thyng, modified.)

(Fig. 103, A). This original diverticulum, in embryos of 5 to 6 mm., has become clearly differentiated into several parts (Figs. 40 and 103, B). A maze of branching and anastomosing cell cords grows out from it ventrally and cephalically. The distal portions of these cords give rise to the secretory tubules of the liver and their proximal portions form the hepatic ducts. Originating where the hepatic ducts become confluent is a dilation which is the primordium of the *gall-bladder*. Closer to the gut tract is a separate outgrowth of cells which constitutes the ventral primordium of the pancreas.

The later changes in the biliary duct region are shown in figures 104 and 105. The gall-bladder elongates very rapidly and its terminal portion becomes distinctly saccular. The narrower proximal portion of this limb of the diverticulum becomes the cystic duct. That portion of the original diverticulum which lies toward the duodenum from the entrance of the hepatic ducts is called the *common bile duct* (ductus choledochus).

The mass of branching and anastomosing tubules which are distal continuations of the hepatic ducts constitute the actively secreting portion of the liver. Their position and extent in embryos of various ages are shown in figures 40, 65, 99, 100, and 106. The organization of these secreting units in the liver is quite characteristic. The hepatic tubules are not packed so closely together in a framework of dense connective tissue as is usually the case in massive glands. Surprisingly

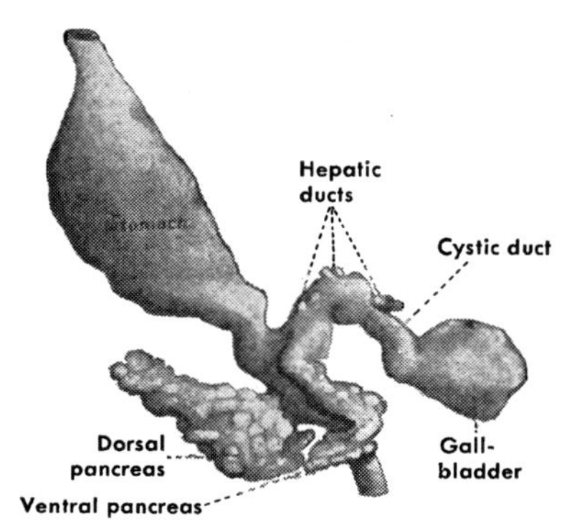

FIG. 104. Reconstruction of gut tract of a 9.4 mm. pig embryo showing pancreatic and hepatic diverticula (× 33). Compare with figure 60.

little connective tissue is formed between them and the intertubular spaces become pervaded by a maze of dilated and irregular capillaries known as sinusoids. This tremendously extensive meshwork of small blood vessels among the cords of liver cells is a condition which we shall find of great importance in the development of the circulatory system in this region.

The Pancreas. The pancreas makes its appearance in the same region and at about the same time as the liver. It is derived from two separate primordia which later become fused. One primordium arises dorsally, directly from the duodenal entoderm; the other[1] arises

[1] The ventral pancreatic diverticulum in a certain number of cases may be paired instead of single. It is probable that the usual unpaired diverticulum seen in mammalian embryos represents originally paired ventro-lateral diverticula.

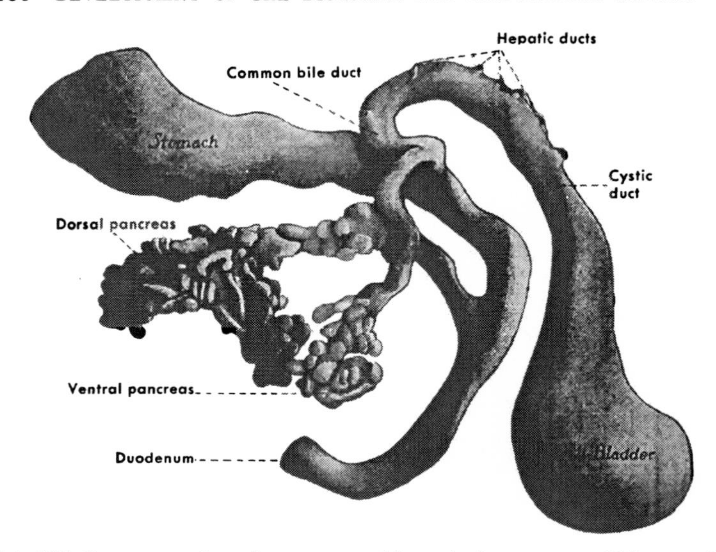

FIG. 105. Reconstruction of pancreas and hepatic duct system of 20 mm. pig embryo. (After Thyng, modified.)

ventrally, from the entoderm of the hepatic diverticulum (Fig. 103). As they increase in size, these two buds approach each other and eventually fuse (Figs. 104 and 105). The glandular tissue of the pancreas is formed by the budding and rebudding of cords of cells derived from this primordial mass. The terminal parts of the cords gradually take on the characteristic configuration of pancreatic acini while their more proximal portions form the duct system draining the acini.

There is, in different forms, considerable variation in the relations of the main pancreatic ducts which persist in the adult. In the horse and dog, for example, there are two ducts, a dorsal one (duct of Santorini) which opens directly into the duodenum, and a ventral one (duct of Wirsung) which opens into the duodenum by way of the common bile duct. These two ducts represent the two original pancreatic buds which appear in mammalian embryos generally. In other forms the two original ducts become confluent within the pancreas and the terminal portion of one duct only is retained. Thus in the sheep and in man the ventral duct persists communicating with the duodenum by way of the common bile duct, while the terminal portion of the dorsal duct usually atrophies. In the pig and the ox the

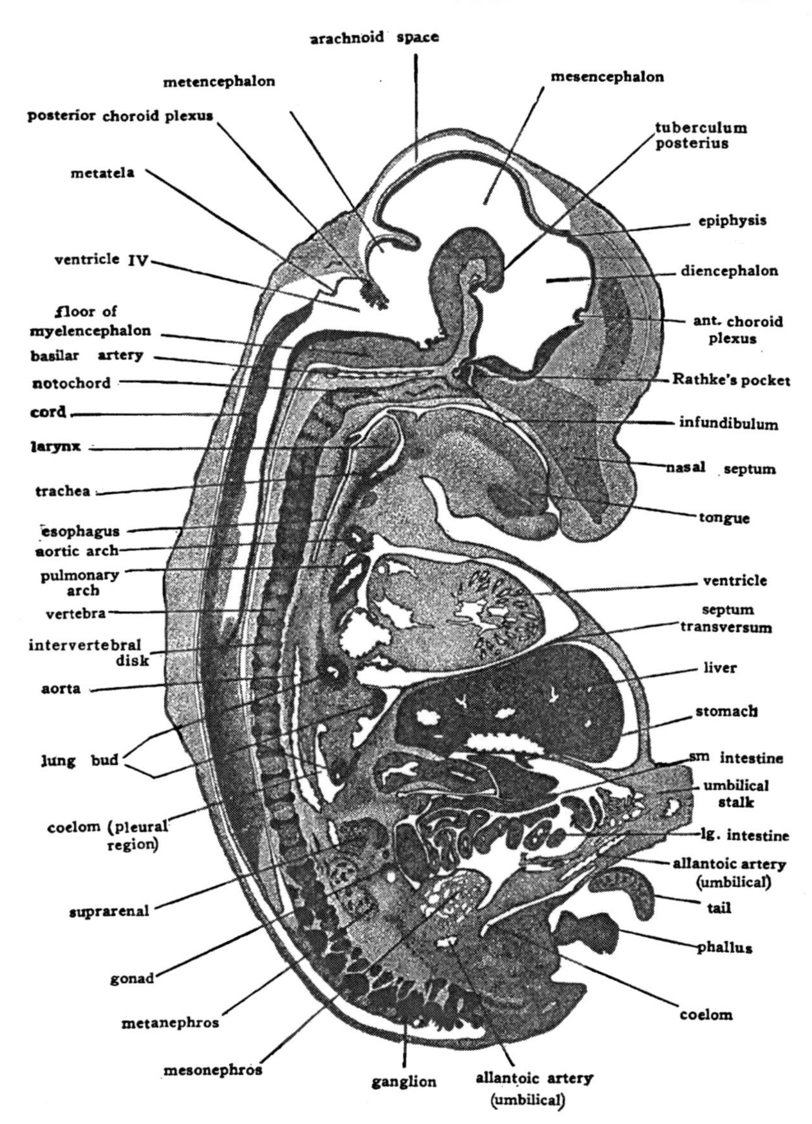

FIG. 106. Sagittal section of 24 mm. pig embryo. (After Minot.)

ventral duct ordinarily disappears and the dorsal one persists as the definitive pancreatic duct.

II. The Respiratory System

The Trachea. The first indication of the differentiation of the respiratory system is the formation of the laryngo-tracheal groove. This mid-ventral furrow in the primitive gut tract appears in embryos of about 4 mm. at the posterior limit of the pharyngeal region. As it becomes deepened it is constricted off from the gut except for a narrowed communication cephalically. Once established as a separate diverticulum, it grows caudad as the trachea, ventral to, and roughly parallel with, the esophagus (Figs. 40, 65, and 103).

The anatomical relations of the trachea in the embryo, even in early stages, are quite similar to adult conditions. We can recognize its communication with the posterior part of the pharynx as the future glottis, and the slightly dilated portion of the embryonic trachea just caudal to the glottis as foreshadowing the larynx (Fig. 106).

Only the epithelial lining of the adult trachea is derived from fore-gut entoderm. The cartilage, connective tissue, and muscle of its wall are formed by mesenchymal cells which become massed about the growing entodermal tube (Figs. 72, 97, D, and 161).

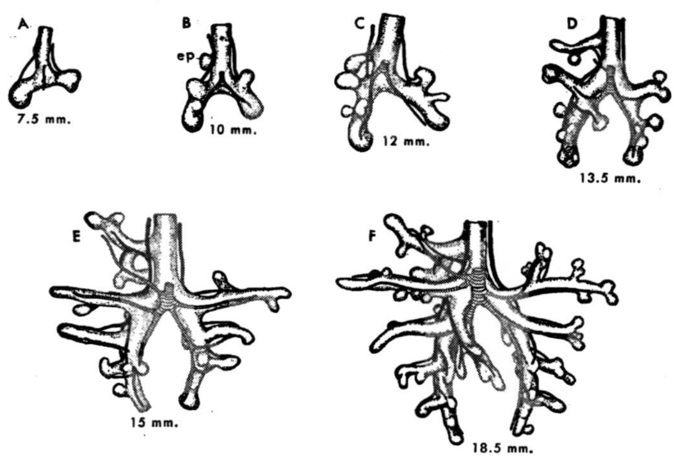

FIG. 107. Stages in the development of the trachea, bronchi, and lungs in the pig. (After Flint.) The pulmonary arteries are shown in black; the veins are cross hatched. Ep, bud of eparterial bronchus.

The Bronchi and Lungs. As the tracheal outgrowth lengthens, it bifurcates at its caudal end to form the two lung buds (Fig. 107, A). These in turn continue to grow and rebranch, giving rise to the bronchial trees of the lungs (Fig. 107). The terminal portions of the branches where cell proliferation is exceedingly active tend to remain somewhat bulbous. Later in development these terminal portions of the bronchial buds become still more dilated, their epithelium thins markedly, and they give rise to the characteristic air sacs of the lungs. As was the case with the trachea, the connective-tissue framework of the lung is derived from mesenchyme, which collects about the entodermal buds during their growth. The entoderm gives rise only to the lining epithelium of the bronchi and the air sacs. The pleural covering of the lungs is derived from splanchnic mesoderm pushed ahead of the lung buds in their growth (Fig. 113).

The lungs do not at first occupy the position characteristic of adult anatomy. In very young embryos they lie dorsal to the heart (Figs. 40 and 110). A little later when they have extended caudad, they are situated dorsal to the heart and liver (Figs. 99 and 138). The changes by which they eventually come to occupy their definitive position in the thorax can best be taken up in connection with the partitioning of the primitive coelom to form the body cavities of the adult.

III. The Body Cavities and Mesenteries

The body cavities of adult mammals are the pericardial cavity containing the heart, the paired pleural cavities containing the lungs, and the peritoneal cavity containing the viscera lying caudal to the diaphragm. All three of these regional divisions of the body cavity are derived from the coelom of the embryo. The general location and extent of the coelom are already familiar from the study of young embryos, but it may be well, nevertheless, to restate some of the more important relations here.

The Primitive Coelom. The coelom arises by the splitting of the lateral mesoderm on either side of the body into splanchnic and somatic layers (Fig. 108, A).

It is, therefore, primarily a paired cavity bounded proximally by splanchnic mesoderm and distally by somatic mesoderm. In forms such as birds and mammals which have highly developed extra-embryonic membranes, the coelom extends between the mesodermal layers of the extra-embryonic membranes beyond the confines of the

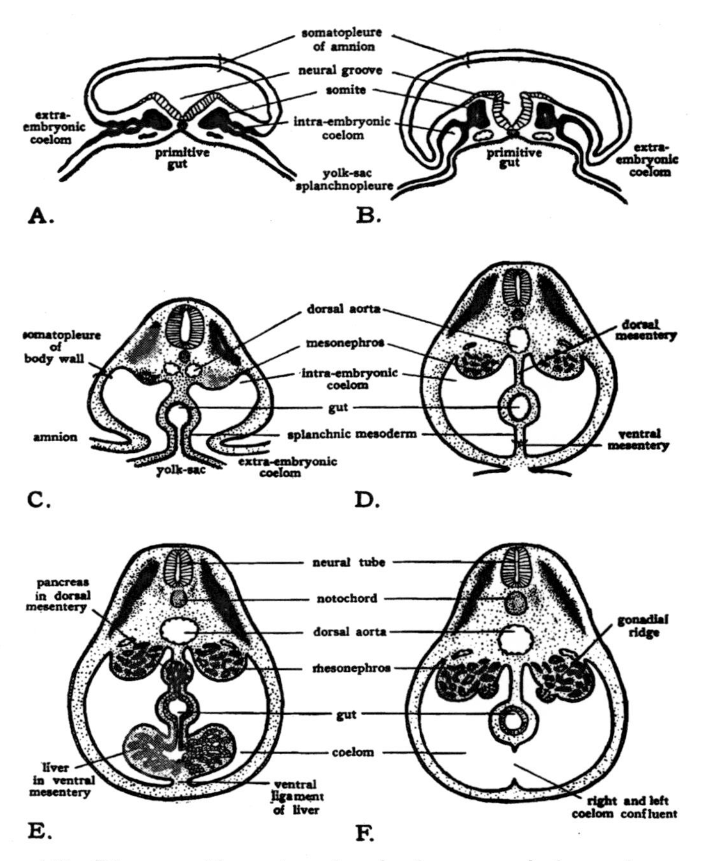

Fig. 108. Diagrams illustrating the development of the coelom and mesenteries.

developing body. In mammalia where the nutrition of the embryo depends on the uterine relations established by the extra-embryonic membranes, they develop exceedingly precociously. It is not surprising, in view of this fact, that the splitting of the mesoderm in mammals occurs first extra-embryonically and progresses thence toward the embryo (Fig. 108, A, B). When the body of the embryo is folded off from the extra-embryonic membranes the extra- and intra-embryonic portions of the coelom are thereby separated from each other, the last place of confluence to be closed off being in the region of the belly-stalk (Fig. 108, C). It is the intra-embryonic portion of the primitive coelom, thus delimited, which gives rise to the body cavities.

It will be recalled that the typical configuration of the mesoderm indicated diagrammatically in figure 108 does not pertain in the cephalic part of the embryo. The mesoderm in the head region consists of mesenchymal cells which wander in from the more definitely organized mesoderm located farther caudally in the body. Thus the intra-embryonic coelom established by the splitting of the lateral mesoderm extends headwards only to the level of the pharynx, and the heart is developed in its most cephalic portion (Figs. 43, 44, 109, and 110).

The Mesenteries. The same folding process that separates the embryo from the extra-embryonic membranes completes the floor of the gut (Figs. 37 and 108). Coincidently the splanchnic mesoderm of either side is swept toward the mid-line enveloping the now tubular

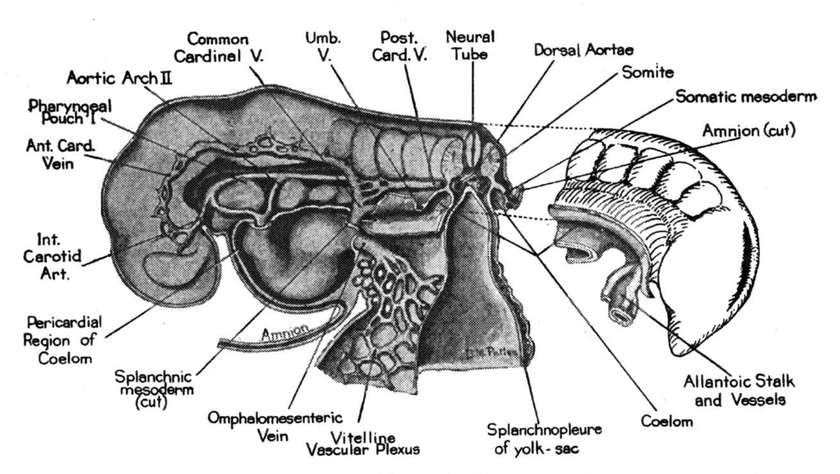

FIG. 109. Schematic plan of lateral dissection of young mammalian embryo to show the relations of the pericardial region of the coelom to the primary paired coelomic chambers caudal to the level of the heart. (Patten: "Human Embryology," The Blakiston Company.) The proportions of the illustration were based in part on Heuser's study of human embryos about 3 weeks old, but all the essential relationships shown are equally applicable to pig embryos of 3 to 4 mm.

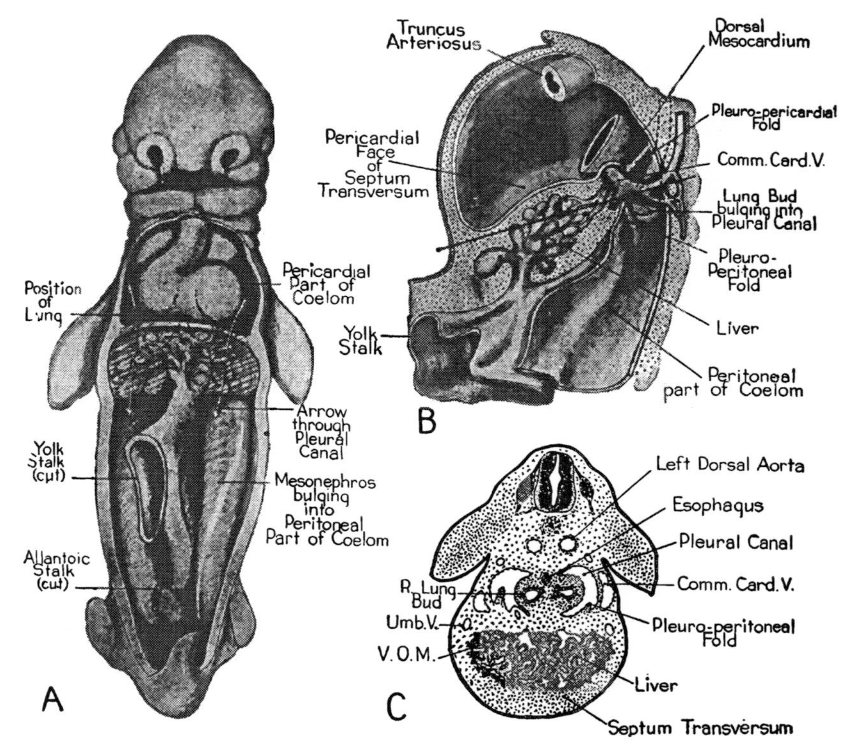

FIG. 110. Diagrams showing the relations of the pericardial, pleural, and peritoneal regions of the coelom while they are still confluent. (Patten: "Human Embryology," The Blakiston Company.)

A, Semischematic frontal plan with body of embryo represented as if it had been pulled out straight. The position of the lungs is indicated by broken lines, and arrows indicate the location of the pleural canals, on either side, dorsal to the liver. (Cf. part C of this figure.)

B, Lateral dissection to show left pleural canal opened with lung bud bulging into it. (Modified from Kollmann.)

C, Schematized section diagonally through body at level of line in B.

digestive tract. The two layers of splanchnic mesoderm which thus become apposed to the gut and support it in the body cavity are known as the *primary or common mesentery.* The part of the mesentery dorsal to the gut, suspending it from the dorsal body-wall, is the *dorsal mesentery.* The part of the mesentery ventral to the gut, attaching it to the ventral body-wall, is the *ventral mesentery* (Fig. 108, D). The primary mesentery, while intact, keeps the original right and left halves of the coelom separate. But the part of the mesentery ventral to the gut breaks through very early, bringing the right and left coelom into confluence and establishing the unpaired condition of the body cavity characteristic of the adult (Fig. 108, F).

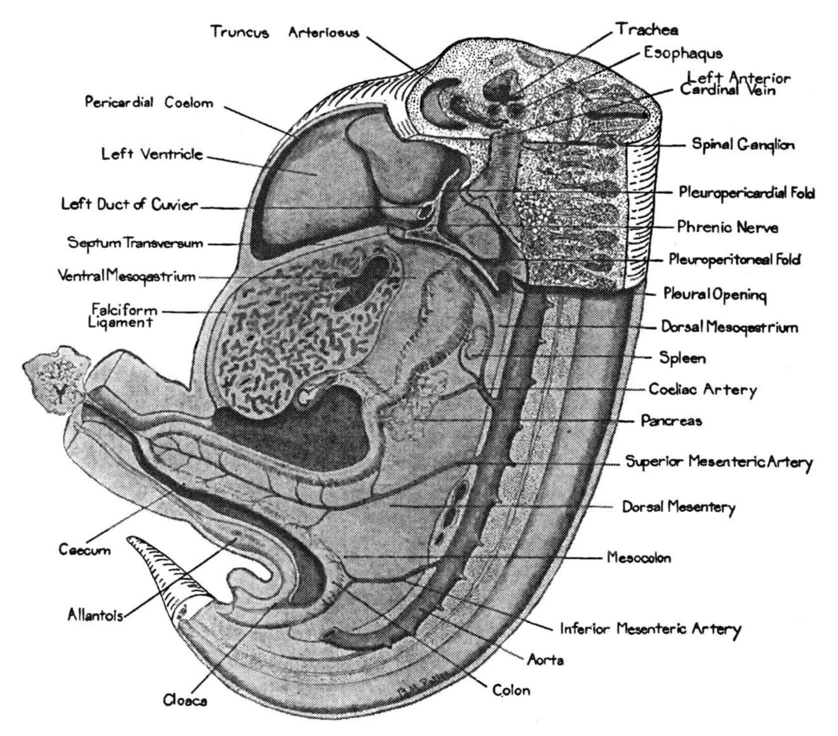

FIG. 111. Semidiagrammatic drawing showing the arrangement of the viscera, body cavities, and mesenteries in young mammalian embryos. (Patten: "Human Embryology," The Blakiston Company.) In all essentials the conditions here represented will be found in pig embryos of 12–15 mm.

In the region of the developing lungs the body is cut parasagittally, well to the left of the mid-line, in order to show the relations of pleuropericardial and pleuroperitoneal folds. Below the developing diaphragm, dissection has been carried to the mid-line.

Abbreviations: G, gall-bladder; Y, yolk-sac.

In the liver region the ventral mesentery does not entirely disappear. The liver arises, as we have seen, from an outgrowth of the gut and in its development pushes into the ventral mesentery (Fig. 108, E). The portion of the ventral mesentery between the liver and the stomach persists as the gastro-hepatic omentum (ventral mesogastrium), and the portion between the liver and the ventral body-wall, although reduced, persists in part as the falciform ligament of the liver (Fig. 111).

While the ventral mesentery, except in the region of the liver, eventually disappears, almost the entire original dorsal mesentery persists. It serves at once as a membrane supporting the gut in the body cavity and a path over which nerves and vessels reach the gut from main trunks situated in the dorsal body-wall. Its different regions are named according to the part of the digestive tube with which they are associated, as, for example, *mesogastrium*, that part of the dorsal mesentery which supports the stomach; *mesocolon*, that part of the dorsal mesentery supporting the colon, etc. (Fig. 111).

The Partitioning of the Coelom. The structure which initiates the division of the coelom into separate chambers is the *septum transversum*. The septum transversum appears very early in development (Figs. 40 and 110) and is already a conspicuous structure in embryos of 9 to 12 mm. (Figs. 64 and 111). Extending from the ventral body-wall dorsad, it forms a sort of semicircular shelf. Fused to the caudal face of the shelf is the liver and on its cephalic face rests the ventricular part of the heart.

The septum transversum is the beginning of the diaphragm. It should be clearly borne in mind, however, that the diaphragm is a composite structure embryologically, and that the septum transversum gives rise only to its ventral portion. The septum transversum itself never grows all the way to the dorsal body-wall. Dorsal to the septum transversum, the region of the coelom occupied by the heart and lungs is confluent with that occupied by the developing gastro-intestinal tract and liver (Fig. 110). Thus, although the division of the coelom into thoracic and abdominal regions is clearly indicated even at this early stage, it is not as yet complete.

The complete isolation from one another of the pericardial, pleural, and peritoneal portions of the coelom is brought about by the growth of the paired pleuroperitoneal and pleuropericardial folds. These folds arise from the dorso-lateral body-walls where the ducts of Cuvier bulge into the coelom as they swing around to enter the sinus venosus of the heart (Figs. 110 and 113, B). The folds thus established rapidly acquire a roughly triangular shape with their bases diagonally along the body-wall and their apices extending toward, and eventually fusing with, the dorsal part of the septum transversum. Because of their different fate and relations the cephalic parts of these primary triangular folds have been called the *pleuropericardial* folds and their caudal parts the *pleuroperitoneal* folds (Fig. 111).

In the growth processes which lead toward the separation of the thoracic from the abdominal region, the dorsal mesentery is caught

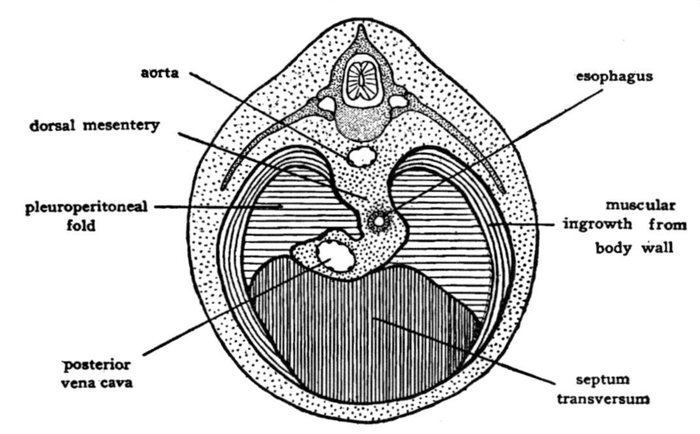

FIG. 112. Diagram indicating the embryological derivation of the various regions of the diaphragm. (Modified from Broman.)

between the converging septum transversum and the pleuroperitoneal folds. Fusions along the lines of contact complete the diaphragm (Fig. 112). The last place to close is near the dorsal body-wall on either side of the mid-line where the pleuroperitoneal folds are bent caudad by the growing lungs (Fig. 111). The manner in which the margins of the pleuroperitoneal folds are forced caudad by the growing lungs is one of the chief factors in establishing the characteristic dome-shaped configuration of the adult diaphragm.

Later in development, the margins of the diaphragm, especially dorso-laterally, are invaded by body-wall tissue which contributes the main part of the diaphragmatic musculature (Fig. 112).

In the thoracic region of the coelom, changes have in the meantime been going on which lead toward its subdivision into a pericardial

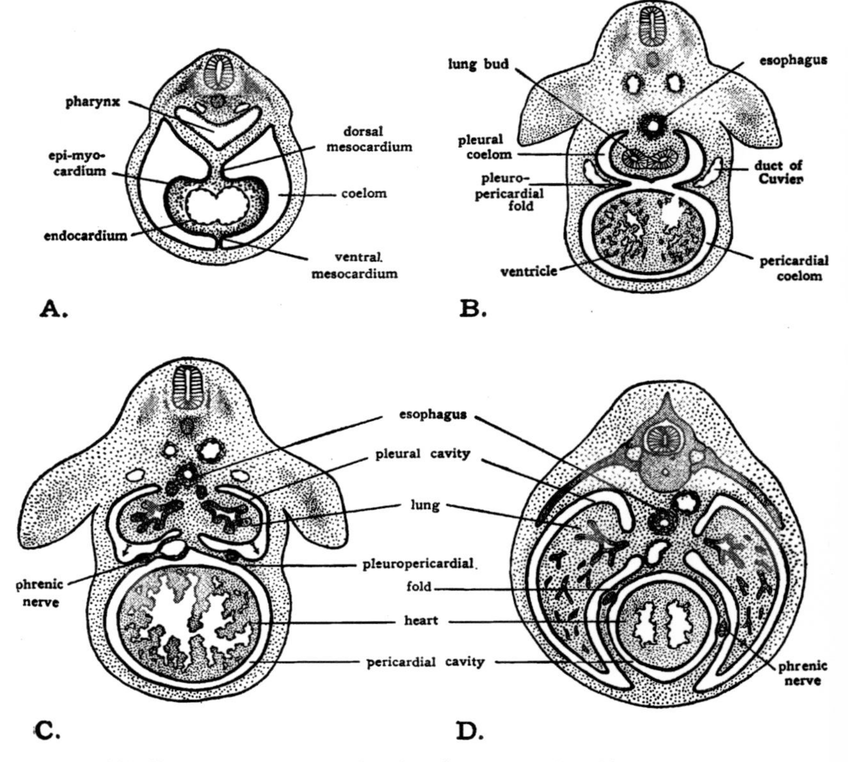

Fig. 113. Schematic diagrams showing the manner in which the pleural and pericardial regions of the coelom become separated.

and paired pleural chambers. The cephalic portions of the primary triangular folds arising about the ducts of Cuvier, constitute, it will be recalled, the pleuropericardial folds (Fig. 111). The convergent growth and ultimate fusion of the pleuropericardial folds isolate the heart from the lungs (Fig. 113, B, C). The pleural cavities thus established lie very far dorsally and are greatly restricted in extent as compared with the pleural cavities of the adult. The schematic diagrams of figure 113 indicate the manner in which, with the growth in mass of the lungs, the pleural cavities are expanded ventralwards on either side of the heart.

CHAPTER 10

The Development of the Urogenital System

The excretory and reproductive systems are so closely related both anatomically and embryologically that they must inevitably be considered together. Neither system is particularly simple in organization and the two of them together present quite a formidable array of structures. Naturally the development of such a composite group involves much of special interest to the embryologist. We shall see organs formed by the association of parts which arise independently at different places. Certain organs appear and then disappear completely without ever having become functional. Other organs fall into disuse in their original capacity and begin to degenerate only to have some part seized upon and salvaged by a new organ for a new function. We have, as it were, in the story of the development of these systems many characters. Each character, individually, is doing things of interest. Sooner or later their activities cross. The method of the novelist in dealing with such a situation would be to switch from one character to another to keep us in confusion and suspense as to what is going to happen next. Our method in dealing with this embryological story must be exactly the reverse. To prevent the various threads of the story from becoming entangled we must, as far as possible, follow one group of structures from their origin to their completion before becoming involved with another. Because the excretory system appears earlier than the reproductive system, we shall take it up first and follow it through. Then we must return to young embryos and pick up the story of the internal reproductive system, watching constantly its relations to that of the excretory system with which we have already become familiar. Yet again we must go back and follow the differentiation of the external genitalia. Any attempt to develop all the threads of the story synchronously would lead only to confusion.

I. The Urinary System

The General Relationships of Pronephros, Mesonephros, and Metanephros. As a preface to the account of the development of the excretory organs in the pig, it is desirable to review certain facts about the structure and development of the excretory organs in the vertebrates generally. Without such information as a background the story of the early stages of the formation of these organs in a mammal seems utterly without logical sequence. With it, the progress of events encountered in mammalian development seems but natural, because it is so clearly an abbreviated recapitulation of conditions which existed in the adult stages of ancestral forms.

There occur in adult vertebrates three distinct excretory organs. The most primitive of these is the pronephros which exists as a functional excretory organ only in some of the lowest fishes. As its name implies, the pronephros is located far cephalically in the body. In all the higher fishes and in the Amphibia the pronephros has degenerated and its functional rôle has been assumed by the mesonephros, a new organ located farther caudally in the body. In birds and mammals a third excretory organ develops caudal to the mesonephros. This is the metanephros or permanent kidney. All three of these organs are paired structures located retroperitoneally in the dorsolateral body-wall. Each consists essentially of a group of tubules which discharge by way of a common excretory duct. In the different nephroi the tubules vary in structural detail but their functional significance is, in all cases, much the same. They are concerned in collecting waste material from the capillary plexuses associated with them and excreting it from the body.

In the development of the urinary system of birds and mammals, pronephros, mesonephros, and metanephros appear in succession, furnishing an excellent epitome of the same evolutionary history which may be learned in more detail from comparative anatomy. In embryos sufficiently young we find only the pronephros established. It consists of a group of tubules emptying into ducts, called the pronephric ducts, which discharge into the cloaca (Fig. 114, A).

A little later in development there arises in close proximity to each pronephric duct a second group of tubules more caudal in position than the pronephros. These are the mesonephric tubules. In their growth they extend toward the pronephric ducts and soon open into them (Fig. 114, B). Meanwhile the pronephric tubules begin to degenerate and the ducts which originally arose in connection with the pronephros are appropriated by the developing mesonephros. After the degeneration of the pronephric tubules these ducts lose their original name and are called mesonephric ducts because of their new associations (Fig. 114, C).

At a considerably later stage outgrowths develop from the mesonephric ducts near their cloacal ends (Fig. 114, C). These outgrowths form the ducts of the metanephroi. They grow cephalo-laterad and eventually connect with a third group of tubules which constitute the metanephros (Fig. 114, D). With the establishment of the metanephroi or permanent kidneys, the mesonephroi begin to degenerate. The only parts of the mesonephric system to persist, except in vestigial form, are some of the ducts and tubules which, in the male, are appropriated by the testis as a duct system (Fig. 114, D, right).

The Pronephros. In the embryos of birds and mammals the pronephros is an exceedingly transitory structure. Very young embryos show rudimentary pronephric tubules arising from the

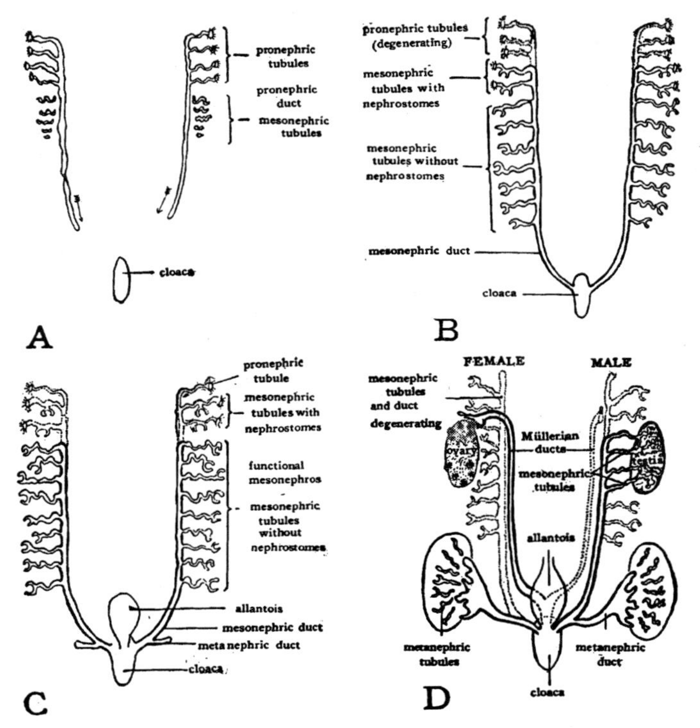

FIG. 114. Schematic diagrams to show the relations of pronephros, mesonephros, and metanephros. (Patten: "Early Embryology of the Chick," The Blakiston Company.)

intermediate mesoderm opposite a few of the somites lying well cephalically in the body. (Birds, usually 5th to 16th somites; mammals, usually 6th to 14th somites.) The significance of these vestigial tubules can readily be understood by comparing them with a plan of the fully developed and functional pronephric tubules of which they are a sketchy recapitulation (Fig. 115, A, B).

The *pronephric duct* arises at the level of the pronephric tubules by the extension caudad of the distal end of each tubule till it meets and fuses with the tubule behind it to form a continuous channel. The

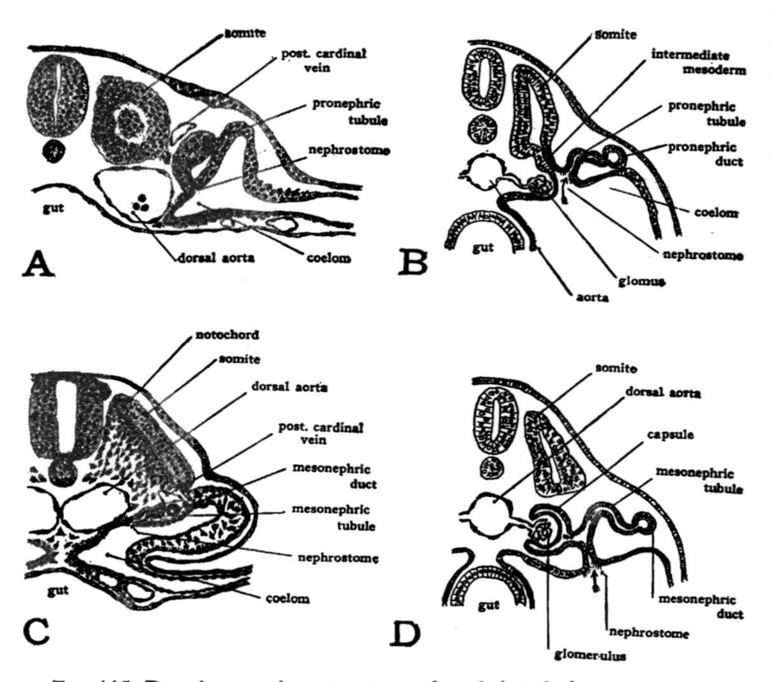

Fig. 115. Drawings to show structure of nephric tubules.

A, Pronephric tubule from section through 12th somite of a 16-somite chick embryo. (After Lillie.)

B, Diagram of functional pronephric tubule. (After Wiedersheim.)

C, Primitive mesonephric tubule with rudimentary nephrostome, from section through 17th somite of 30-somite chick embryo.

D, Schematic diagram of functional mesonephric tubule of the primitive type which retains the nephrostome. (After Wiedersheim.)

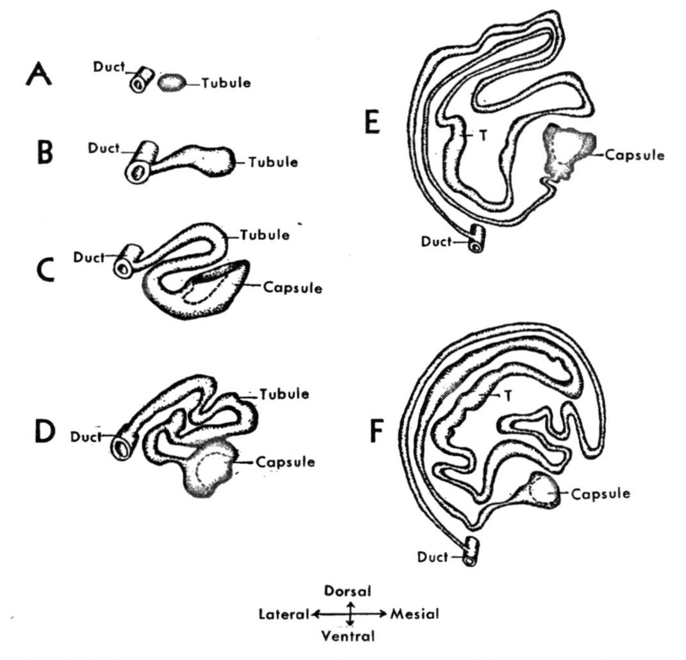

FIG. 116. Drawings showing the development of the mesonephric tubules in the pig. (Based on figures by McCallum and Lewis.) Abbreviation: T, mesonephric tubule.

duct thus established continues to grow caudad beyond the level of the tubules until it eventually opens into the cloaca (Fig. 114). Since the pronephric tubules never become functional in mammalian embryos, we need give them no further consideration. The pronephric duct, however, becomes of importance through its subsequent relations to the mesonephros.

The Mesonephros. The mesonephros in young mammalian embryos attains a high degree of development. In the pig it is especially large, being one of the most conspicuous organs in the embryo (Figs. 59 and 60). Its tubules become highly differentiated and, pending the development of the metanephros, are believed to play an active part in the embryo's elimination of nitrogenous waste.

As was the case with the pronephric tubules, the mesonephric tubules are derived from the intermediate mesoderm. At the time the tubules arise from it the intermediate mesoderm shows no trace of segmentation. When viewed in reconstructions or dissections showing its longitudinal extent, it appears as a continuous band connecting the somites with the lateral mesoderm. For this reason it is sometimes spoken of as the *nephrogenic cord*. When the mesonephric tubules are first budded off from the intermediate mesoderm they appear as cell clusters very close to, but not in contact with, the mesonephric (old pronephric) duct (Fig. 116, A). Once the process of tubule formation starts, the nephrogenic tissue is soon completely converted into young tubules, three or four tubules being formed opposite each somite from about the 14th to the 32nd.

The newly formed tubules grow rapidly, extending toward the mesonephric duct with which they soon attain connection (Fig. 116, B). In birds a few of the more cephalic tubules show a rudimentary *nephrostome* opening to the coelom, a condition comparable to that in some of the lower forms in which the mesonephros is the adult functional kidney (Fig. 115, C, D). Most if not all of the tubules in the mammalian mesonephros slur over this phase in recapitulation

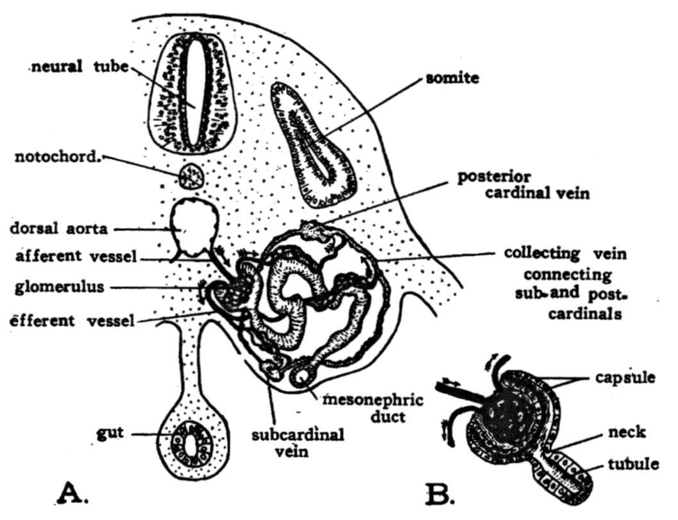

Fig. 117. Diagrams showing the relations of the blood vessels to a mesonephric tubule. (Based on figures by McCallum.)

and develop without a nephrostome (Figs. 116 and 117). Having no ciliated nephrostome capable of drawing in coelomic fluid, such tubules obtain their liquid content from the glomerular capillaries (Fig. 117). This fluid serves to carry off by way of the mesonephric duct waste materials from the blood stream. The discarding of the nephrostome by the more specialized of the mesonephric tubules is an interesting step toward the still more highly differentiated tubule we shall encounter in the metanephros.

After they have attained connection with the duct, the mesonephric tubules elongate rapidly. Starting from a simple S-shaped configuration, their pattern is complicated by a series of secondary bendings (Fig. 116, C, F). This growth in length greatly increases their surface exposure, thereby enhancing their capacity for interchanging materials with the blood in the adjacent capillaries.

The relations of the mesonephric tubules to the vascular system are indicated schematically in figure 117. The mesonephros is fed by many small arteries arising ventro-laterally from the aorta. Each of these arterial twigs pushes into the dilated free end of a developing tubule, forming from it a double-walled cup called a *glomerular (Bowman's) capsule* (Fig. 117, B). Within the capsule the artery breaks up into a knot of capillaries known as the *glomerulus*. Blood from the glomerulus leaves the capsule over one or more vessels (efferent with reference to the glomerulus) which again break up into capillaries. This time the capillaries form a plexus in close relation to the body of the tubule in its tortuous course from glomerulus to duct. From these capillaries the blood passes to collecting veins which are for the most part peripherally located in the mesonephros and more or less circularly disposed about it (Fig. 117, A). These collecting veins form a freely anastomosing system connecting both with the posterior cardinals and the subcardinals through which the blood is eventually returned to the general circulation.

Although it is relatively more conspicuous earlier in development, the mesonephros does not attain its greatest actual bulk until the embryo has reached a size of about 60 mm. When the metanephros becomes well developed, the mesonephros undergoes rapid involution and ceases to be of importance in its original capacity. In dealing with the reproductive system, however, we shall see that its ducts and some of its tubules still persist and give rise to structures of vital functional importance.

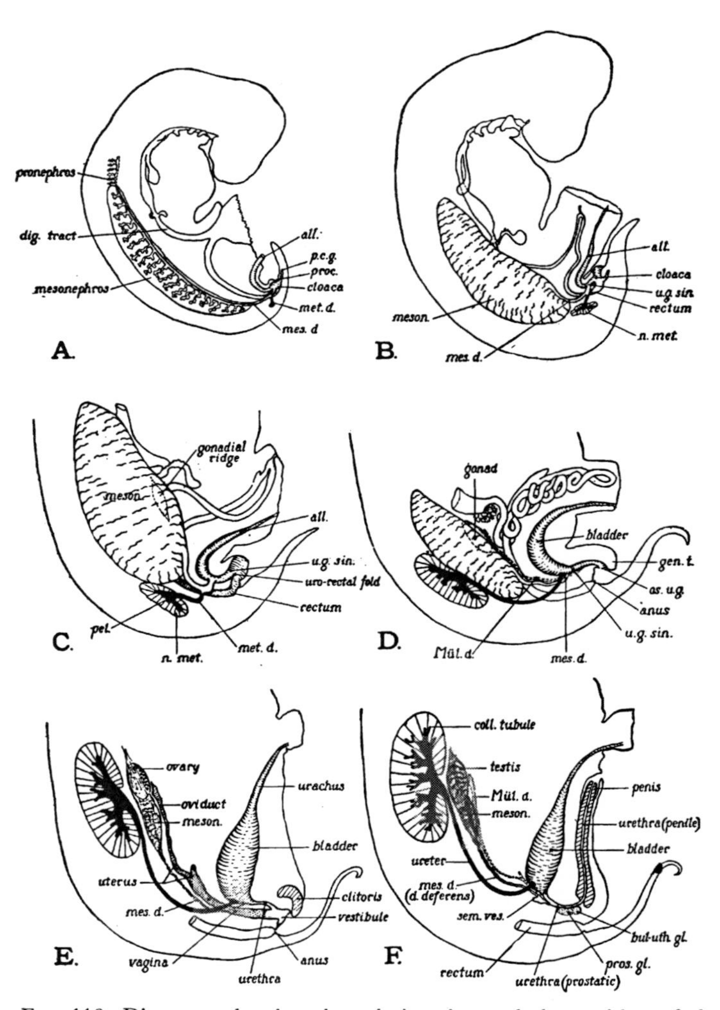

FIG. 118. Diagrams showing the relative size and the position of the nephroi of the pig at various stages of development. A, 6 mm.; B, 10 mm.; C, 15 mm.; D, 35 mm.; E, 85 mm. female; F, 85 mm. male.

The Metanephros. The metanephros has a dual origin. It arises in part from the mesonephric duct, and in part from the intermediate mesoderm. Of these separate primordia the diverticulum arising from the mesonephric duct is the first to appear. In embryos as small as 5 or 6 mm. this *metanephric diverticulum* can usually be identified as a tiny bud-like outgrowth just cephalic to the point where the mesonephric duct opens into the cloaca (Fig. 118, A). Almost from its first appearance the blind end of the metanephric diverticulum is dilated, foreshadowing its subsequent enlargement to form the lining of the *pelvis* of the kidney. The portion of the diverticulum near the mesonephric duct remains slender, presaging its eventual fate as the duct draining the kidney (*ureter*).

As the metanephric diverticulum pushes out, it collects about its distal end mesoderm which has arisen from the nephrogenic cord of intermediate mesoderm caudal to the mesonephros. The original relations of this mass of mesoderm are soon entirely lost because it becomes closely massed about the pelvic end of the metanephric diverticulum and pushed farther and farther away from its point of origin as the diverticulum continues to grow cephalad (Figs. 118, 119, and 120). This mesoderm gives rise to the secretory tubules of the metanephros or permanent kidney, and is, therefore, often designated as *metanephrogenous tissue.*

While the metanephric primordium is being pushed cephalad, it is increasing rapidly in size and encroaching on the space occupied by the mesonephros. Coincidently, rapid internal differentiation is progressing. The pelvic end of the diverticulum expands within its investing mass of mesoderm and takes on a shape suggestive of the pelvic cavity of the adult kidney (Figs. 119 and 120). From this primitive pelvic dilation arise numerous outgrowths which push radially into the surrounding mass of nephrogenic mesoderm (Fig. 138). These outgrowths become hollow, forming ducts which branch and rebranch as they extend toward the periphery. These are the collecting ducts of the kidney (*straight collecting tubules*) (Fig. 119, E).

The first changes in the mesoderm which presage the formation of the uriniferous tubules, occur near the distal ends of terminal branches of the collecting ducts. The mesodermal cells become arranged in small vesicular masses which lie in close proximity to the blind end of the collecting duct (Fig. 121, A). Each of these vesicular cell masses is destined to become a uriniferous tubule draining into the duct near which it arises. (In figure 121, A, this condition is

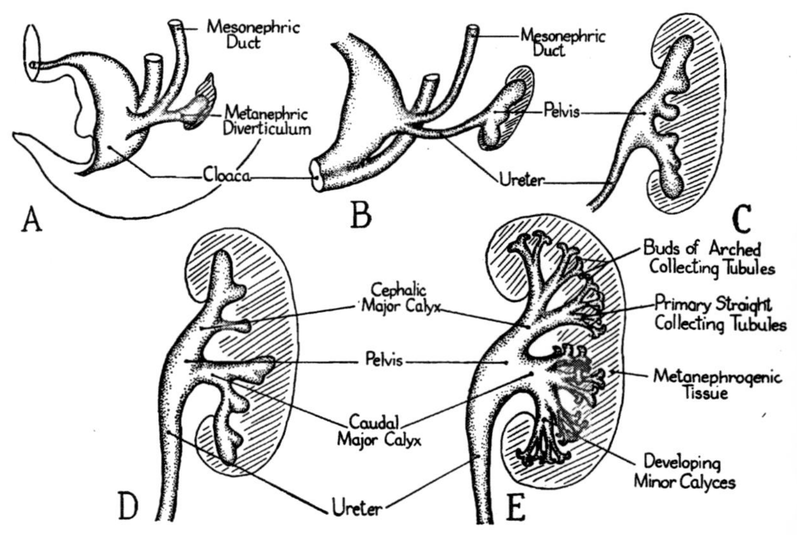

FIG. 119. Diagrams showing a series of stages in the growth and differentiation of the metanephric diverticulum. (Patten: "Human Embryology," The Blakiston Company.)

represented schematically, only two uriniferous tubules being shown in relation to the end of the collecting ducts whereas there are actually several. The tubule on the right is represented as slightly further differentiated than that on the left.) As the developing tubules extend toward the end of the collecting duct, the bud-like tips of the duct itself grow out to meet them (Fig. 121, B). Soon the two become confluent (Fig. 121, C). In this stage the metanephric tubules are very similar to young mesonephric tubules (cf. Fig. 116). In their later development the metanephric tubules become much more elaborately convoluted than the mesonephric but their functional significance is the same.

As the kidney grows in mass additional generations of tubules are formed in its peripheral zone. New orders of straight collecting tubules arise from buds, called ampullae, which appear at about the point where the excretory tubules become confluent with the straight collecting tubules of the previous order (Fig. 121, D). At the tips of the new straight collecting tubules a new order of excretory tubules is formed from the metanephrogenous tissue in the same manner that

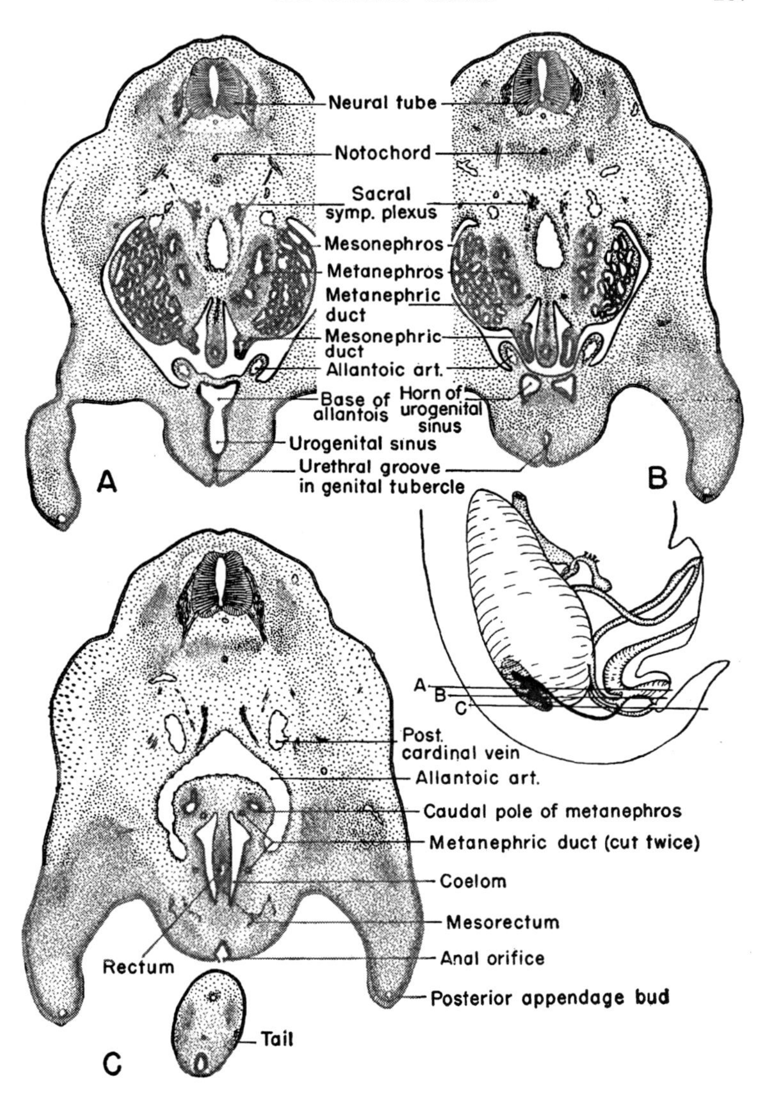

FIG. 120. Drawings (× 15) of transverse sections through the pelvic region of a 15 mm. pig embryo. The level of each section is indicated on the inset lateral plan.

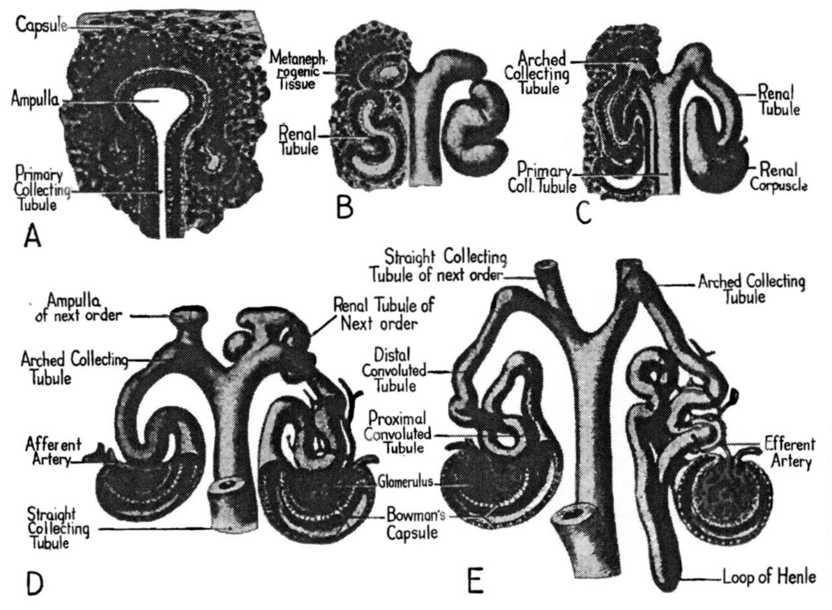

FIG. 121. Diagrams showing the development of the metanephric tubules of mammalian embryos. (After Huber, from Kelly and Burnam: "Diseases of Kidneys, Ureters, and Bladder," courtesy, D. Appleton-Century Co.)

the previous group was formed. This process is repeated many times during the growth of the kidney, about 12 to 14 generations of tubules usually having made their appearance by the time of birth. Some additional generations of tubules may be formed in the period of rapid growth immediately following birth but most of the postnatal growth of the kidneys, by which they keep pace with increased body mass, is due to the growth of the tubules rather than to further increase in their number.

The blood supply to the metanephros, instead of coming directly from the aorta by numerous small branches as is the case in the mesonephros, is brought in from the aorta through the renal artery and thence distributed by a very elaborate system of smaller vessels. Nevertheless the relations of the smaller vessels to the tubules are essentially alike in the two organs. An arterial twig breaks up into a glomerulus within a capsule at the distal end of each tubule. An efferent vessel leaves the glomerulus to break up again in a meshwork of capillaries in close relation to the tortuous tubule. Collecting veins return the blood to the general circulation, freed of the nitrogenous waste matter which is a constant by-product of metabolism.

Formation of the Bladder and Early Changes in the Cloacal Region. In dealing with the development of the extra-embryonic membranes we have already taken up the formation of the allantois as an evagination from the caudal end of the primitive gut (Fig. 37). Shortly after this occurs, the gut caudal to the point of origin of the allantois becomes enlarged to form the cloaca (Fig. 118, A). When the cloacal dilation is first formed, the hind-gut still ends blindly, but there is an ectodermal depression under the root of the tail which has sunk in toward the gut until the tissue separating the gut from the outside is very thin (Fig. 37, D). This ectodermal depression is known as the proctodaeum and the thin plate of tissue still closing the hind-gut is called the *cloacal membrane*. Eventually the cloacal membrane ruptures, establishing a caudal outlet for the gut. This rupture is similar to the rupture of the oral plate which has previously established communication between the stomodaeum and the cephalic end of the primitive gut.

Before this occurs important changes take place internally. The cloaca begins to be divided into two parts, a dorsal part which forms the *rectum* and a ventral part, the urogenital sinus (Fig. 122). This division is effected by the growth of the *urorectal fold*, a crescentic fold which cuts into the cephalic part of the cloaca where the allantois and the gut meet (Fig. 122). The two limbs of the fold bulge into the lumen of the cloaca from either side, eventually meeting and fusing with each other. The progress of this partitioning fold toward the proctodaeal end of the cloaca makes it difficult to keep track of the original limits of the allantois, since as the urogenital sinus is lengthened, it is, in effect, added onto the allantois (cf. Fig. 118, A–D). The point of entrance of the mesonephric ducts, however, affords a landmark which is sufficiently accurate for all practical purposes. Before the urorectal fold has changed the relations, the mesonephric ducts open from either side into the cephalic part of the cloaca. After the urorectal fold has divided the cloaca, the mesonephric ducts appear to empty into the allantois (Fig. 122). This gives us our bearings, for the mesonephric ducts are actually opening into the newly established urogenital sinus which is continuous with the allantois.

Before the cloacal membrane ruptures, separation of the cloaca is

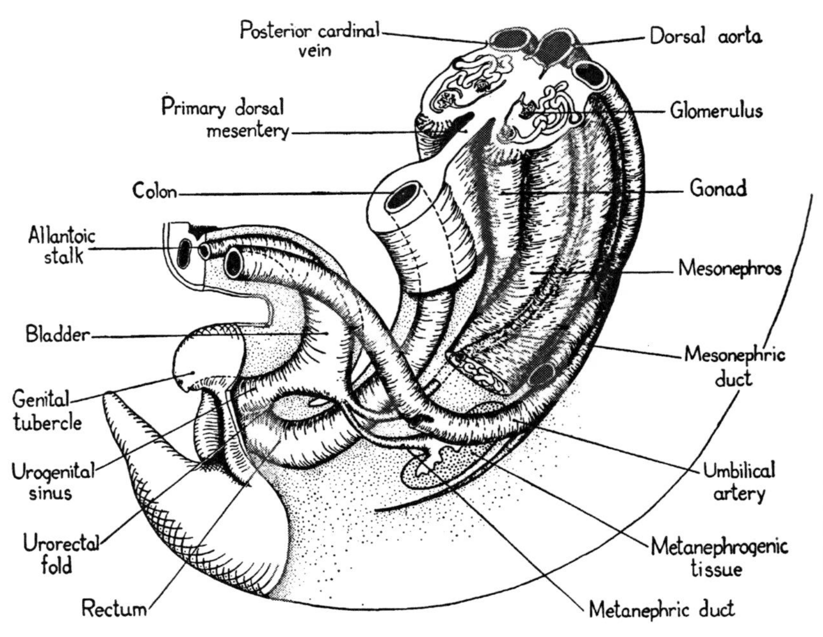

FIG. 122. Schematic ventro-lateral view of the urogenital organs of a young mammalian embryo. (Redrawn from Kelly and Burnam, "Diseases of Kidneys, Ureters and Bladder," courtesy, D. Appleton-Century Co.) The figure as originally drawn was based on human embryos of 12 to 14 mm. In all essentials the relations shown are applicable to 14 to 15 mm. pig embryos.

complete and its two parts open independently. The opening of the rectum is the *anus* and that of the urogenital sinus is the *ostium urogenitale* (Fig. 118, D).

Meanwhile the proximal part of the allantois has become greatly dilated and may now quite properly be called the *urinary bladder*. We should remember, however, that the neck of the bladder has been formed largely from tissue which was originally part of the cloaca.

In the growth of the bladder the caudal portion of the mesonephric duct is absorbed into the bladder wall. This absorption progresses until the part of the mesonephric duct caudal to the point of origin of the metanephric diverticulum has disappeared. The end result of this process is that the mesonephric and metanephric ducts open independently into the urogenital sinus. The metanephric duct, possibly due to traction exerted by the kidney in its migration headwards, acquires its definitive opening somewhat laterally and cephalically to that of the mesonephric duct. It then discharges into the part of the urogenital sinus which was incorporated in the bladder. The mesonephric ducts open into the part of the urogenital sinus which remains narrower and gives rise to the *urethra* (Fig. 118, D–F). The urethra acquires quite different relations in the two sexes. It is, therefore, desirable to defer consideration of it and take it up in connection with the external genitalia.

II. The Development of the Internal Reproductive Organs

The Indifferent Stage. One of the striking things in the development of the reproductive system is the condition which at first exists as to sexual differentiation. One might expect that reproductive mechanisms as totally unlike as those of adult males and females would be sharply differentiated from one another from their earliest appearance. Such is not the case. Young embryos exhibit gonads which at first give no evidence as to whether they are destined to develop into testes or ovaries. Along with these neuter or indifferent gonads there are present in an undeveloped state two different duct systems. If the individual develops into a female, one of these duct systems forms the oviducts, uterus, and vagina, and the other remains rudimentary. If the individual is destined to become a male, the potentially female ducts remain rudimentary and the other set gives rise to the duct system of the testes. In dealing with the embryology of the reproductive organs, therefore, conditions as they exist in the indifferent stage (Fig. 123) form a common starting point for the consideration of the later developmental changes in either sex.

Origin of the Gonads. From their earliest appearance the gonads are intimately associated with the nephric system. While the mesonephros is still the dominant excretory organ, the gonads arise as ridge-like thickenings (gonadial ridges, germinal ridges) on its ventromesial face (Fig. 108, F, and 138). Histologically the gonadial ridge consists essentially of a mesenchymal thickening covered by mesothelium. The mesothelial coat of the developing gonad is directly continuous with the mesothelium covering the mesonephros—is in fact merely a part of it stretched over the mesenchymal thickening. It soon, however, begins to show characteristics which differentiate it from the adjacent mesothelium. It grows markedly thicker and its cells round out and increase in size. Some of the cells in the germinal

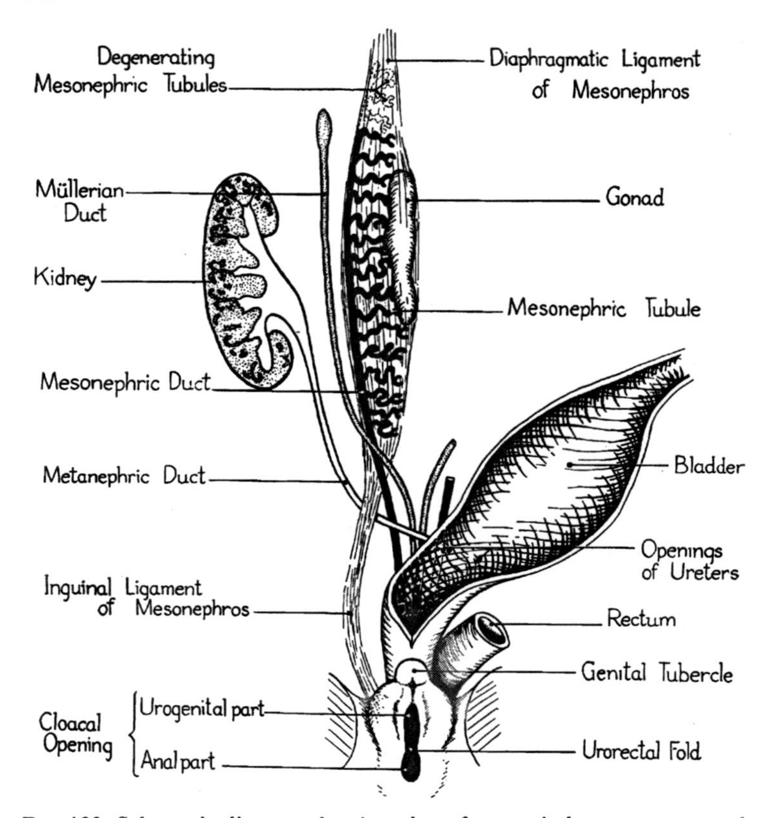

FIG. 123. Schematic diagram showing plan of urogenital system at an early stage when it is still sexually undifferentiated. (Modified from Hertwig.)

epithelium, as this modified layer of mesothelium is now termed, are conspicuously larger than their neighbors. These large cells are the primordial germ cells of the gonad. Considerable evidence has been adduced of late that these germ cells are not formed in situ by the differentiation of mesothelial cells. It is maintained that they can be identified elsewhere in the body before they appear in the germinal epithelium, and that they migrate from their place of origin (yolk-sac entoderm) to settle down in the germinal epithelium and there rear their families. Whatever their previous history may be, they are clearly recognizable in the germinal epithelium and it is not difficult to follow their differentiation from then on, through succeeding generations, to give rise, finally, to the gametes.

If the gonad is to develop into a testis the cells of the germinal epithelium grow into the underlying mesenchyme and form cord-like masses. These cords eventually become differentiated into the seminiferous tubules in which the spermatozoa are formed. In case the gonad develops into an ovary the primordial germ cells grow into the mesenchyme and there become differentiated into ovarian follicles containing the ova. (See Chap. 2.)

The Sexual Duct System in the Male. The ducts which in the male convey the spermatozoa away from the testis are, with the exception of the urethra, appropriated from the mesonephros—a developmental opportunism facilitated by the proximity of the growing testes to the degenerating mesonephros (Fig. 128). The mesonephric structures which are taken over by the testes are shown schematically in figure 124.

Epididymis. Some of the mesonephric tubules which lie especially close to the testes are retained as the efferent ductules (Figs. 3 and 124). They, together with that part of the mesonephric duct into which they empty, become the epididymis. Cephalic to the tubules which are converted into efferent ductules a few mesonephric tubules sometimes persist in vestigial form as the *appendix of the epididymis*. Caudal to the efferent ductules a cluster of mesonephric tubules almost invariably persists in rudimentary form as the *paradidymis*.

Ductus Deferens, Seminal Vesicle, and Ejaculatory Duct. Caudal to the epididymis the mesonephric duct receives a thick investment of smooth muscle and becomes the ductus (vas) deferens. A short distance before the vasa deferentia enter the urethral part of the urogenital sinus, local dilations appear in them which become elaborately sacculated and form the seminal vesicles (Fig. 3). The short part of the mesonephric duct between the seminal vesicles and the urethra constitutes the ejaculatory duct. From this point on, the spermatozoa traverse the urethra which thus serves as a common passageway to the exterior for both the sexual cells and the renal excretion.

Prostate and Cowper's Glands. From the urethral epithelium the prostate and bulbo-urethral glands develop. The prostate surrounds the urethra near the neck of the bladder; the bulbo-urethral (Cowper's) glands lie adjacent (Fig. 124). Their secretions, discharged

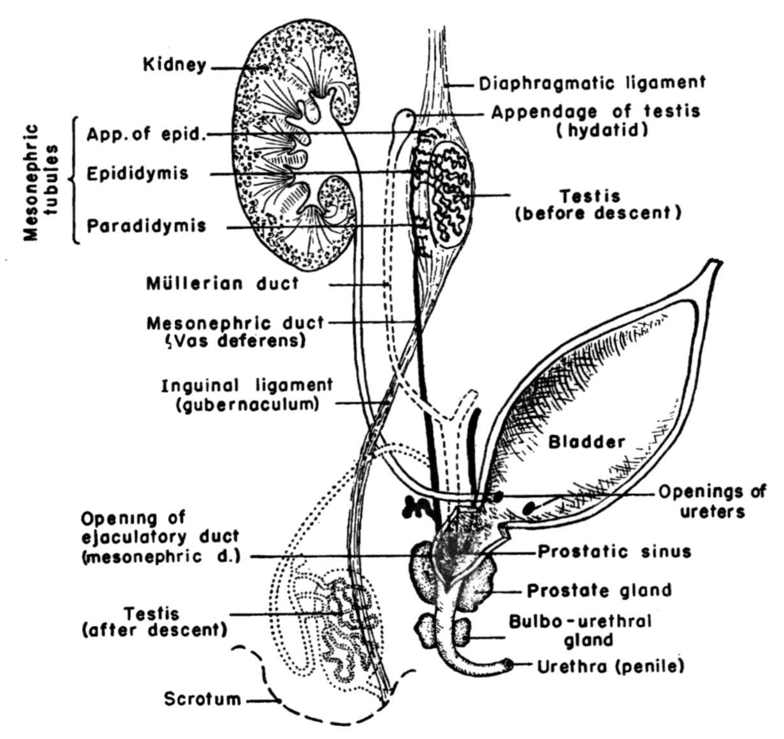

FIG. 124. Diagram of the male sexual duct system in mammalian embryos. (Modified from Hertwig.)

into the urethra with that of the seminal vesicles, serve as a conveying fluid for the spermatozoa.

The Female Duct System. The Müllerian ducts first appear close beside and parallel to the mesonephric ducts. They are the primordial structures from which the uterine tubes (oviducts), uterus, and vagina arise in the female. It is possible that phylogenetically the Müllerian ducts arose directly from the mesonephric ducts. Ontogenetically in the mammals any such process of splitting has been slurred over and they seem to arise side by side from the same parent tissue. The mesonephric ducts become well developed earlier than the Müllerian ducts and it is very easy to overlook the Müllerian ducts

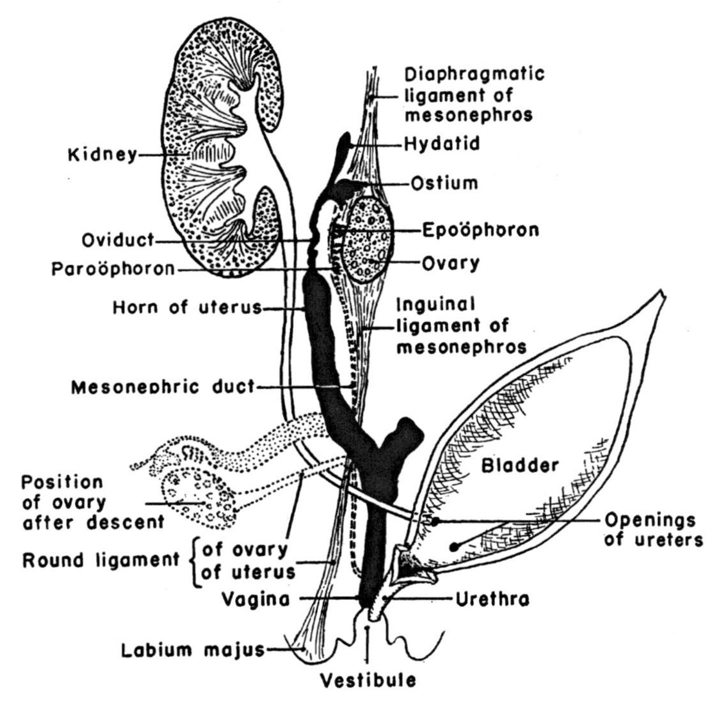

FIG. 125. Diagram of the female sexual duct system in mammalian embryos. (Modified from Hertwig.)

altogether in young specimens. By the time embryos have reached 30 or 40 mm. in length, however, it should be possible to locate them readily in sections or, with care, by dissection.

Vagina. When the Müllerian ducts first appear they are paired throughout their entire length. At their cloacal ends the right and left ducts lie close to the mid-line. In this region they soon approach and fuse with each other in the mid-line to form the vagina (Fig. 125).

Uterus. In some of the primitive mammals fusion of the Müllerian ducts does not progress cephalad beyond the vagina. Such animals have paired uteri formed by enlargement of the Müllerian ducts cephalic to their entrance into the vagina (Fig. 126, A). In all the higher mammals, fusion of the Müllerian ducts involves the caudal end of the uterus so that it opens into the vagina in the form of an

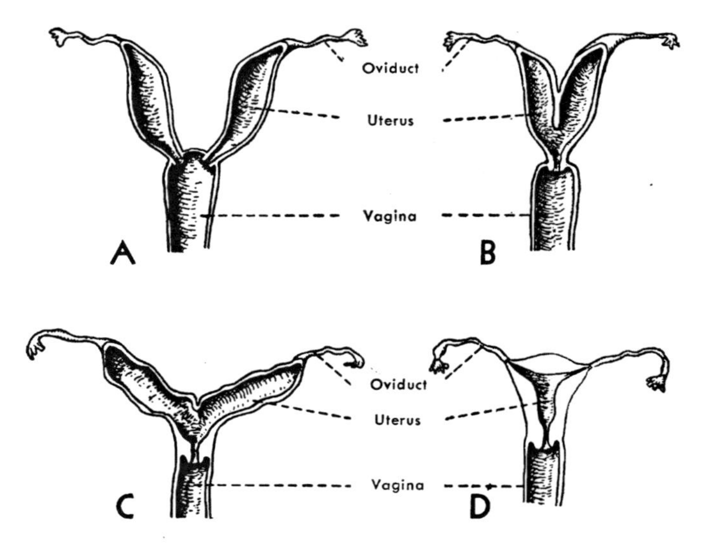

FIG. 126. Four types of uteri occurring in different groups of mammals. (After Wiedersheim.)

A, Duplex, the type found in marsupials.

B, Bipartite, the type found in certain rodents.

C, Bicornate, the type found in most ungulatse, and carnivores.

D, Simplex, the type characteristic of the primates.

unpaired neck or cervix. Toward the ovary from the cervix there is great variation in the degree of fusion encountered in the different groups (Fig. 126, B, C, D). In the sow the fusion is carried only a short way beyond the cervix to form a typical bicornate uterus (Fig. 126, C).

Uterine Tubes. The part of the Müllerian duct between the uterus and the ovary remains slender and forms the uterine tube (oviduct). Near its cephalic end, but not usually at the extreme tip, a more or less funnel-shaped opening develops (*ostium tubae abdominale*). In different forms the detailed configuration of the ostium and its relation to the ovary are quite variable. Conditions range all the way between a pouch-like dilation which almost completely invests the ovary (sow), and an elaborately fringed, funnel-shaped ostium which opens in the general direction of the ovary (man). Whatever the morphological eccentricities of the ostium may be, they apparently make less difference in its efficiency in picking up the discharged ovum than one might suppose. Even in forms where the relation of the ostium to the ovary is least intimate, abdominal pregnancies resulting from the fertilization of an ovum which the ostium failed to catch and start on its way to the uterus are comparatively uncommon.

Vestigial Structures in the Genital Duct System. In the conversion of the primordial duct systems to their definitive conditions, some of the parts which are not utilized in the formation of functional structures persist in vestigial form even in the adult. Mention has already been made of the rudimentary mesonephric tubules which persist in the male as the paradidymis and the appendix of the epididymis. Traces of the old Müllerian duct system also can usually be found in the male. Attached to the connective tissue investing the testis there is sometimes a well-marked vesicular structure called the *appendix of the testis* (*hydatid*) which represents the cephalic end of the Müllerian duct. These ducts also leave a vestige at their opposite ends in the form of a minute diverticulum (*prostatic sinus, vagina masculina*) which persists where the fused Müllerian ducts opened into the urogenital sinus (Fig. 124).

In the female the ostium of the oviduct does not ordinarily develop at the extreme cephalic end of the Müllerian duct. The tip of the duct is likely to persist in rudimentary form as a stalked vesicle (hydatid) attached to the oviduct (Fig. 125).

The mesonephric tubules and ducts may remain recognizable to a variable extent. Usually there is embedded in the mesovarium a cluster of blind tubules and traces of a duct, corresponding to the part of the mesonephric duct and tubules which in the male form the epididymis. These vestiges are called the *epoöphoron.* Less frequently the more distal portion of the mesonephric duct (the part which in the male forms the vas deferens) leaves traces known as the canals of Gartner in the broad ligament close to the uterus and vagina.

Changes in Position of the Gonads. Neither the testes nor the ovaries remain located in the body at their place of origin. The excursion of the testes is particularly extensive. Many factors are involved in their descent from the mesonephric region, where they first appear, to their definitive position in the scrotal sac. We can only sketch very briefly the course of events.

The urogenital organs arise in the dorsal body-wall, covered by the mesothelial lining of the coelom. Later when the coelomic mesothelium of the abdominal region is reinforced by connective tissue,

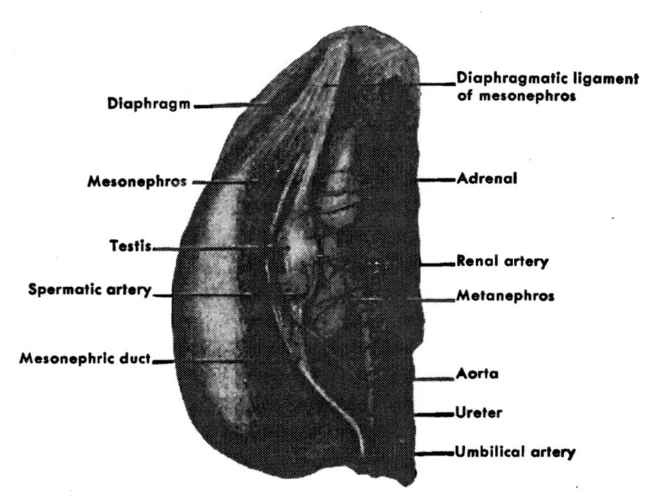

FIG. 127. Dissection showing the relative position of mesonephros, metanephros, and testis in a 33 mm. pig embryo. (Modified from Hill.)

the two layers together constitute the *peritoneum*. As to position of origin with reference to the body cavity, the urogenital organs may, therefore, be briefly characterized as retroperitoneal. This primary positional relationship is already familiar but it is emphasized again here because it is involved in many phases of the change in position and relations undergone by the reproductive organs.

Descent of the Testes. When the mesonephros begins to grow rapidly in bulk, it bulges out into the coelom, pushing ahead of itself a covering of peritoneum. At either end of the mesonephros the peritoneum is, in this process, thrown into folds. One of them extends cephalad to the diaphragm and is known as the *diaphragmatic ligament of the mesonephros* (Figs. 124 and 127). The other, which extends to the extreme caudal end of the coelom, becomes fibrous and is then known as the *inguinal ligament of the mesonephros* (Fig. 124). The inguinal ligament is destined to play an important part in the descent of the testes.

We have already seen that when the testis develops it causes a local expansion of the peritoneal covering of the mesonephros to accommodate its increasing mass. As the testis grows, the mesonephros decreases in size and the testis takes to itself more and more of the peritoneal coat of the mesonephros (Figs. 127 and 128). In this process it becomes closely related to the inguinal ligament of the mesonephros. In effect the inguinal ligament extends its attachment to include the growing testis as well as the shrinking mesonephros. With this change the ligament is spoken of as the *gubernaculum* (Figs. 124, 129, and 131, A).

In the meantime a pair of coelomic evaginations are formed, one in the inguinal region of each side of the pelvis where the caudal end of the gubernaculum is attached. These are the *scrotal pouches*. Due perhaps in part to traction exerted by the gubernaculum, the testes and the mesonephric structures which give rise to the epididymis begin to shift their relative position progressively farther caudad (cf. Figs. 127–130). Eventually they come to lie in the scrotal pouches. It would be more direct and vivid to say that the gubernaculum "pulls the testis down." Although the end results of their association very nearly justify such a phrase, it would not be strictly correct. We would be overlooking the more important factor of differential growth. Failure

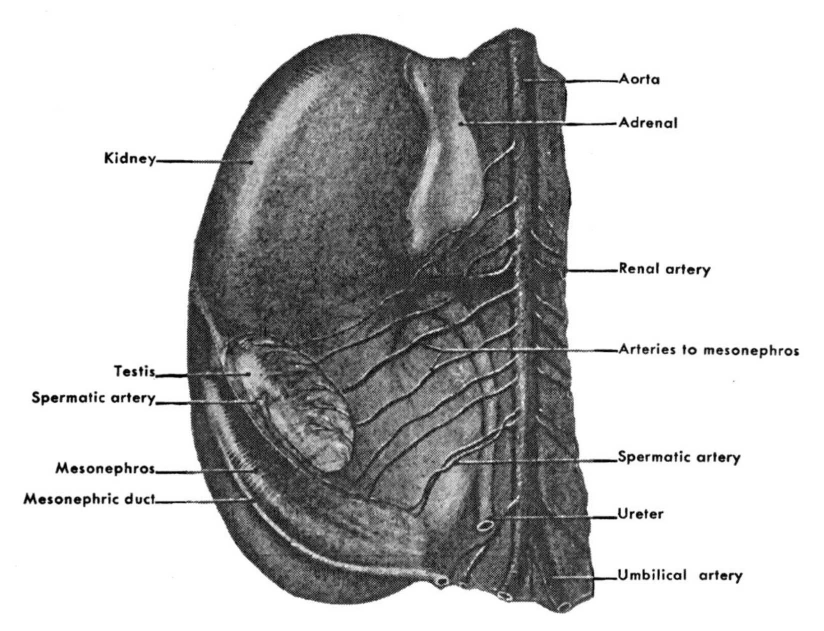

FIG. 128. Dissection showing the relative size and position of mesonephros, metanephros, and testis in an 87 mm. pig embryo. (Modified from Hill.)

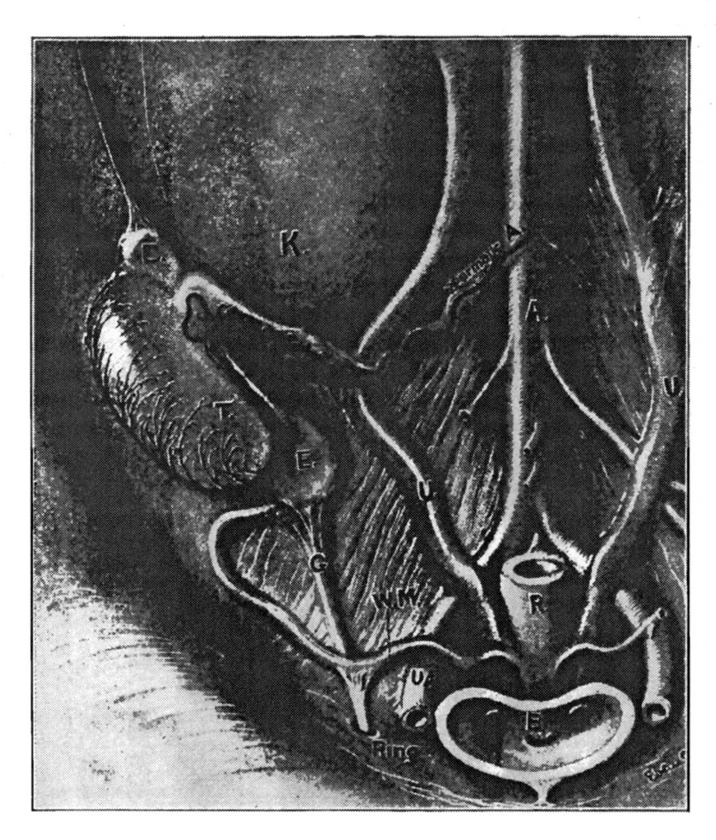

Fig. 129. Dissection of 128 mm. embryo showing an early stage in the descent of the testis. (After Hill.) Abbreviations: A., aorta; B., urinary bladder; E., epididymis (the retrogressing mesonephros); G., gubernaculum; K., kidney (metanephros); R., rectum; Ring, "inguinal ring," the fibrous tissue surrounding the opening (inguinal canal) into the scrotal sac; T., testis; U., ureter; U. A., umbilical artery; W. M., mesonephric (Wolffian) duct and Müllerian duct. The common investment of connective tissue largely conceals the smaller Müllerian duct. In this figure the line of demarcation between the two is most clearly shown just to the left of the gubernaculum.

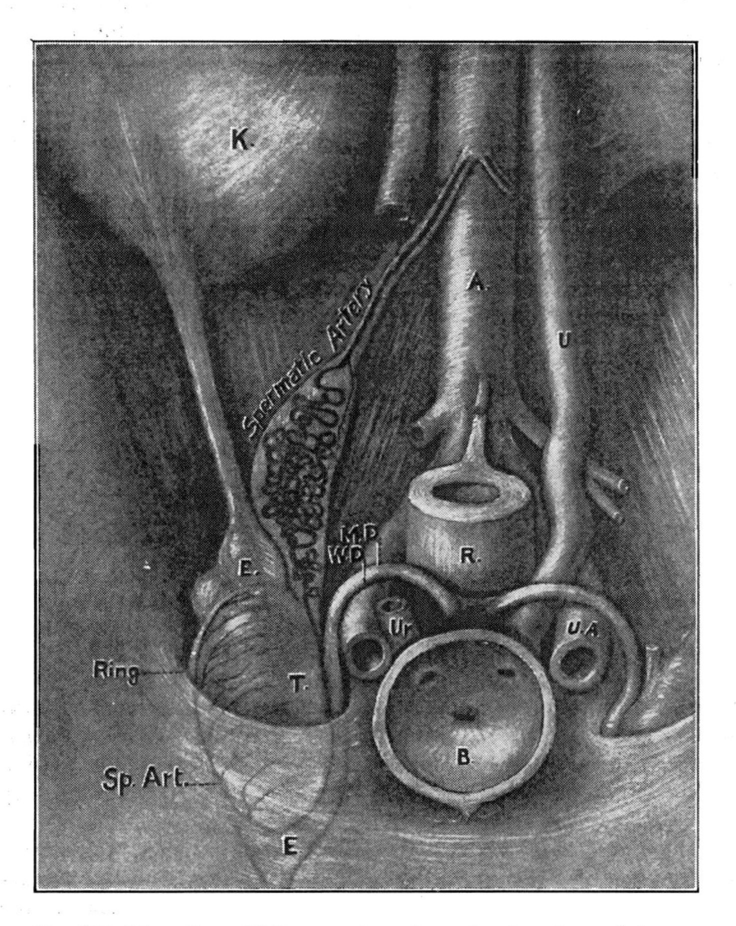

FIG. 130. Dissection of 210 mm. pig embryo showing the testis just entering the inguinal canal. (After Hill.) Abbreviations: A., aorta; B., bladder; E., epididymis; K., kidney; M. D.; Müllerian duct; R., rectum; Ring, inguinal ring; Sp. Art., spermatic artery; T., testis; U. A., umbilical artery; U., left ureter; Ur., right ureter (cut); W. D., mesonephric (Wolffian) duct.

of the gubernaculum to elongate in proportion to the growth of surrounding pelvic structures is more responsible for the traction exerted on the testis than actual shortening on its part.

In its entire descent, the testis moves caudad beneath the peritoneum. It does not, therefore, enter the lumen of the scrotal pouch directly but slips down under the peritoneal lining and protrudes into the lumen, reflecting a peritoneal layer over itself (Fig. 131). This layer of reflected peritoneum is known anatomically as the *visceral tunica vaginalis*. In most mammals when the testis has come to rest in the scrotal sac, the canal connecting the sac with the abdominal cavity becomes closed. In some of the rodents, however, it remains patent and the testes descend into the scrotum only during the breeding

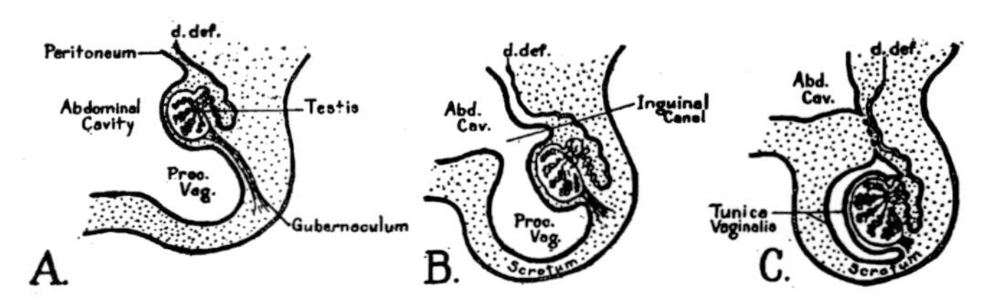

FIG. 131. Schematic diagrams illustrating the descent of the testis as seen from the side. Abbreviations: d. def., ductus deferens; Proc. Vag., processus vaginalis (the diverticulum of the peritoneum pushed into the scrotal sac).

season, to be retracted again into the abdominal cavity until the next period of sexual activity. Even in those forms normally exhibiting complete closure of the inguinal canal the obliteration of the opening is not uncommonly incomplete or structurally weak as evidenced by the not infrequent occurrence of inguinal hernias.

Descent of the Ovaries. Although the ovaries move through far less distance than the testes their change in position is quite characteristic and definite. As they increase in size, both the gonads and the Müllerian ducts sag progressively farther into the body cavity. In so doing they pull with them peritoneal folds comparable to the mesenteries of the intestinal tract. As these folds are stretched out they allow the ovaries, uterine tubes, and uterus to move caudally, laterally, and somewhat ventrally (Fig. 125). The peritoneal folds remain attached to the dorsal and lateral body-walls, become reinforced by fibrous tissue, and constitute the *broad ligaments*. The inguinal ligament of the mesonephros which in the male forms the gubernaculum, in the female is caught in the peritoneal folds which form the broad ligaments. When the ovaries move caudad and laterad the inguinal ligament is bent into angular form. Cephalic to the bend it becomes the *round ligament of the ovary*, caudal to it, the *round ligament of the uterus* (Fig. 125). Thus the changes in position of the female reproductive organs are carried out in a manner quite different from those in the male. In both sexes the organs arise retroperitoneally, but in the male the testes slide along close to the body-wall beneath the peritoneum, while in the female the ovaries, oviducts, and uterus stretch the peritoneum into a mesentery-like structure which permits a certain latitude of positional change and at the same time serves as a supporting ligament and a path of ingress for blood vessels and nerves.

The Adrenal Glands. The adrenal glands and the accessory chromaffin bodies are endocrine organs which are in no way part of the urogenital system. But the close proximity of the adrenal glands to the kidneys (Figs. 127 and 128) makes it convenient to give them a word of comment at this point.

Certain cells which migrate ventrally from the neural crest at the time the sympathetic ganglia are formed become, not nerve cells, but gland cells active in the production of a specific hormone. Due presumably to the presence of this internal secretion in their cytoplasm they exhibit a characteristic reaction with chromic acid salts which has led to their designation as *chromaffin cells*. Clusters of these chromaffin cells become located in close proximity to each sympathetic ganglion. These clusters are called the *paraganglionic chromaffin bodies*. Other masses of chromaffin tissue from the same source appear in various places beneath the mesoderm lining the coelom. There is usually a considerable amount of chromaffin tissue present in the region of the abdominal sympathetic plexus. This mass constitutes the *aortic chromaffin body* (organ of Zuckerkandl). The largest mass of extra-sympathetic chromaffin tissue appears just cephalic to the kidney and becomes converted into the *medulla of the adrenal*.

The cortical portion of the adrenal gland appears very early in development. Even in embryos of 9–12 mm. there is accelerated local proliferation of cells from the splanchnic mesoderm around the notch on either side of the base of the primary dorsal mesentery adjacent to the cephalic pole of the mesonephros. These cells push into the underlying mesenchyme and begin to show a tendency to become arranged in cords. By the 15–17 mm. stage the aggregation

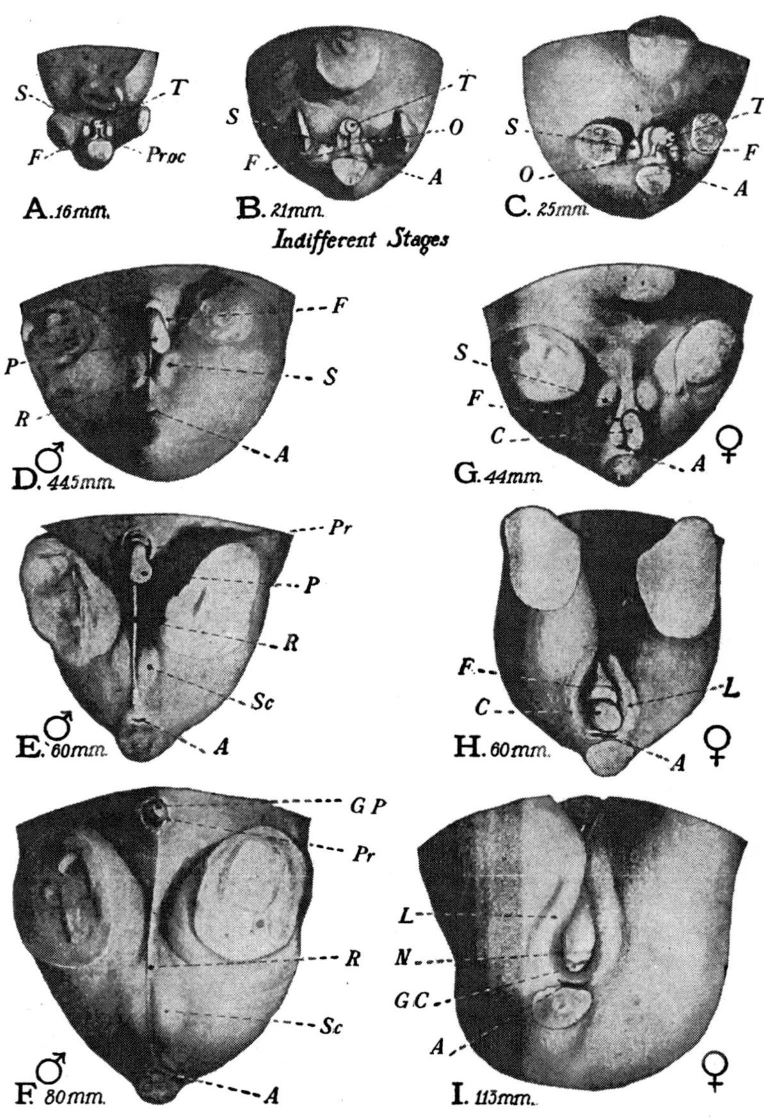

FIG. 132. Photographs (× 5) of the external genitalia of a series of pig embryos. Abbreviations: A, anus; C, clitoris; F, genital fold; GC, glans clitoridis; GP, glans penis; L, labia majora; N, labia minora (nymphae); O, urogenital orifice; P, penis; Pr, prepuce; Proc, proctodaeum; R, raphe; S, genital swelling; Sc, scrotum; T, genital tubercle.

of cell cords which constitutes the primordium of the adrenal cortex is quite conspicuous (Fig. 138). Later in development the migrating cells which give rise to the medulla of the adrenal invade the cortical primordium and become encapsulated within it.

III. The External Genitalia

Indifferent Stages. Still another thread in the story which has to be picked up separately is the development of the external genitals of either sex by divergent differentiation from a common starting point.

In very young embryos there is formed in the mid-line just cephalic to the proctodaeal depression, a vaguely outlined elevation known as the *genital eminence*. This is soon differentiated into a central prominence (*genital tubercle*) closely flanked by a pair of folds (*genital folds*) extending toward the proctodaeum. Somewhat farther to either side are rounded elevations known as the *genital swellings* (Figs. 132, A, B, C). Between the genital folds is a longitudinal depression which attains communication with the urogenital sinus to establish the urogenital orifice (ostium urogenitale). This opening is separated from the anal opening by the urorectal fold (Figs. 118, 122, and 132 A–C).

The Male Genitalia. If the individual develops into a male the genital tubercle becomes greatly elongated to form the penis, the genital folds ensheath the penis as the prepuce, and the genital swellings become enlarged to form the scrotal pouches (Fig. 132, C, D, E, F). During the growth of the penis there develops on its caudal face a groove extending throughout its entire length. Posteriorly the groove is continuous with the slit-like opening of the urogenital sinus. This groove in the penis later becomes closed over by a ventral fusion of its margins, establishing the penile portion of the male urethra. That portion of the urogenital sinus between the neck of the bladder and the original opening of the urogenital sinus becomes the prostatic urethra (Fig. 118, F). Since the margin of the slit-like urogenital orifice closes coincidently with the closure of the urethral groove in the penis, the prostatic urethra and the penile urethra become continuous and the urogenital orifice is projected to the tip of the penis. The line of fusion in the urogenital sinus region and along the caudal surface of the penis is clearly marked by the persistence of a ridge-like thickening known as the *raphe* (Fig. 132, D, E, F).

Female Genitalia. In the female the genital tubercle becomes the clitoris, the genital folds become the labia minora and the genital swellings the labia majora (Fig. 132, C, G, H, I). The original opening of the urogenital sinus undergoes no such changes as occur in the male but persists nearly in its original position. Its orifice, enlarged and flanked by the labia, becomes the vestibule into which open the vagina and the urethra (Fig. 118, E). The urethra in the female is derived entirely from the urogenital sinus, being homologous with the prostatic portion of the male urethra.

CHAPTER 11

The Development of the Circulatory System

I. The Interpretation of the Embryonic Circulation

The embryonic circulation is difficult to understand only when the meaning of its arrangement is overlooked. If one bears in mind certain fundamental conceptions as to the significance of the circulatory system in organic economics, and the basic morphological principle that any embryo must go through certain ancestral phases of organization before it can arrive at its adult structure, the changes in the arrangement of vascular channels during the course of development form a coherent and logical story.

In the embryo as in the adult the main vascular channels lead to and from the centers of metabolic activity. The circulating blood carries food from the organs concerned with its absorption to parts of the body remote from the source of supplies; oxygen to all the tissues of the body from organs which are especially adapted to facilitate the taking of oxygen into the blood; and waste materials from the places of their liberation to the organs through which they are eliminated. One of the primary reasons the arrangement of the vessels in an embryonic mammal differs so much from that in the adult, is the fact that the embryo lives under conditions totally unlike those which surround its parents. Its centers of metabolic activity are, therefore, different; and, since the course of its main blood vessels is determined by these centers, the vascular plan is different. No such profound changes occur between the embryonic and the adult stages in the circulation of a fish where embryo and adult are both living under similar conditions.

The organs which in the adult mammal carry out such functions as digestion and absorption, respiration, and excretion are extremely complex and highly differentiated structures. They are for this reason slow to attain their definitive condition and are not ready to become functional until toward the close of the embryonic period. Moreover the conditions which surround certain of the developing organs during intra-uterine life absolutely prevent their becoming functional even were they sufficiently developed so to do. Suppose the lungs, for example, were functionally competent at an early stage of development. The fact that the embryo is reliving ancestral conditions in its private amniotic aquarium renders its lungs as incapable of functioning as those of a man under water. Likewise the developing digestive organs of the embryo are inaccessible to raw food materials. Further examples are not necessary to make it obvious that were the embryo dependent on the same organs which carry on metabolism in the adult, development would be at an impasse.

An embryo must, nevertheless, solve the problem of existence during the protracted time in which it is building a set of organs similar to those of its parents. In the absence of a dowry of stored food in the form of yolk, the mammalian embryo draws upon the uterine circulation of the mother. Utilization of this source of supplies depends on the development of a special organ which serves through fetal life and is then discarded. The embryo takes food not into its slowly developing gastro-intestinal tract but into its chorion, a membrane projected outside its own body and applied to the uterine wall to form, together with it, the placenta. The nutritive materials there absorbed from the maternal blood must be transported to the body of the growing embryo by its own blood stream.

The use of food materials to produce the energy expressed in growth depends on the presence of oxygen. For growth there must be a means of securing oxygen and carrying it, as well as food, to all parts of the body. Nor can continued growth go on unless the waste products liberated by the developing tissues are eliminated. The blood of the embryo cannot be relieved of its carbon dioxide and acquire a fresh supply of oxygen in the primordial cell clusters which will later become its lungs. It cannot excrete its nitrogenous waste products through undeveloped kidneys. Its respiration and excretion, like its absorption of food, are carried out in the rich plexus of small blood vessels in the chorion. Here the fetal blood is separated from the maternal, by tissues so thin that it can readily give up its waste materials to, and receive food and oxygen from, the maternal blood stream, just as the mother's own tissues constantly carry on this interchange with the circulating blood. The placenta is thus the temporary alimentary system, lung, and kidney of the mammalian embryo. The large size of the umbilical blood vessels to the placenta is not a surprising thing—it is the entirely logical, the inevitable, expression of the conditions under which the embryo develops.

The enormous chorionic blood supply during fetal life, with the entire disappearance of this special arc of the circulation when the organism assumes adult methods of living, is a striking example of the determination of vascular channels by the location of functional centers. We must not, however, overlook the fact that there are many other centers of activity in the growing embryo less conspicuous but equally important for its continued existence. Each developing organ in the embryonic body is a center of intense metabolic activity. During fetal life it must be supplied by vascular channels adequate to care for its growth. But that is not all. Up to the time of birth each organ has been drawing on blood furnished with food and freed of waste materials by the activities of the maternal organism. At birth all this must change. Each organ essential to metabolism must be ready to assume its own active share in the process. Their vessels must be adequate to take care not only of the needs of these organs themselves but also of the functions these organs must now take over in maintaining the metabolism of the organism as a whole.

While the functional significance of the arrangement of the blood vessels is always of importance, especially in understanding the progressive changes in vascular plan, there is another factor which we cannot overlook. This factor is conservative, having to do with the things we inherit from our forebears. The goal of the embryonic period is the attainment of a bodily structure similar to that of the parents. Because it is so familiar, we accept with complaisance the remarkable fact that this goal is attained with absolute regularity. Accidents there may be, leading to defective development or malformation—but the fertilized ovum of a pig never gives rise to a cow. The new individual will show detailed differences from its parents, differences which are capitalized in the slow march of evolution; but in a single generation these differences are never radical. We say that the offspring has inherited the structure of its parents. It does more. It inherits the tendency to arrive at its adult condition by passing through the same sort of changes which its ancestors underwent in the countless millions of years it took their present structure to evolve.

Applied to the development of the circulatory system of mammals this means that the earliest form in which it appears will not be a miniature of the adult circulation. The simple tubular heart pumping blood out over aortic arches to be distributed over the body and returned to the posterior part of the heart by a bilaterally symmetrical venous system, in short the vascular plan which we see in young mammalian embryos (Fig. 45), is essentially the plan of the circulation in fishes. When we realize this, we are not puzzled either by the appearance of a full complement of aortic arches, or by their subsequent disappearance to make way for a new respiratory circulation in the lungs. We see the march of progress from a logical beginning in ancestral conditions toward the consummation of fetal life with an organization like that of the parent.

In addition to the fundamental ground plan of the circulation of the mammalian embryo, recapitulations account for many transitory peculiarities. The formation of a conspicuous though empty yolk-sac with a complement of blood vessels almost as well developed as the vitelline vessels of animals well endowed with yolk, is clearly a recapitulation of ancestral conditions. So also is the highly developed system of venous channels in the mesonephros. If the organ itself appears it brings with it its quota of vessels, no matter whether or not the organ is destined to degenerate later in development.

Whatever peculiarities may be impressed on the course of the circulation by the appearance of ancestral structures or by the development of special fetal organs such as the yolk-sac and the placenta, the main blood currents will at any time be found concentrated at the centers of activity. Changes of these main currents as one center retrogresses and another becomes dominant, must take place gradually. Large vessels become smaller, what was formerly an irregular series of small vessels becomes excavated to form a new main channel, but the circulation of blood to all parts of the body never ceases. Even slight curtailment of the normal blood supply to any region would stop its growth; any marked local decrease in the circulation would result in local atrophy or malformation; complete interruption of any important circulatory channel, even for a short time, would inevitably mean the death of the embryo.

II. The Arteries

The Derivatives of the Aortic Arches. In vertebrate embryos six pairs of aortic arches are formed connecting the ventral with the dorsal aorta. The portions of the primitive paired aortae which bend around the anterior part of the pharynx constitute the first (i.e., the most anterior) of these aortic arches. In its course around the pharynx the first aortic arch is embedded in the tissues of the mandibular arch (Fig. 136, A). The other aortic arches develop later, in sequence, one aortic arch in each branchial (gill) arch posterior to the mandibular

(Fig. 136, C–F). But in mammalian embryos we never find the entire series of aortic arches well developed at the same time. The two most anterior arches degenerate as main channels before the posterior arches have been established (Fig. 136, D–F). It should be noted, however, that their disappearance is neither abrupt nor complete as the schematic diagrams (Fig. 133) summarizing the changes in the aortic arches might lead one to believe. They break down as main

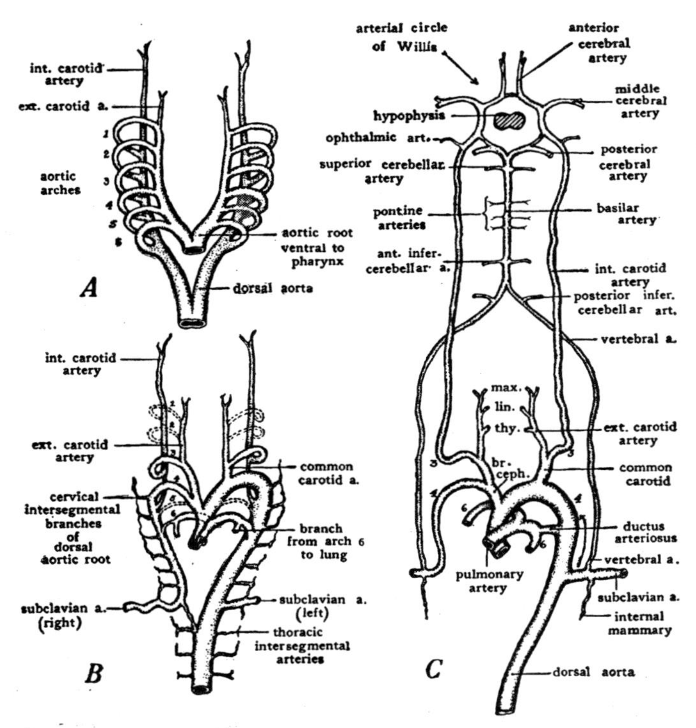

FIG. 133. Diagrams illustrating the changes which occur in the aortic arches of mammalian embryos. (Adapted from several sources.) A, Ground plan of complete set of aortic arches. B, Early stage in modification of arches. C, Derivatives of aortic arches. Abbreviations: br. ceph., brachio-cephalic (innominate) artery; lin., lingual; max., maxillary; thy., thyroid arteries. Arrow in C indicates later forward shift of left subclavian. Figures 134–137 show in detail the changes schematically summarized in this figure.

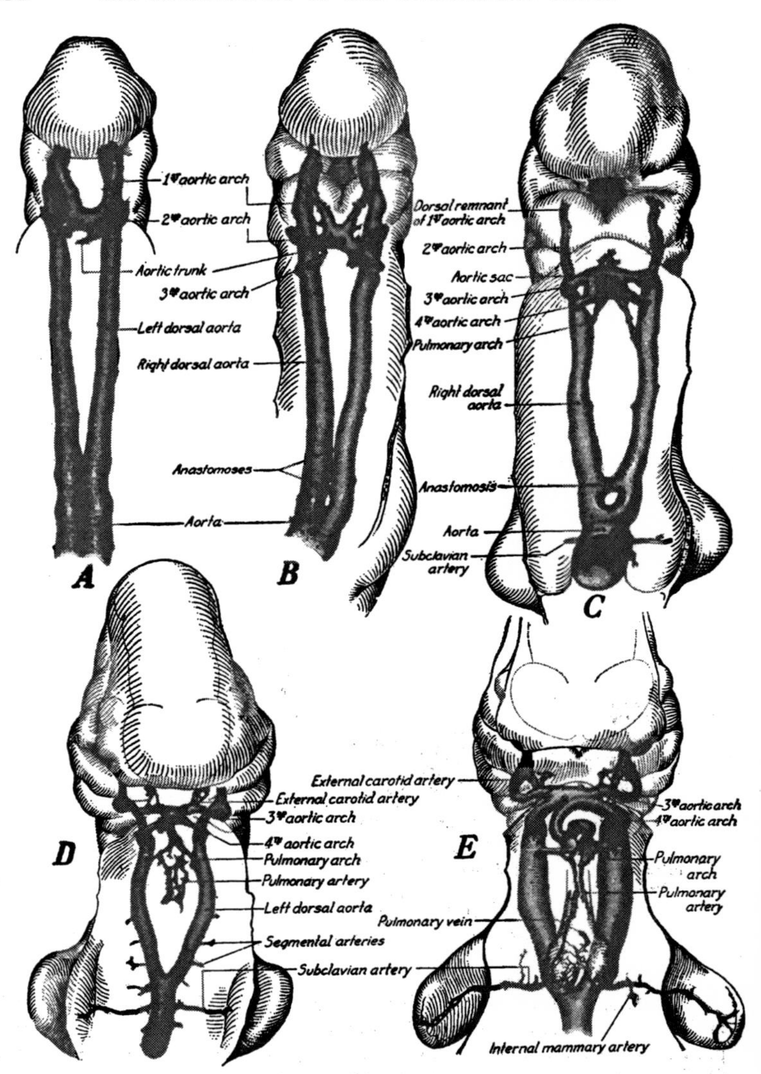

FIG. 134. Ventral aspect of vessels in the branchial region of pig embryos of various ages. (After Heuser.) The drawings in this and the three following figures were made directly from injected specimens rendered transparent by treatment with wintergreen oil. A, 24 somites; B, 4.3 mm.; C, 6 mm.; D, 8 mm.; E, 12 mm.

channels, but leave behind small vessels appropriated by the local tissues as their source of nutrition (Fig. 136, F).

The early degeneration of the first two aortic arches and the fact that the fifth arch never appears in mammalian embryos except transitorily as a vestigial vessel appended to one of the neighboring

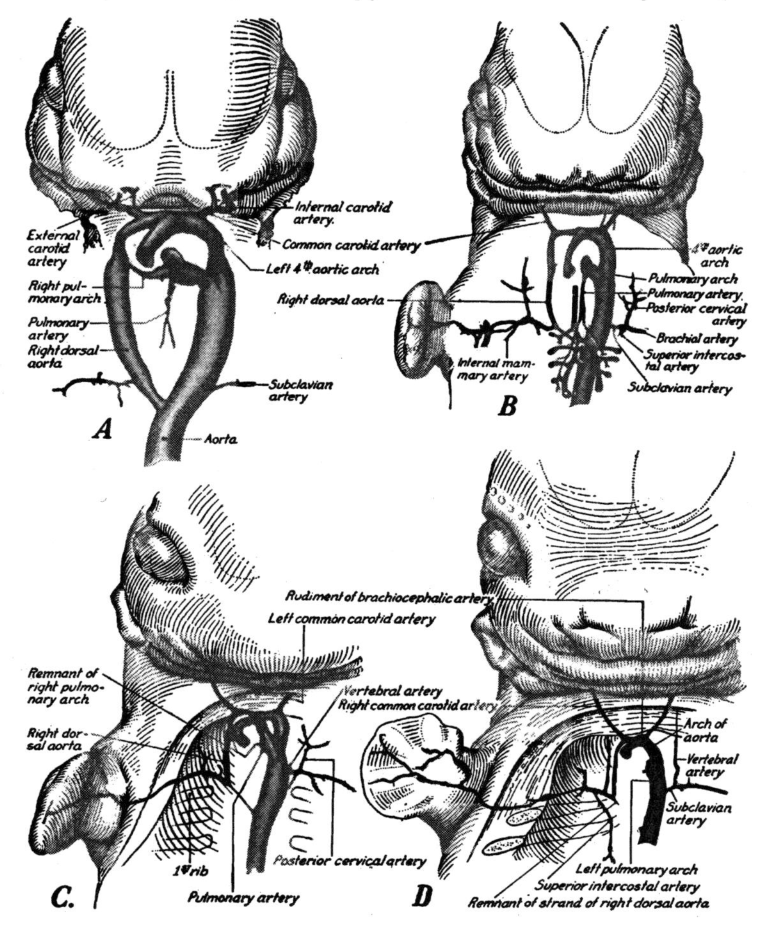

FIG. 135. Continuation of the series of injected embryos shown in figure 134. A, 14 mm.; B, 17 mm.; C, 19.3 mm.; D, 20.7 mm.

arches, leaves only the ventral and dorsal aortic roots and the third, fourth, and sixth arches to play an important rôle in the formation of adult vessels.

In dealing with embryos of 9 to 12 mm. we have already seen how the portions of the ventral aortic roots which formerly acted as feeders to the first two arches were retained as the external carotid arteries. These vessels, in part through the small channels left by the disintegration of the aortic arches with which they were originally associated, and in part through the formation of new branches to subsequently formed structures, nourish the oral and cervical regions (Figs. 67, 133 and 137, A, B).

The internal carotid arteries, also, are familiar as vessels which arise as prolongations of the dorsal aortic roots and extend to the brain (Fig. 67). When the portion of the dorsal aortic root which lies between arch 3 and arch 4 dwindles and drops out, the third arch is left constituting the curved proximal part of the internal carotid artery (Figs. 133, B, C, and 137, A–C). The part of the ventral aortic root which, from the first, has fed the third aortic arch becomes somewhat elongated and persists as the common carotid artery (Figs. 133, 137).

The fourth aortic arch has a different fate on opposite sides of the body. On the left it is greatly enlarged and persists as the arch of the adult aorta (Figs. 133 and 134–137). On the right the fourth arch forms the root of the subclavian artery. The short section of the right ventral aortic root proximal to the fourth arch persists as the innominate (brachiocephalic) artery from which both the right subclavian and the right common carotid artery arise (Fig. 133, C).

The sixth aortic arch changes its original relationships somewhat more than the others. At an early stage of development branches extend from its right and left limbs toward the lungs (Fig. 134, D, E). After these pulmonary vessels have been established[1] the right side

[1] The details of the formation of the pulmonary arteries differ somewhat in different mammals. In most of the forms which have been carefully studied (man, cat, dog, sheep, cow, opossum) the pulmonary arteries maintain their original paired condition throughout their entire length. In these forms part of the right sixth arch is retained as the proximal portion of the adult right pulmonary artery (Fig. 133). The pig is unusual in having its pulmonary branches fuse with each other proximally, forming a median vessel ventral to the trachea (cf. Figs. 134, E, and 135, A). Distal to this short median trunk the pulmonary vessels retain their original paired condition, each running to the lung on its own side of the body. Proximally the median trunk becomes associated with the left sixth arch and the right sixth arch drops out altogether.

of the sixth aortic arch loses communication with the dorsal aortic root and disappears (Fig. 135, A, B). On the left, however, the sixth arch retains its communication with the dorsal aortic root. The portion of it between the point where the pulmonary trunk is given off and the dorsal aorta is called the *ductus arteriosus* (Figs. 133, C, and 138). During the fetal period when the lungs are not inflated the ductus arteriosus shunts the excess blood from the pulmonary circulation directly into the aorta. The functional importance of this channel will be more fully appreciated when we have given it further consideration in connection with the development of the heart and the changes which take place in the circulation at the time of birth.

While these changes have been taking place in the more peripheral part of the vascular channels which lead to the lungs, a fundamental alteration has occurred in the main ventral aortic stem. Formerly a single channel leading away from the undivided ventricle of the primitive tubular heart, the ventral aorta now becomes divided lengthwise into two separate channels. This division begins in the aortic root just where the sixth arches come off, and progresses thence toward the heart. Meanwhile, as we shall see when we take up the development of the heart, the ventricle has become divided into right and left chambers. The final result of these two synchronous partitionings is the establishment of a channel leading from the right ventricle to the lungs by way of the sixth aortic arches, and another separate channel leading from the left ventricle to the dorsal aorta by way of the left fourth aortic arch.

The Derivatives of the Intersegmental Branches of the Aorta. In dealing with the structure of 9–12 mm. embryos comment was made on the importance of the small intersegmental branches from the dorsal aorta (Figs. 67 and 136, F). At that time, too, we became familiar with some of the vessels which are derived from these branches. The anterior appendage bud first appears at the level of the seventh cervical intersegmental, and it is this artery which becomes enlarged to form the subclavian. With the enlargement of the left fourth aortic arch to form the main channel leading from the heart to the dorsal aorta, the dorsal aortic root on the right side becomes much reduced (Fig. 135, A–D). Caudal to the level of the subclavian it drops out entirely. It will be recalled that the sixth aortic arch also drops out on this side. This leaves the right subclavian communicating with the dorsal aorta by way of a considerable section of the old dorsal aortic root and the fourth aortic arch. In the adult, both the distal part of

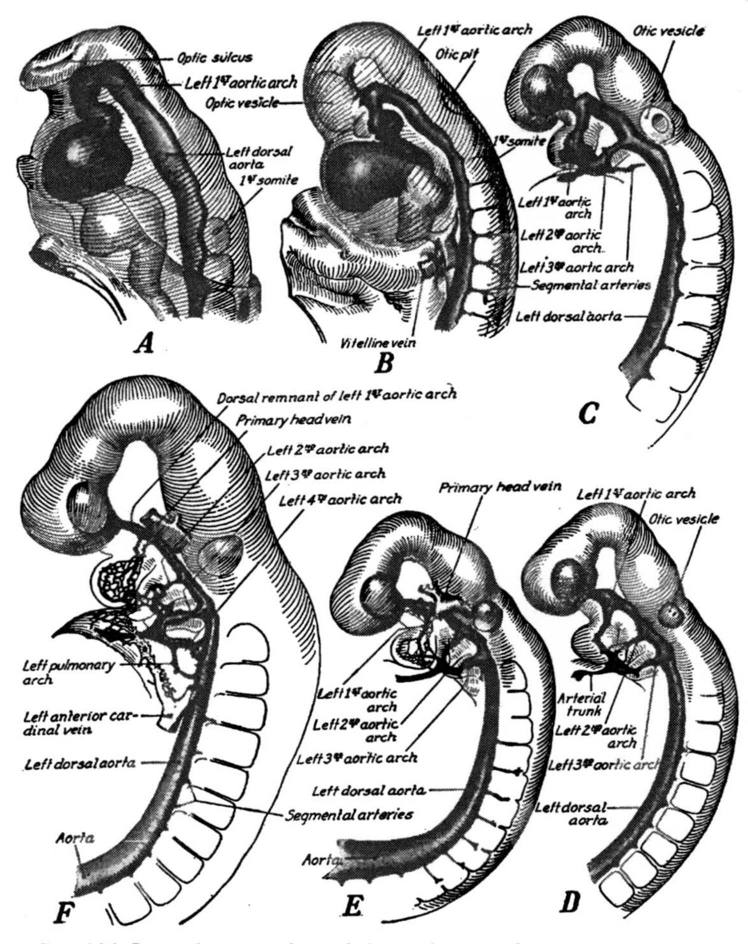

FIG. 136. Lateral aspect of vessels in the branchial region of pig embryos of various ages. (After Heuser.) A, 10 somites; B, 19 somites; C, 26 somites; D, 28 somites; E, 30 somites; F, 36 somites (6 mm.).

this vessel (formed by the intersegmental artery) and its proximal portion (appropriated from the old aortic arch system) pass under the name subclavian. This accounts for the striking dissimilarities of origin between the right and left subclavian arteries in the adult.

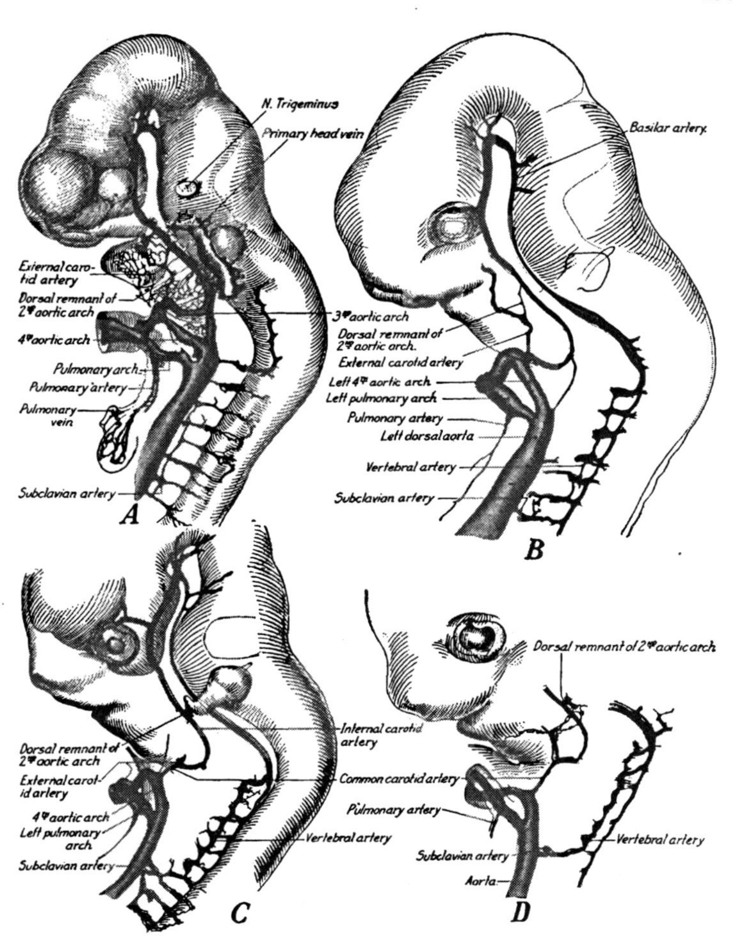

FIG. 137. Continuation of the same series of lateral views of injected embryos. A, 14 mm.; B, 17 mm.; C, 19.3 mm.; D; 20.7 mm.

Cephalic to the subclavian arteries, a series of longitudinal anastomoses appear connecting the cervical intersegmentals to form the vertebral arteries (Fig. 137). When the vertebral arteries are thus established, all the intersegmental roots back to the subclavian drop out, leaving the vertebral as a branch of the subclavian (Figs. 133, C, and 137). The manner in which the vertebrals swing in to the mid-line rostrally and become confluent with each other to form the basilar artery, and the anastomosis between the internal carotids and the basilar artery in the region of the hypophysis, are already familiar.

Caudal to the subclavian, the internal mammary artery is formed by longitudinal anastomosing of the more cephalic of the thoracic intersegmental arteries. Subsequent dropping out of the proximal parts of the other intersegmentals leaves it arising from the subclavian. Thus the steps in its origin are strikingly similar to the processes by which the vertebral artery was established cephalic to the subclavian (Fig. 133). Still farther caudally in the body, the intersegmental arteries retain their original independent condition as paired branches extending from the aorta dorsad on either side of the neural tube and the developing spinal column (Fig. 67). Even in the adult these vessels appear with little change in their original relations.

The Enteric Arteries. The first of the three enteric arteries to appear is the *anterior* (*superior*) *mesenteric*. We have already followed its origin as a pair of arteries originally called the omphalomesenterics which, in young embryos, extended to the surface of the yolk-sac (Fig. 45). When the yolk-sac degenerates and the ventral part of the body closes in, these paired channels fuse with each other to form a median vessel situated in the mesentery and extending to the gut loop in the belly-stalk (Fig. 66). This is now called the anterior mesenteric artery. With the elongation and coiling of the intestine in the region fed by it, the anterior mesenteric artery acquires many radiating terminal branches. Its primary relations, however, remain unchanged.

The *celiac artery* arises from the aorta in a manner basically similar to the origin of the anterior mesenteric artery, but at a slightly later stage of development. The embryonic body has, therefore, become more nearly closed ventrally and the primary dorsal mesentery has been established. As a result the primary paired condition we expect to see in the early stages of the formation of all of the main enteric arteries is greatly abbreviated in the case of the celiac artery, and almost from its first appearance it is a median vessel extending in the mesentery toward the gastric region of the gut (Fig. 67). As development progresses it becomes extensively branched, being the main artery which feeds the gastro-hepato-pancreatic region of the digestive system and also the spleen which arises in its territory (Fig. 111).

The *inferior mesenteric artery* has an origin similar to that of the celiac. It is established caudal to the anterior mesenteric artery slightly

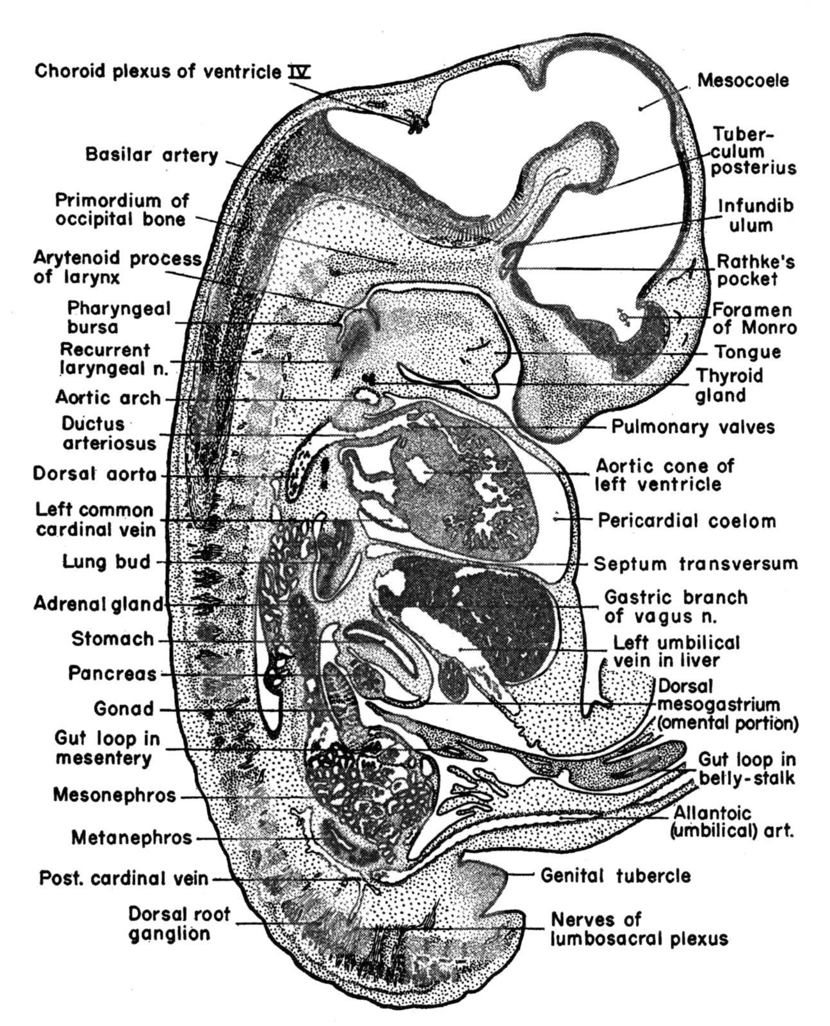

FIG. 138. Drawing (× 10) of parasagittal section of 15 mm. pig embryo. The section is in a plane slightly to the left of the mid-line and passes through the ductus arteriosus, lung bud, stomach, gonad, and metanephros.

later in development than the time at which the celiac appears and is the main vessel to the posterior part of the intestinal tract (Fig. 111).

The Renal Arteries. The mesonephros is supplied by many small arteries which arise ventro-laterally from the aorta. While the metanephroi or permanent kidneys are still very small, they lie in close proximity to the mesonephroi and are fed by small arteries which arise from the aorta along with the mesonephric vessels (Fig. 127). The local vessels associated with the kidneys are progressively enlarged as the kidneys themselves grow in bulk and become the renal arteries of the adult (Fig. 128).

The Arteries Arising from the Caudal End of the Aorta. The main aortic trunk decreases abruptly in size where the large umbilical (allantoic) arteries (Fig. 138) turn off into the belly-stalk. Beyond this point the aorta is continued toward the tail as a slender median vessel called the *caudal artery* (Fig. 67).

The posterior appendage buds arise some time after the placental circulation has been established. The umbilical arteries are consequently of considerable size, and the small vessels which branch off from them to feed the appendage buds are by comparison quite insignificant. As the appendage buds grow, these small vessels grow with them to become the *external iliac arteries*. When, at birth, the placental circulation stops, the umbilical arteries are reduced to small vessels nourishing the local tissues between their point of origin and the umbilicus. We then know their proximal portions as the *internal iliac*, or *hypogastric arteries*, and the fibrous cords which still mark their course along the wall of the bladder (the old allantoic stalk) as the obliterated branches of the hypogastric arteries. Thus the tables are turned between fetal and adult life. In the fetus the external iliac artery to the leg appears as a branch of the dominating umbilical artery. After birth the reduced umbilical arteries under their new name of internal iliacs, or hypogastrics, appear as branches of the now larger external iliacs. The original umbilical root proximal to the origin of the external iliac is called the *common iliac* (Fig. 150).

III. The Veins

There is a natural grouping of the veins according to their relationships which it is convenient to follow in discussing their development. Under the term systemic veins we can include all the vessels which collect the blood distributed to various parts of the body in the routine of local metabolism. In young embryos these would be the cardinal veins and their tributaries, that is, the return channels of the primitive intra-embryonic circulatory arc. In older embryos and adults the systemic veins would include the anterior (superior) caval system which is evolved from the anterior cardinals, and the posterior (inferior) caval system which takes the place of the postcardinals and their tributaries.

We can set apart from the general systemic circulation three special venous arcs: the umbilical, returning the blood from the placenta; the pulmonary, returning the blood from the lungs; and the hepatic portal, carrying blood from the intestinal tract to the liver. The specialized nature of the placental and pulmonary circulations is obvious. The peculiarities of the hepatic portal system call, perhaps, for a word of explanation. Ordinarily veins[2] collect blood from local capillaries and pass it on directly to the heart. Their blood stream is away from the organ with which they are associated; once collected within a vein the blood is not redistributed in capillaries until it has again passed through the heart. The portal vein arises in typical fashion by collecting the blood from capillaries in the digestive tube. But then, contrary to the usual procedure, its blood flows, not directly to the heart, but to the liver where it enters a second capillary bed and is returned by a second set of collecting vessels to the heart. With reference to the plexus of capillaries in the liver this vein is afferent. Hence its designation as a portal (translated = carrying to) vein, setting it apart from other veins which carry blood only away from the organ with which they are associated.

---

[2] There is a tendency among those who have done but little work on the circulation to regard any vessel which carries oxygenated blood as an artery, and any vessel which carries blood poor in oxygen and high in carbon dioxide content as a vein. This is not entirely correct even for the circulation of adult mammals on which the conception is based. In comparative anatomy and especially in embryology it is far from being the case. It is necessary, therefore, in dealing with the circulation of the embryo, to eradicate this not uncommon misconception.

The differentiation between arteries and veins which holds good for all forms, both embryonic and adult, is based on the structure of their walls, and on the direction of their blood flow with reference to the heart. An artery is a vessel carrying blood away from the heart under a relatively high, fluctuating pressure due to the pumping of the heart. Correlated with the pressure conditions in it, its walls are heavily reinforced by elastic tissue and smooth muscle. A vein is a vessel carrying blood toward the heart under relatively low and constant pressure from the blood welling into it from capillaries. Correlated with the pressure conditions characteristic for it, the walls of a vein have much less elastic and muscle tissue than artery walls, and more non-elastic connective-tissue fibers.

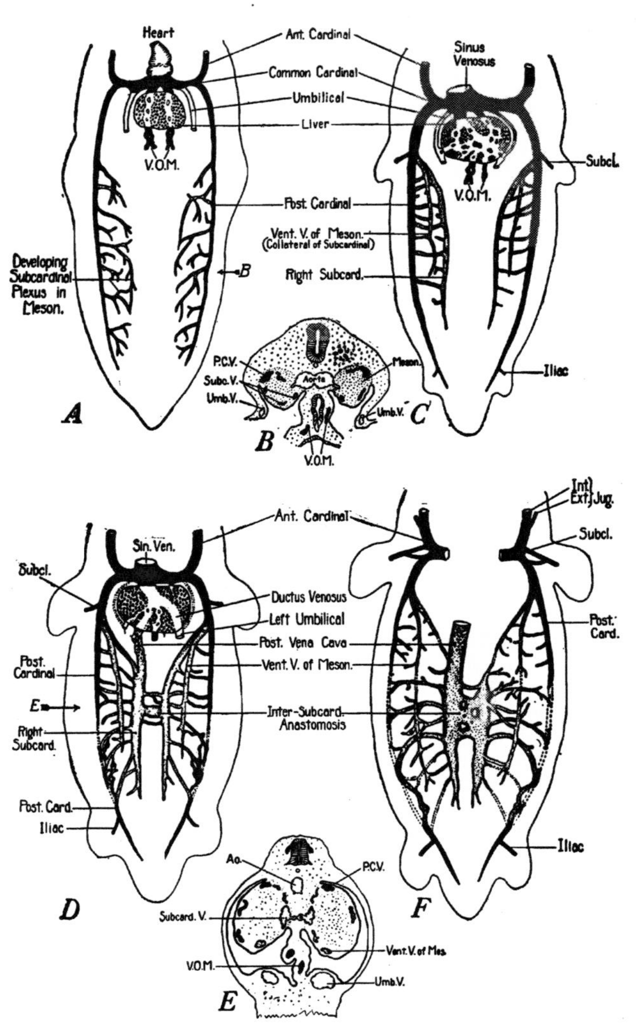

FIG. 139. Diagrams illustrating stages in the development of the systemic veins of the pig. (After Butler.) The cardinal and omphalomesenteric veins are shown in black, the subcardinal system is stippled, the supracardinals are horizontally hatched, and vessels arising independently of these three systems are indicated by small crosses.

A, Ground plan of the veins of a young mammalian embryo (cf. Fig. 45).

B, Cross-section (at level of arrow in A) showing dorso-ventral relations of the various veins.

C, Diagrammatic plot of veins of 5–6 mm. pig embryos.

D, Arrangement of veins in 6–7 mm. pig embryos.

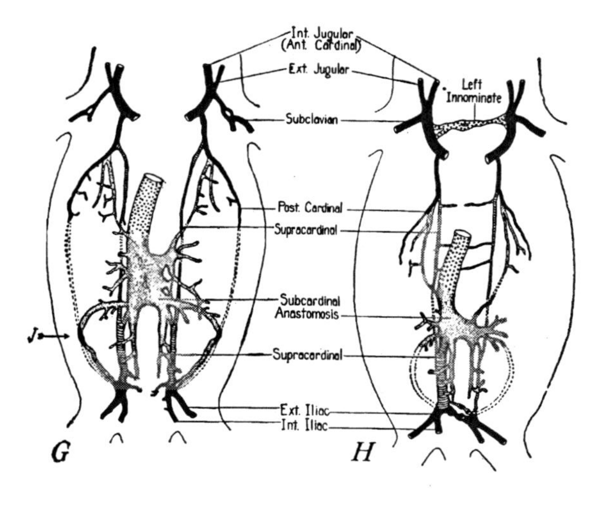

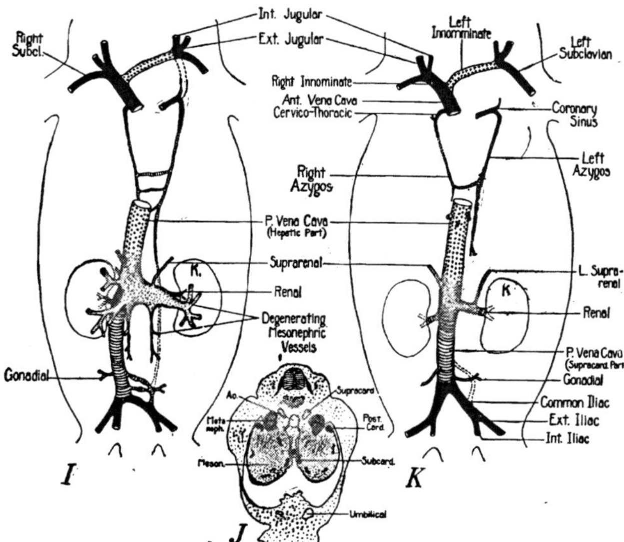

FIG. 139—(*Continued*)

E, Cross-section (at level of arrow in D) showing dorso-ventral relations of vessels.

F, Veins in 12–13 mm. embryos.

G, Veins in 16–19 mm. embryos.

H, Veins in 22–24 mm. embryos.

I, Veins in 30–35 mm. embryos.

J, Cross-section of 17 mm. embryo at level of arrow in G.

K, Plan of veins in adult pig.

Changes in the Anterior Systemic Veins Resulting in the Establishment of the Anterior (Superior) Vena Cava. The main tributaries draining the anterior parts of the adult body are the external and internal jugulars and the subclavians. In 12 mm. embryos we saw all these vessels laid down. The internal jugular is merely the original anterior cardinal under a new name. The external jugular develops from the small branch draining the mandibular region, and the subclavian as an enlargement of one of the segmental tributaries at the level of the anterior appendage bud (Fig. 68).

When the appendage buds first appear, the heart lies far forward in the body. As development progresses it is carried caudad. With this change in the position of the heart, the common cardinal veins (ducts of Cuvier) change their relative position in the body and come to lie caudal to the anterior appendage buds. As a result of this altered relation the subclavian veins from the anterior appendages, which early in development drain into the posterior cardinals (Fig. 139, C) eventually empty into the anterior cardinals (Fig. 139, G).

The outstanding characteristic of the systemic venous plan of a young embryo is its bilateral symmetry. Paired vessels from the anterior and from the posterior parts of the body become confluent to enter the sinus region of a simple tubular heart (Figs. 139, A and 144, A–C). The rerouting of the blood to enter the right side of the heart is the all-important end toward which the mammalian venous system is progressing throughout its development. In dealing with local changes this basic trend should never for a moment be forgotten.

In the anterior systemic channels this shift to the right is accomplished very simply and directly. A new vessel forms between the right and left anterior cardinals and shunts the left anterior cardinal blood stream across to the right (Fig. 139, H). With the establishment of this new channel, the part of the left anterior cardinal toward the heart drops out (Fig. 139, I). We have now but to apply the familiar adult names (Fig. 139, K): the new connecting vessel is the left innominate; the old anterior cardinal between the union of the subclavian with the jugulars and the new transverse connection is the right innominate; from the confluence of the innominates to the heart is the anterior vena cava. The anterior vena cava is thus composed of the most proximal part of the right anterior cardinal and the right common cardinal vein (duct of Cuvier). The small azygos (cervicothoracic) vein, which is the reduced posterior cardinal, indicates the old point of transition from anterior cardinal to common cardinal.

Changes in the Posterior Systemic Veins Resulting in the Establishment of the Posterior (Inferior) Vena Cava. The changes in the systemic veins of the posterior part of the body are much more radical than they are anteriorly. The posterior cardinal veins which are the primitive systemic drainage channels are associated primarily with the mesonephroi. When the mesonephroi degenerate, it is but natural that the posterior cardinals should degenerate with them. The posterior vena cava which replaces the cardinals is a composite vessel which gradually takes shape by the enlargement and straightening of small local channels which are, as it were, pressed into service as the posterior cardinals degenerate.

The subcardinal veins initiate the diversion of the postcardinal blood stream. The subcardinals are established as vessels lying along the ventro-mesial border of the mesonephros parallel with, and ventral to, the postcardinals. Taking origin from an irregular plexus of small vessels emptying from the mesonephroi into the posterior cardinals, the subcardinals from their first appearance have many channels connecting them with the postcardinals (Fig. 139, A–C). As the mesonephroi increase in size and bulge toward the mid-line the subcardinals are brought very close together. In the mid-mesonephric region they anastomose with each other to form a large median vessel, the subcardinal sinus, or intersubcardinal anastomosis (Figs. 60, 77, and 139, D–F). When this sinus is established the small vessels connecting sub- and postcardinals drain into the capacious sinus rather than toward the posterior cardinals. The result of this change very soon becomes apparent in the disappearance of the posterior cardinals at the level of this sinus. The blood from the posterior part of the body is still collected by the distal ends of the postcardinals but it returns to the heart by way of the subcardinal sinus. Consequently the anterior portions of the posterior cardinals, although they persist, are much reduced in size (Fig. 139, F).

Meanwhile the increased volume of blood entering the subcardinal sinus is finding a new and more direct route to the heart. The cephalic pole of the right mesonephros lies close to the liver. A fold of dorsal body-wall tissue, just to the right of the primary dorsal mesentery, early makes a sort of bridge between these two organs. This fold is known as the *caval plica* (*caval mesentery*) (Fig. 140). In it, as everywhere in the growing body, are numerous small vessels. Connection of these small vessels with the plexus of channels in the liver cephalically (Fig. 140, A), and the mesonephros caudally (Fig. 140, B),

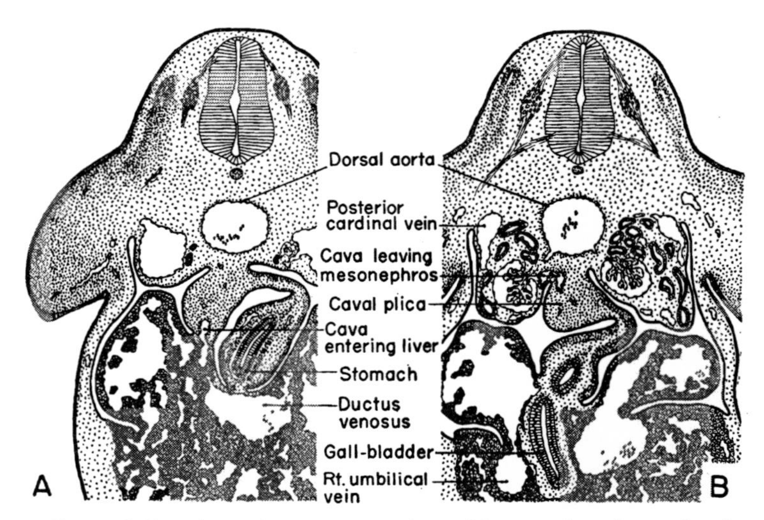

Fig. 140. Drawings of transverse sections of 5 mm. pig embryos showing the relations of the caval fold of the mesentery (caval plica). A, Section at level where cephalic end of caval plica merges with liver. Here the small vessel which is the primordium of the mesenteric portion of the posterior vena cava enters the sinusoidal circulation of the liver. B, Section showing merging of caudal end of caval plica with right mesonephros. At this level the developing mesenteric portion of the posterior vena cava anastomoses with the right subcardinal vein (cf. Fig. 139, D).

provides the entering wedge. Once a current of blood finds its way from the mesonephros to the liver through these small vessels, enlargement of the channel proceeds with great rapidity. This new channel becomes the *mesenteric part of the inferior vena cava.* (See relations of posterior vena cava in figures 75 and 76, and portion of cava indicated with small crosses in figure 139, D.)

Within the liver this new blood stream at first finds its way by devious small channels eventually entering the sinus venosus along with the omphalomesenteric circulation. As its volume of blood increases it excavates through the liver a main channel which gradually becomes walled in. As this new vessel becomes more and more definitely organized it gradually crowds toward the surface and eventually appears as a great vein lying in a notch along the dorsal side of the liver. This is the *hepatic part of the inferior vena cava.*

From the subcardinal sinus the most direct route to this new outlet is by way of the right subcardinal vein. Thus in embryos as young as 9 to 12 mm., the formation of the posterior vena cava is well started with its proximal portion consisting of subcardinal sinus, a portion of the right subcardinal vein, and the new channels through the mesentery and through the liver. (Follow this part of the cava through figure 139 from D to K.)

Posterior to the level of the subcardinal sinus still another set of veins enters into the formation of the post-cava. These are the supracardinal veins which appear, relatively late in development, as paired channels draining the dorsal body-wall (Fig. 139, G, J).

At the mid-mesonephric level, the supracardinals are diverted into the subcardinal sinus just as happened with the postcardinals earlier in development. Cephalic to the sinus, parts of the supracardinals persist as the azygos group draining in a somewhat variable manner into the reduced proximal part of the postcardinals. Caudal to the anastomosis with the subcardinal sinus, the right supracardinal becomes the principal drainage channel of the region. Its appropriation of the tributary vessels from the posterior appendages establishes it as the *postrenal portion of the inferior vena cava* and is the last step in the formation of that composite vessel (Fig. 139, G–K).

The Coronary Sinus. The ultimate fate of the left common cardinal vein (duct of Cuvier) is a result of the shift in the course of the systemic blood so that it all enters the right side of the heart. Formerly returning a full half of the systemic blood stream to the heart, the left common cardinal vein is finally left almost without a tributary from the body. In the pig a small amount of blood usually does continue to enter it over the left azygos (Fig. 139, K). Occasionally in the pig, and normally in most other mammals, even this is cut off and the azygos drainage is by way of the right side to the superior cava (dotted line in Fig. 139, K). Nevertheless the proximal part of the old left cardinal channel is utilized. Pulled around the heart in the course of the migration of the sinus venosus toward the right, the left common cardinal vein lies close against the heart wall for a considerable distance (Fig. 144, D). As the heart muscle grows in bulk it demands a greater blood supply for its metabolism. The small returning veins of this circulation find their way into this conveniently located main vessel (Fig. 144, E). Thus even when its peripheral circulation is cut off the left common cardinal vein still persists as the coronary sinus into which the vessels of the cardiac wall drain (Fig. 144, F).

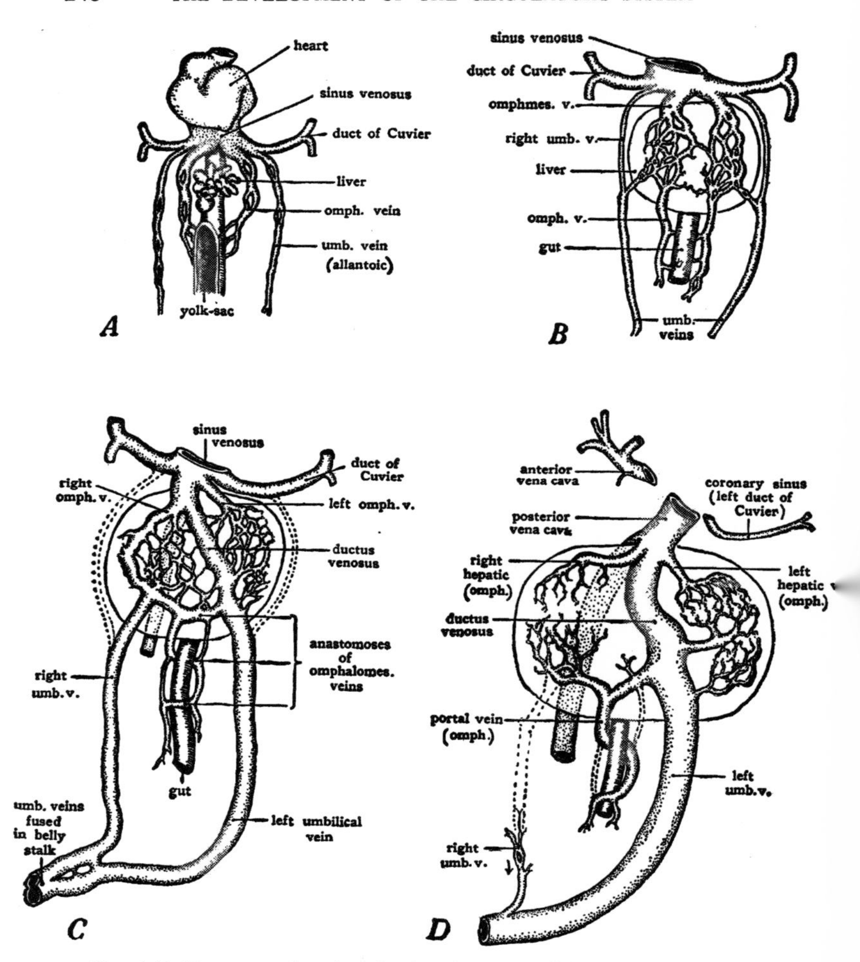

FIG. 141. Diagrams showing the development of the hepatic portal circulation from the omphalomesenteric veins, and the relations of the umbilical veins to the liver. (Adapted from several sources.)

A is based on conditions in pig embryos of 3–4 mm.; B, on embryos of about 6 mm.; C, on embryos of 8–9 mm.; D, on embryos of 20 mm. and older.

The Pulmonary Veins. Phylogenetically the lungs are relatively new structures. It is not surprising, therefore, that we find the pulmonary veins arising independently and not by the conversion of old vascular channels. They originate as vessels which drain the various branches of the lung buds and converge into a common trunk (Fig. 107), entering the left atrium dorsally. In the growth of the heart this trunk vessel is gradually absorbed into the atrial wall and two or more of its original branches open directly into the left atrium as the main pulmonary veins of the adult (Fig. 144, D–F).

The Portal Vein. The blood supply to the intestines is first established through the omphalomesenteric arteries which later become modified to form the anterior mesenteric artery. Likewise the drainage of the intestinal tract is provided for by vessels which were originally the return channels of the primitive omphalomesenteric circulatory arc (Fig. 45). In dealing with 9 to 12 mm. embryos we have already seen the primary changes which occur in these vessels. The growing cords of hepatic tissue break up the proximal portion of the omphalomesenteric veins into a maze of small channels ramifying through the substance of the liver (Fig. 141, A, B). But the stubs of the omphalomesenterics persist and drain this plexus. Distal to the liver, the original veins are for a time retained, bringing blood from the yolk-sac and intestines to the liver. With the disappearance of the yolk-sac and the growth of the intestines, the omphalic (yolk-sac) portions of these veins necessarily disappear, but the mesenteric branches persist and become more extensive concomitantly with the increased length and complexity of the intestinal tract.

The original omphalomesenteric trunks into which these tributaries converge become the unpaired portal vein by forming transverse anastomoses and then abandoning one of the original channels. The curious spiral course of the portal vein is due to the dropping out of the original left channel cephalic to the middle anastomosis and the original right channel caudal to the anastomosis (Fig. 141, D).

The Umbilical Veins. When they are first established the umbilical (allantoic) veins are embedded in the lateral body-walls throughout their course from the belly-stalk to the sinus venosus (Fig. 45). As the liver grows in bulk, it fuses with the lateral body-wall. Where this fusion occurs vessels develop connecting the umbilical veins with the plexus of vessels in the liver (Fig. 141, B). Once these connections are established the umbilical stream tends more and more to pass by way of them to the liver, and the old channels to the sinus venosus gradually degenerate (Fig. 141, C).

Meanwhile the umbilical veins distal to their entrance into the body become fused with each other so that there comes to be but a single vein in the umbilical cord (Fig. 141, C). Following this fusion in the cord, the intra-embryonic part of the umbilical channel also loses its original paired condition. The right umbilical vein is abandoned as a route to the liver and all the placental blood is returned over the left umbilical vein. It is interesting to note that in spite of its ceasing to be a through channel, part of the right umbilical vein persists, draining the body-wall. The small blood stream it then carries is reversed in direction, flowing back into the left umbilical (Fig. 141, D).

When first diverted into the liver, the umbilical blood stream passes through by way of a meshwork of small anastomosing sinusoids. As its volume increases it excavates a main channel through the substance of the liver which is known as the *ductus venosus* (Figs. 138 and 141, B–D). Leaving the liver, the ductus venosus becomes confluent with the hepatic veins (omphalomesenteric stubs) which drain the maze of small sinusoids in the liver. At this point, also, the vena cava joins the others. Thus the blood streams from the posterior systemic circulation, from the portal circulation, and from the placental circulation all enter the heart together. Embryologically this great trunk vessel represents the fused proximal parts of the old omphalomesenteric veins enlarged by the placental blood from the ductus venosus and by the systemic blood from the vena cava (Fig. 141). During its early developmental phases it is often called the *common revehent hepatic vein*. In the adult or in older fetuses it is more convenient to regard it as a part of the vena cava because, with the cessation of the placental circulation at the time of birth, the caval blood stream becomes the dominant one.

IV. The Lymphatic System

Important and interesting as the subject is, it has seemed expedient to omit any account of the development of the lymphatic vessels. Those interested in this field will find that an unusual amount of careful work has been done on the development of the lymphatics in pig embryos. References to some of the more important recent papers have been included in the bibliography appended at the end of the book.

V. Blood Corpuscles

The first blood corpuscles which appear in the circulation are produced extra-embryonically in the blood islands of the yolk-sac (Fig. 48). Later in development there are many blood-forming centers within the embryo. Concerning the establishment of these centers we find the same controversy that was commented on in connection with the origin of blood vessels. It is maintained by some authorities that these centers always arise from cells originally produced in the yolk-sac blood islands. According to this interpretation some of these blood cells are believed to remain sufficiently undifferentiated to retain their power of active proliferation. As such cells are carried by the circulation to various parts of the body they settle in favorable locations and raise new families of blood corpuscles.

According to the local origin idea it is not necessary to account for all the centers of blood corpuscle formation on the basis of blood-mother-cells wandering in from the yolk-sac and settling down in new locations within the embryo. It is maintained that mesodermal cells arising in the body have the same capacity of becoming differentiated into blood-mother-cells as mesodermal cells which arise in the yolk-sac. On this interpretation blood-forming centers arise in various parts of the body from blood-mother-cells differentiated in situ from local mesoderm. The recent experimental work tends to indicate that such "local origin" does occur. That this same evidence proves that the origin of blood-forming centers from migrating cells never occurs is by no means so clear.

Whatever the source of the original blood-mother-cells may be we always find them establishing their centers of proliferation in places where the current of the circulation is sluggish. In very young embryos the centers of corpuscle formation are located in the maze of small channels in the yolk-sac and the allantois. When the yolk-sac degenerates and the allantois becomes highly specialized as part of the chorion, these centers cease to be active and new ones are established in connection with such rich vascular plexuses as those in the mesonephros and the liver. Still later other centers appear in the lymphoid organs, and last of all in the bone marrow (Fig. 152).

The histological details of the processes involved in the production of red blood corpuscles and the various types of white blood corpuscles are exceedingly complex. Moreover there is by no means agreement as to the exact manner in which these processes occur, nor as to the genetic relations of one type of blood corpuscle to another. This whole subject, other than the recognition of the multiplicity of blood-forming centers and their shifting locations in the embryo at various phases of development, is a special field of histogenesis entirely beyond the scope of an elementary text.

VI. The Heart

To appreciate the significance of the changes which occur in the growing heart one must have in mind the exigencies under which it develops. Starting as a simple tube with the blood passing through

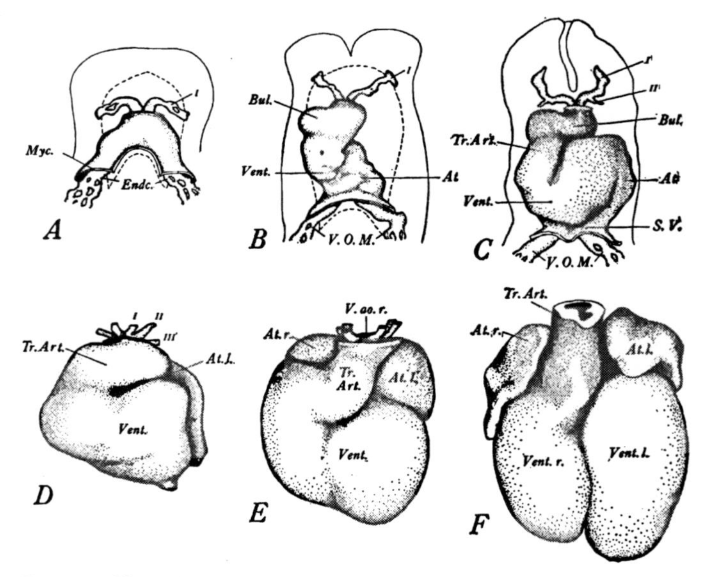

FIG. 142. Ventral aspect of the heart of the pig at various stages showing the formation of the cardiac loop and the establishment of the primary regional divisions of the heart. Drawn (A–E, × 30; F, × 20) from reconstructions made from series in the Carnegie Collection (A–D) and in the Western Reserve University Collection (E, F).

A, 7 somites; B, 13 somites; C, 17 somites; D, 25 somites; E, 3.7 mm. (after flexion); F, 6 mm.

Abbreviations: At., atrium (r., right; l., left); Bul., aortic bulb (bulbus arteriosus); Endc., endocardial tubes; Myc., cut edge of epi-myocardium; S.V., sinus venosus; Tr. Art., truncus arteriosus; V. ao. r., ventral aortic roots; Vent., ventricle (r., right; l., left); V.O.M., omphalomesenteric veins.

it in an undivided stream, it must become converted into an elaborately valved, four-chambered organ, partitioned in the mid-line and pumping from its right side a pulmonary stream which is returned to the left side and pumped out again as the systemic blood stream. And the heart cannot cease work for alteration; there can be no interruption in the current of blood it pumps to the growing embryo. This is but one phase of the matter. At the end of gestation the vascular mechanism must be prepared to function under conditions radically different from those surrounding the embryo. In spite of the impossibility of the developing lungs being effectively exercised under air-breathing conditions, they themselves, their blood vessels,

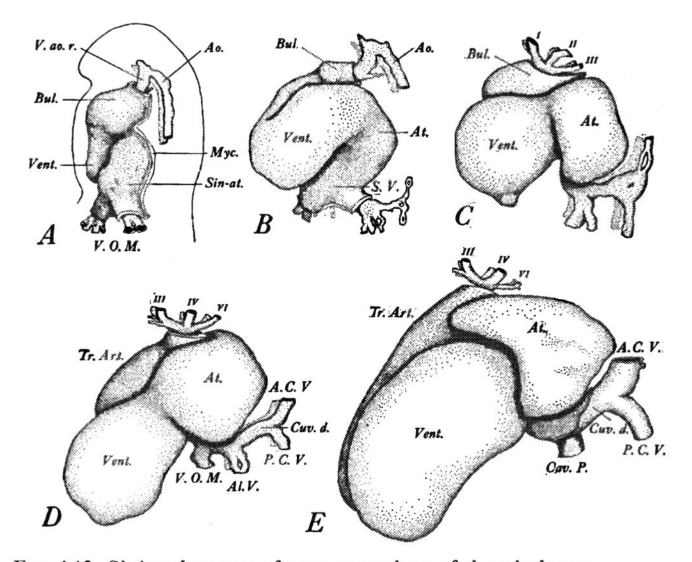

FIG. 143. Sinistral aspect of reconstructions of the pig heart.

A, 13 somite embryo (cf. Fig. 142, B); B, 17 somite embryo (cf. Fig. 142, C); C, 25 somite embryo (cf. Fig. 142, D); D, 3.7 mm. embryo (cf. Fig. 142, E); E, 6 mm. embryo (cf. Fig. 142, F).

Abbreviations: A.C.V., anterior cardinal veins; Al.V., allantoic (umbilical) vein; Ao., aorta; At., atrium; Bul., aortic bulb; Cav. P., posterior vena cava; Cuv. d., duct of Cuvier (common cardinal vein); Myc., cut edge of epi-myocardium; P.C.V., posterior cardinal vein; S.V., sinus venosus; Sin-at., sino-atrial region of heart; Tr. Art., truncus arteriosus; V. ao. r., ventral aortic roots; Vent., ventricle; V.O.M., omphalomesenteric veins.

and the right ventricle which pumps blood to them, must at the moment of birth be ready to take over the entire responsibility of oxygenating the blood. And the systemic part of the circulation as well as the pulmonary must be prepared. Throughout intra-uterine development the left side of the heart receives less blood from the pulmonary veins than the right side of the heart receives from the venae cavae. Yet after birth the left ventricle is destined to carry a greater load than the right ventricle. It must pump through the myriad peripheral vessels of the systemic circulation, sufficient blood to care for the active metabolism and continued growth of the entire body. These are some of the situations which must be faced before the heart can arrive at its adult condition. The manner in which they are met is doubly interesting because they seem at first sight so difficult.

The Formation of the Cardiac Loop and the Establishment of the Regional Division of the Heart. In dealing with the establishment of the circulatory system in very young embryos we saw how the tubular heart was formed by the fusion of paired primordia (see Chap. 5 and especially Figs. 43 and 44). The primary factor which brings about its regional differentiation is the rapid elongation of this primitive cardiac tube. The heart increases in length so much faster than the chamber in which it lies that it is first bent to the right and then twisted into a loop. Since the anterior end of the heart is anchored in the body by the aortic roots, and the posterior end by the great veins, it is the mid-portion of the heart-tube which, in this process, undergoes the most extensive changes in position. This is facilitated by the early disappearance of the dorsal mesocardium which leaves the heart entirely free in its mid-region.

During the period in which the cardiac loop is being formed, the primary regional divisions of the heart become clearly differentiated. The *sinus venosus* is the thin-walled chamber in which the great veins become confluent to enter the heart at its primary posterior end (Fig. 144). The atrial region is established by transverse dilation of the heart-tube just cephalic to the sinus venosus (Fig. 144).

The ventricle is formed by the bent mid-portion of the original cardiac tube. As this *ventricular loop* becomes progressively more extensive, it at first projects ventrally beneath the attached aortic and sinus ends of the heart (Fig. 143, A–C). Later it is bent caudally so that the ventricle, formerly situated cephalic to the atrium, is brought into its characteristic adult position caudal to the atrium (Fig. 143, D, E). Between the atrium and ventricle the heart remains

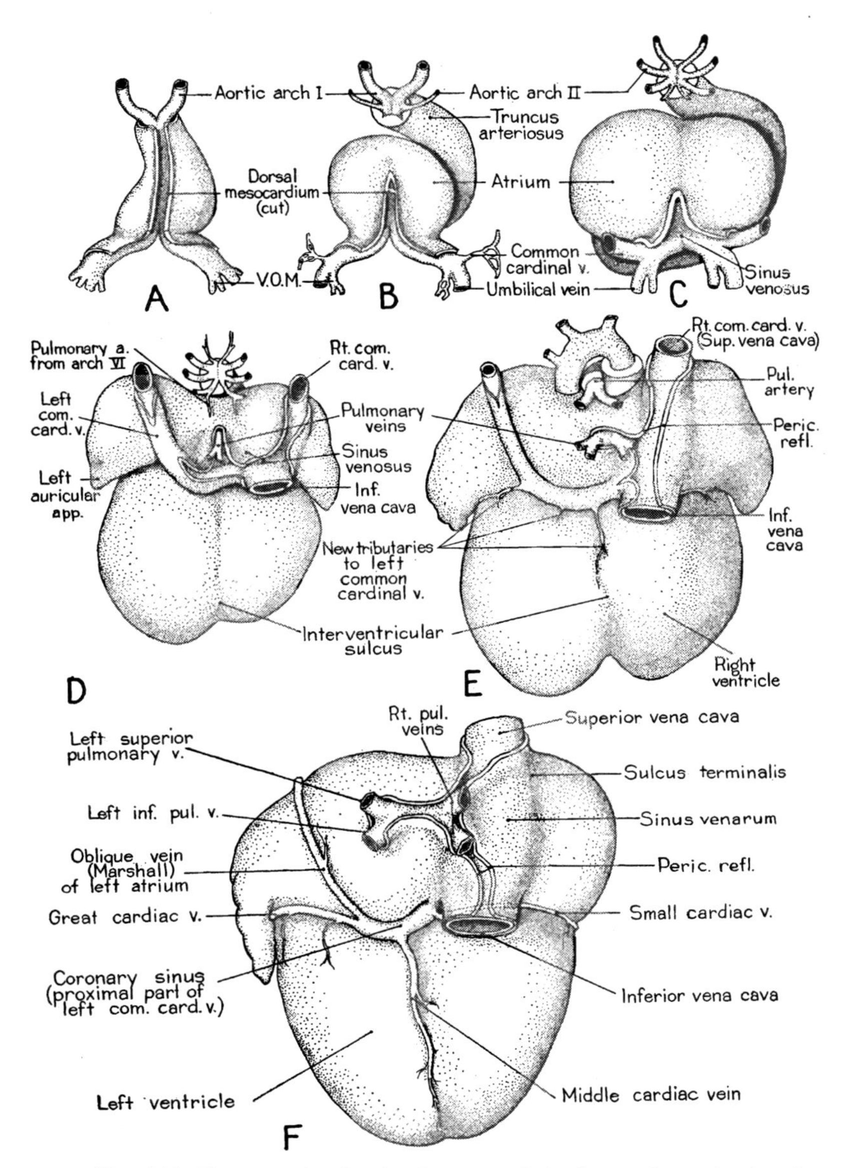

FIG. 144. Six stages in the development of the heart, drawn in dorsal aspect to show the changing relations of the sinus venosus and great veins entering heart. (Patten: "Human Embryology," The Blakiston Company.)

relatively undilated. This narrow connecting portion is the *atrio-ventricular canal*.

The most cephalic part of the cardiac tube undergoes least change in appearance, persisting as the *truncus arteriosus* connecting the ventricle with the ventral aortic roots (Fig. 142). In very young embryos there is a conspicuous bulge where the truncus arteriosus swings toward the mid-line to break up in the aortic arches. This sharply bent and somewhat dilated region is called the *aortic bulb* (Fig. 142, B, C). Its location is of interest as being the place at which the paired endocardial primordia first fused with each other (Figs. 43 and 44) and the place at which, later in development, the division of the truncus arteriosus into separate aortic and pulmonary roots will first become apparent. The bulge itself soon merges into the rest of the truncus arteriosus without giving rise to any special structure.

Almost from their earliest appearance the atrium and the ventricle show external indications of the impending division of the heart into right and left sides. A distinct median furrow appears at the apex of the ventricular loop (Figs. 142, E, F, and 144, D, E). The atrium meanwhile has undergone rapid dilation and bulges out on either side of the mid-line (Fig. 144). Its bilobed configuration is emphasized by the manner in which the truncus arteriosus compresses it mid-ventrally (Fig. 142, E, F).

The Partitioning of the Heart. These superficial features suggest the more important changes going on internally. As the wall of the ventricle increases in thickness it develops on its internal face a meshwork of interlacing muscular bands, the *trabeculae carneae*. Opposite the external furrow in the ventricle these muscular bands become consolidated as a partition which appears to grow from the apex of the ventricle toward the atrium. This is the *interventricular septum* (Fig. 147).

Meanwhile two conspicuous masses of peculiar, loosely organized mesenchyme (called endocardial-cushion tissue) develop in the walls of the narrowed portion of the heart between the atrium and the ventricle. One of these so-called *endocardial cushions of the atrio-ventricular canal* is formed in its dorsal wall (Fig. 147, A) and the other is formed opposite, on the ventral wall. These two masses nearly occlude the central part of the canal and thus initiate its separation into right and left channels.

At the same time a median partition appears in the cephalic wall of the atrium. Because another closely related partition is destined to form here later, this one is called the first interatrial septum or *septum primum.* In shape it is crescentic with its concavity directed toward the ventricle and the apices of the crescent extending, one along the dorsal wall, and one along the ventral wall of the atrium, all the way to the atrio-ventricular canal where they merge, respectively, with the dorsal and ventral endocardial cushions (Fig. 147). This leaves the atria separated from each other except for an opening called the *interatrial foramen primum.*

While these changes have been occurring, the sinus venosus has been shifted out of the mid-line so that it opens into the atrium to the right of the interatrial septum (Fig. 147). The heart is now in a critical stage of development. Its simple tubular form has been altered so that the four chambers characteristic of the adult heart are clearly recognizable. Partitioning of the heart into right and left sides is well under way. But there is as yet no division of the blood stream because there are still open communications from the right to

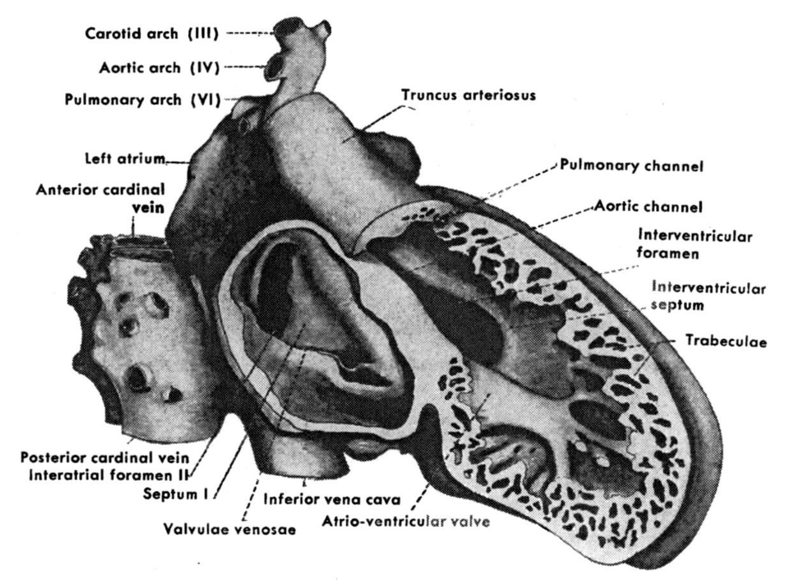

FIG. 145. Reconstruction of the heart of a 9.4 mm. pig embryo cut open somewhat to the right of the mid-line to show its internal structure (cf. Fig. 146).

the left side in both atrium and ventricle. A little further progress in the growth of the partitions, however, and the two sides of the heart would be completely separated. Were this to occur now, the left side

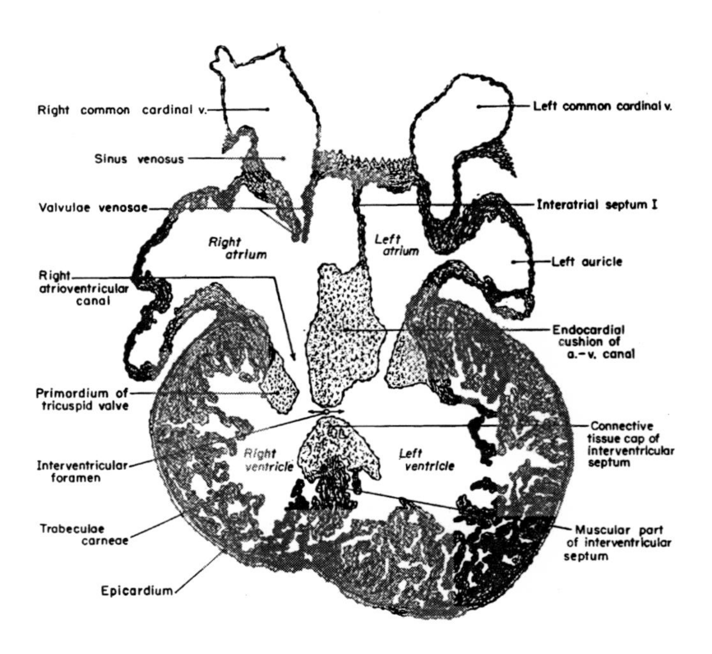

FIG. 146. Section (× 45) through the heart of a 9.4 mm. pig embryo, at the level of the atrio-ventricular canals. (From the series used in making the reconstruction appearing as figure 145. This section may be oriented as passing horizontally through Fig. 145 at the level of the most caudal portion of the interventricular foramen.)

of the heart would become almost literally dry. For the sinus venosus, into which systemic, portal, and placental currents all enter, opens on the right of the interatrial septum, and not until much later do the lungs and their vessels develop sufficiently to return any considerable volume of blood to the left atrium. The partitions in the ventricle and in the atrio-ventricular canal do progress rapidly to completion (Figs. 147, 148, and 149), but an interesting series of events takes place at the interatrial partition which assures an adequate supply of blood reaching the left atrium and thence the left ventricle.

Just when it appears that the septum primum is going to fuse with the endocardial cushions of the atrio-ventricular canals, closing the interatrial foramen primum and isolating the left atrium, a new opening is established. The more cephalic part of the septum primum ruptures to form the *interatrial foramen secundum*, thus keeping a route open from the right to the left atrium (Fig. 148).

At about this time the second interatrial partition makes its appearance just to the right of the first. Also crescentic in form, this *septum secundum* extends its apices along the dorsal and ventral walls of the atrium to fuse with the septum primum near the endocardial cushion mass which now completely divides the atrio-ventricular

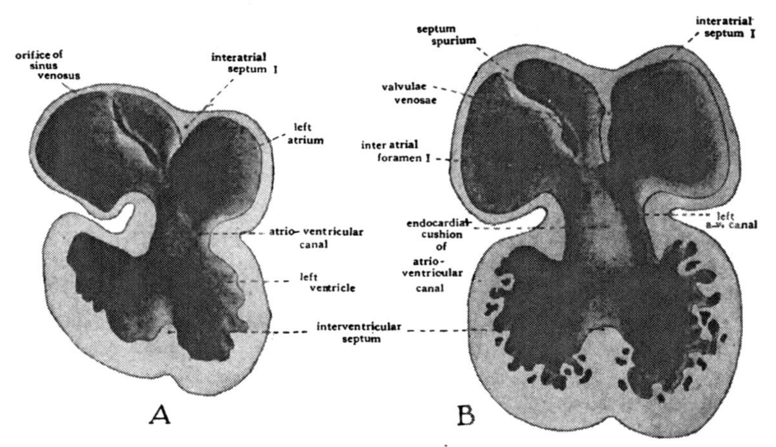

Fig. 147. Drawings showing the initial steps in the partitioning of the heart.

A, Slightly schematized drawing from reconstruction of heart of 3.7 mm. pig embryo. The heart has been opened by a diagonally frontal cut in the plane which can be indicated by drawing a line through the labels At. and Vent. in figure 143, D. The dorsal portion of the heart is shown viewed from the ventral side.

B, Slightly schematized drawing from a reconstruction of the heart of a 6 mm. pig embryo. The heart has been opened in a plane which can be indicated by drawing a line through the labels At. and Vent. in figure 143, E. The dorsal portion of the heart is shown viewed from the ventral side.

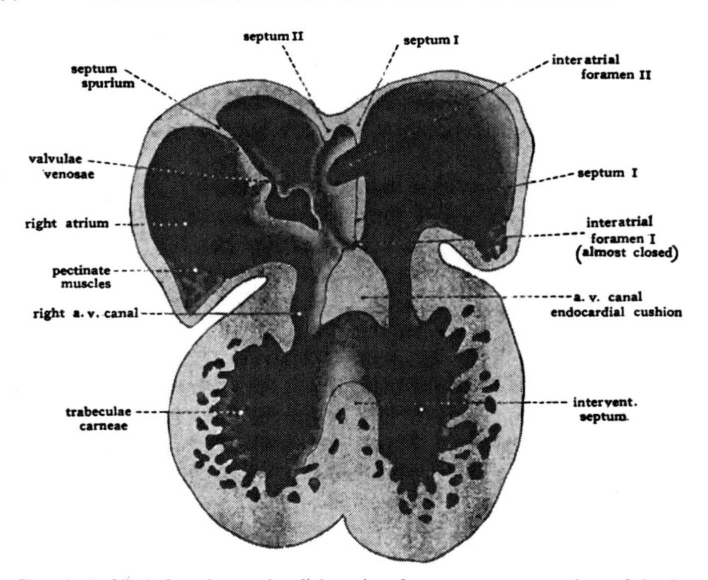

FIG. 148. Slightly schematized drawing from a reconstruction of the heart of a 9.4 mm. pig embryo. Dorsal part of heart, interior view. The plane in which the heart is opened can be indicated by drawing a line through the center of the interatrial and the interventricular foramina in figure 145.

canals. But the septum secundum never becomes a complete partition. An oval opening of considerable size persists in its center. This is the *foramen ovale* (Fig. 149).

The newly established septum secundum and the flap-like remains of the septum primum constitute an efficient valvular mechanism between the two atria. When the atria are filling, some of the blood discharged from the venae cavae into the right atrium can pass freely through the foramen ovale by merely pushing aside the flap of the septum primum. The inferior caval entrance lies adjacent to, and is directed straight into, the orifice of the foramen ovale (Fig. 149). Consequently it is primarily—some think exclusively—blood from the inferior vena cava which passes through the foramen ovale into the left atrium. When the atria start to contract, pressure of the blood within the left atrium forces the flap of the septum primum against the foramen ovale, effectively closing it against return flow into the right atrium. Without some such mechanism affording a supply of blood

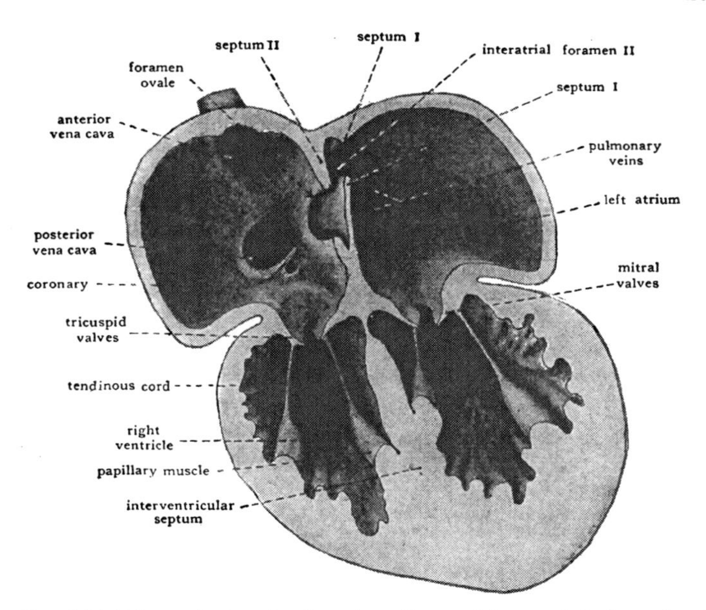

FIG. 149. Schematic drawing based on dissected heart of pig fetus shortly before birth. Interior aspect of dorsal part of heart to show valvular mechanism at foramen ovale.

to its left side, the developing heart could not be partitioned in the mid-line ready to assume its adult function of pumping two separate blood streams.[3]

While all these changes have been going on in the main part of the heart, the truncus arteriosus has been divided into two separate channels. Reference has already been made to the start of this process in the aortic root between the fourth and sixth arches (Fig. 133). Continuing toward the ventricle, the division is effected by the formation of longitudinal ridges of plastic young connective tissue of the same type as that making up the endocardial cushions of the atrioventricular canal. These ridges, called *truncus ridges*, bulge progres-

[3] There are on record a few cases of a peculiar cardiac malformation in which this valvular communication was prematurely closed and the left atrium thereby shut off from the right. In all these cases the left side of the heart has not been sufficiently developed to support life for any length of time after birth.

sively farther into the lumen of the truncus arteriosus and finally meet to separate it into aortic and pulmonary channels. (Note shape of lumen in Figs. 142, F, and 145.) The semilunar valves of the aorta, and of the main pulmonary trunk (Fig. 138), develop as local specializations of these truncus ridges. Toward the ventricles from the site of formation of the semilunar valves the same ridges are continued into the conus of the ventricles. The fact that proximal to the valves these ridges are, for descriptive purposes, called *conus ridges* should not be allowed to obscure their developmental and functional continuity with the truncus ridges. The truncus and conus ridges follow a spiral course such that where the conus ridges extend down into the ventricles they meet and become continuous with the interventricular septum. The right ventricle then leads into the pulmonary channel and the left into the aorta. With this condition established the heart is completely divided into right and left sides except for the interatrial valve which must remain open throughout fetal life, until, after birth, the lungs attain their full functional capacity and the full volume of the pulmonary stream passes through them to be returned to the left atrium.

This leaves but one of the exigencies of heart development still to be accounted for. If, during early fetal life, before the lungs were well developed, the pulmonary channel were the only exit from the right side of the heart, the right ventricle would have an outlet inadequate to develop its pumping power. For it is only late in fetal life that the lungs and their vessels develop to a degree which prepares them for assuming their postnatal activity, and the power of the heart muscle must be built up gradually by continued functional activity. This situation is met by the ductus arteriosus leading from the pulmonary trunk to the aorta (Fig. 138). The right ventricle is not unprepared for its adult function because it pumps its full share of the blood throughout fetal life. Instead of all going to the lungs, however, part of the blood pumped by the right ventricle passes by way of the ductus arteriosus into the aorta (Fig. 150). As the lungs increase in size, relatively more blood goes to them and relatively less goes through the ductus arteriosus. By the time of birth enough blood is passing through the lungs to support life, and within a short time after birth, under the stimulus of functional activity, the lungs are able to take all the blood from the right side of the heart and the ductus arteriosus is gradually obliterated. The ductus arteriosus, therefore, serves during intra-uterine life as what might be called a "compensated exercising channel" for the right ventricle.

Changes in the Sinus Region. In following the story of the development of the heart from the functional standpoint, many things of less striking significance have been passed over. Some of these should have a word of comment. Cephalic to the sinus orifice the valvulae venosae which guard it against return flow fuse and are prolonged onto the dorsal wall of the atrium in the form of a ridge called the *septum spurium* (false septum). This septum is, for a time, too conspicuous to ignore, but it is of little importance in the partitioning of the heart and soon undergoes retrogression.

As the heart grows it absorbs the sinus venosus into its walls so that eventually the anterior and posterior venae cavae and the coronary sinus all open separately into the right atrium (cf. Figs. 147, 148, and 149). Portions of the right valvula venosa are retained as the valves of the caval and coronary orifices. In the adult heart a small external sulcus can usually be found between the entrance of the anterior and the posterior vena cava which records the old line of demarcation between sinus venosus and atrium.

The Atrio-ventricular Valves and the Papillary Muscles. At the point where the atrio-ventricular canals open into the ventricles there are early indications of the establishment of valves. From the partition in the atrio-ventricular canal and from the outer walls on either side, masses of tissue in the shape of thick, blunt flaps project toward the ventricle (Figs. 146 and 148). It is these masses of a primitive type of connective tissue, similar to that in the endocardial cushions of the canal, which later become differentiated into the flaps of the adult valves (Fig. 149). The papillary muscles and tendinous cords which, in the adult, act as stays to these valves, arise by modification of some of the related trabeculae carneae (cf. Figs. 146, 148, and 149).

The Aortic and Pulmonary Valves. The valves which guard the orifices of the aorta and of the pulmonary artery arise from mesenchymal pads (Fig. 138) already mentioned as developing in connection with the truncus septum. In early histological appearance and in the manner in which this loose tissue gradually becomes organized into the exceedingly dense fibrous tissue characteristic of adult valves, they are similar to the atrio-ventricular valves. They do not, however, develop any supporting strands comparable to the papillary muscles and tendinous cords.

Course and Balance of Blood Flow in the Fetal Heart. All the steps in the partitioning of the embryonic heart lead gradually toward the final adult condition in which the heart is completely divided into right and left sides. Yet from the nature of its living conditions it is not possible for the fetus in utero fully to attain the adult type of circulation. The plan of the completely divided circulation is predicated on lung breathing. In the adult the right side of the heart receives the blood returning from a circuit of the body and pumps it to the lungs where it is relieved of carbon dioxide and acquires a fresh supply of oxygen. The left side of the heart receives the blood that has just passed through the lungs and pumps it again through ramifying channels to all the tissues of the body. In the fetus the function of respiration is carried out in the placenta by interchange with the maternal blood circulating through the uterus. The lungs, although in the last part of fetal life they are fully formed and ready to function, cannot actually begin their work until after birth. The radical change which must inevitably take place immediately following birth in the manner in which the blood is oxygenated has led to a widespread belief that there must be revolutionary changes in the routing of blood through the heart and great vessels. However, as the embryology of the circulatory system has been studied more closely from a functional angle it is becoming increasingly clear that the heart and the major vascular channels develop in such a manner that the pumping load on the different parts of the heart remains balanced at all times during fetal life. Moreover, the very mechanisms which maintain this cardiac balance during intra-uterine life are perfectly adapted to rebalance the circulatory load on the new postnatal basis without involving any sudden overloading of previously inactive parts of the vascular system.

To understand the changes in circulation which are so smoothly accomplished at the time of birth it is necessary to have clearly in mind the manner in which the way for them has been prepared during intra-uterine life. In the foregoing account of the development of the interatrial septal complex, emphasis was laid upon the fact that at no time were the atria completely separated from each other. This permits the left atrium, throughout prenatal life, to receive a contribution of blood from the inferior cava and the right atrium by a transseptal flow which compensates for the relatively small amount of blood entering the left atrium of the young embryo by way of the pulmonary circuit, and maintains an approximate balance of intake into the right and left sides of the heart.

The precise manner in which this transseptal flow occurs, and where, and to what extent the various blood streams of the fetal circulation are mixed has long been a controversial subject. The recent brilliant work of Barcroft and Barron and their co-workers has gone far toward putting some of these old controversies into proper perspective. Their first approach was through the quantitative analysis of blood samples drawn from various critical parts of the fetal circulation. The oxygen content of such samples has given important evidence as to what mixing of the currents is actually taking place in the living fetus. Later work involving the collaboration of Barclay and Franklin utilized serial x-ray photography following the injection of opaque material into the blood stream at various points. This method has given further direct evidence as to the course followed by some of the important blood currents. Synthesizing the most significant of the anatomical evidence with the newer experimental evidence, the course followed by the blood in passing through the fetal heart may be summarized somewhat as follows. The inferior caval entrance is so directed with reference to the foramen ovale that a considerable portion of its stream passes directly into the left atrium (Figs. 149 and 150). Under fluctuating pressure conditions—say following uterine contractions which send a surge of placental blood through the umbilical vein—the placental flow may temporarily hold back any blood from entering the circuit by way of either the portal vein or the inferior caval tributaries (Fig. 150). For a time, under these conditions, the left atrium would be charged almost completely with fully oxygenated blood. Such conditions, however, would be but temporary and would be counterbalanced by periods when the portal and systemic veins poured enough blood into the common channels to load the heart for a time with mixed or depleted blood. The important thing physiologically is not the fluctuations, but the maintenance of the average oxygen content of the blood at adequate levels.

Compared with conditions in adult mammals, the mixing of oxygenated blood freshly returned from the placenta with depleted blood returning from a circuit of the body may seem inefficient. But this is a one-sided comparison. The fetus is an organism in transition. Starting with a simple ancestral plan of structure and living an aquatic life, it attains its full heritage but slowly. It must be viewed as much in the light of the primitive conditions from which it is emerging as in comparison with the definitive conditions toward which it is progressing. Below the bird-mammal level circulatory mechanisms with partially divided and undivided hearts and correspondingly un-

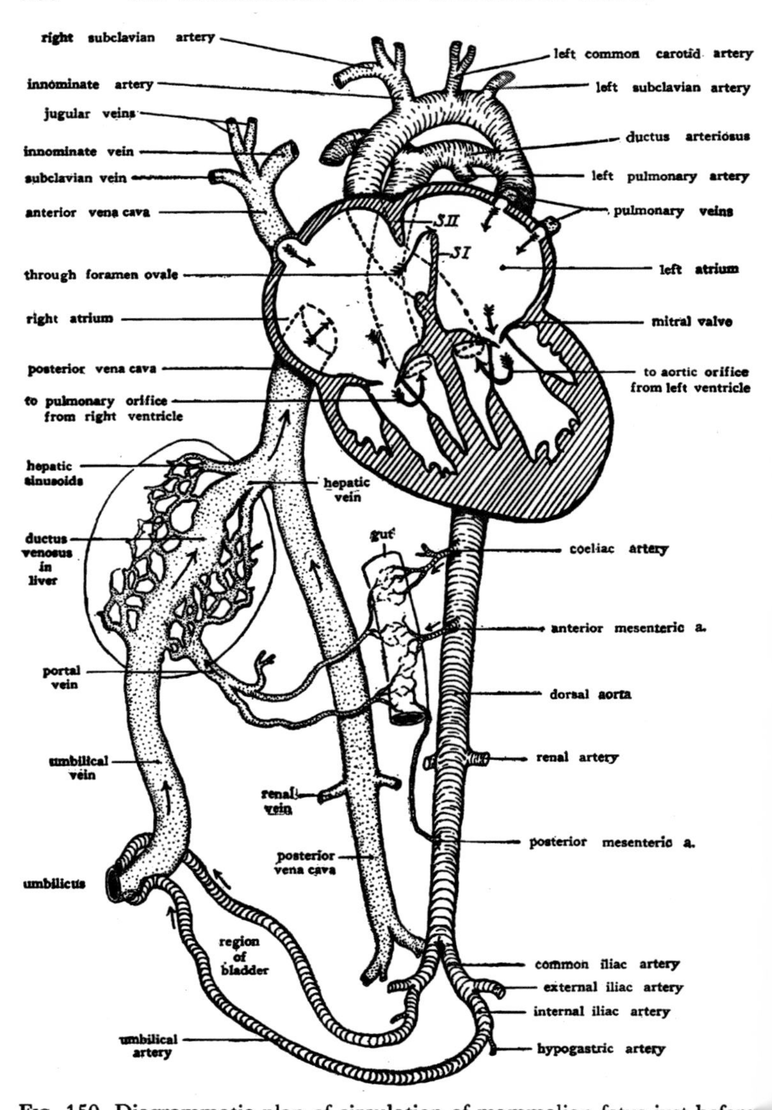

FIG. 150. Diagrammatic plan of circulation of mammalian fetus just before birth.

separated blood streams meet all the needs of metabolism and growth. Maintenance of food, oxygen, and waste products at an average level which successfully supports life does not depend on "pure currents," although such separated currents undoubtedly make for higher efficiency in the rate of interchange of materials. From a comparative viewpoint, the fact that the mammalian fetus is supported by a mixed circulation seems but natural.

Another significant fact is that careful measurements have shown that the interatrial communication in the heart of the fetus at term is considerably smaller than the inferior caval inlet. This would mean that the portion of the inferior caval stream which could not pass through this opening into the left atrium would eddy back and mix with the rest of the blood in the right atrium.

One of the most important inferences as to the fetal circulation based on vessel size is that the circulation through the lungs in a fetus which is sufficiently mature to be viable is of considerable volume. This too has now been supported by experimental work. From the standpoint of smooth postnatal circulatory readjustments, the larger the fetal pulmonary return becomes the less will be the balancing transatrial flow, and the less will be the change entailed by the assumption of lung breathing. Very early in development, before the lungs have been formed, the pulmonary return is negligible and the flow from the right atrium through the interatrial ostium primum constitutes practically the entire intake of the left atrium. After the ostium primum is closed and while the lungs are but little developed, flow through the interatrial ostium secundum must still be the major part of the blood entering the left atrium. During the latter part of fetal life the foramen ovale in septum secundum becomes the transseptal route. As the lungs grow and the pulmonary circulation increases in volume, a progressively smaller proportion of the left atrial intake comes by way of the foramen ovale and a progressively larger amount from the vessels of the growing lungs.

The balanced atrial intake thus maintained implies a balanced ventricular intake, and this in turn implies a balanced ventricular output. Although not in the heart itself, we have seen that there is in the closely associated great vessels a mechanism which affords an adequate outlet from the right ventricle during the period when the pulmonary circuit is developing. When the pulmonary arteries are formed from the sixth pair of aortic arches, the right sixth arch soon loses its original connection with the dorsal aorta. On the left, however, a portion of the sixth arch persists as a large vessel connecting the pulmonary artery with the dorsal aorta (Figs. 138 and 150). This vessel, already familiar to us as the ductus arteriosus, remains open throughout fetal life and acts as a shunt, carrying over to the aorta whatever excess of blood the pulmonary vessels at any particular phase of their development are not prepared to receive from the right ventricle. As has already been pointed out, the ductus arteriosus can be called the "exercising channel" of the right ventricle because it makes it possible for the right ventricle to carry its full share of work throughout development and thus to be prepared for pumping all the blood through the lungs at the time of birth.

VII. The Changes in the Circulation Following Birth

The two most obvious changes which occur in the circulation at the time of birth are the abrupt cutting off of the placental blood stream and the immediate assumption by the pulmonary circulation of the function of oxygenating blood. One of the most impressive things in embryology is the perfect preparedness for this event which has been built into the very architecture of the circulatory system during its development. The shunt at the ductus arteriosus, which has been one of the factors in balancing ventricular loads throughout development, and the valvular mechanism at the foramen ovale, which has at the same time been balancing atrial intakes, are perfectly adapted to effect the postnatal rebalancing of the circulation. The closure of the ductus arteriosus is the primary event and the closure of the foramen ovale follows as a logical sequel.

It has long been known that the lumen of the ductus arteriosus is gradually occluded postnatally by an overgrowth of its intimal tissue. This process in the wall of the ductus is as characteristic and regular a feature of the development of the circulatory system as the formation of the cardiac septa. Its earliest phases begin to be recognizable in the fetus as the time of birth approaches, and postnatally the process continues at an accelerated rate to terminate in complete anatomical occlusion of the lumen of the ductus about six to eight weeks after birth.

Barcroft, Barclay, and Barron have conducted an extensive series of experiments on animals delivered by Cesarean section which indicate that the ductus arteriosus closes functionally far sooner than it does anatomically. Following birth there appears to be a contraction of the circularly disposed smooth muscle in the wall of the ductus which promptly reduces the flow of blood through it. This reduction in the shunt from the pulmonary circuit to the aorta, acting together with the newly assumed respiratory activity of the lungs themselves, aids in raising the pulmonary circulation promptly to full functional level. At the same time the functional closure of the ductus by muscular contraction paves the way for the ultimate anatomical obliteration of its lumen by overgrowth of intimal connective tissue. This concept of the immediate functional closure of the ductus is so appealing on theoretical grounds that a little extra caution in evaluating the evidence is indicated. It should be borne in mind that an initial tendency on the part of the circular smooth muscle of the ductus to contract does not necessarily imply a contraction sufficiently strongly and steadily maintained to shut off all blood flow during the six to eight weeks occupied by morphological closure. The dramatic quality of an immediate muscular response should not cause us to forget the importance of the slower but more positive structural closure.

The results of increased pulmonary circulation with the concomitant increase in the direct intake of the left atrium are manifested secondarily at the foramen ovale. Following birth, as the pulmonary return increases, compensatory blood flow from the right atrium to the left decreases correspondingly, and finally ceases altogether. This is indicated anatomically by a progressive reduction in the looseness of the valvula foraminis ovalis and the consequent diminution of the interatrial communication to a progressively narrower slit between the valvula and the septum. When equalization of atrial intakes has occurred, the compensating one-way valve at the foramen ovale falls into disuse, and the foramen ovale may be regarded as functionally closed.

Anatomical obliteration of the foramen ovale follows leisurely in the wake of its functional abandonment. There is a considerable interval following birth before the septum primum fuses with the septum secundum to seal the foramen ovale. This delay is, however, of no import because as long as the pulmonary circuit is normal and pressure in the left atrium does not fall below that in the right, the orifice between them is functionally inoperative. It is not uncommon to find the fusion of these two septa incomplete in the hearts of individuals who have, as far as circulatory disturbances are concerned, lived uneventfully to maturity. Such a condition can be characterized as "probe patency" of the foramen ovale. When, in such hearts, one inserts a probe under the margin of the fossa ovalis and pushes it toward the left atrium one is, so to speak, prying behind the no longer used, but still unfastened, interatrial door.

With birth and the interruption of the placental circuit there follows the gradual fibrous involution of the umbilical vein and the umbilical arteries. The flow of blood in these vessels, of course, ceases immediately with the severing of the umbilical cord, but obliteration of the lumen is likely to take from three to five weeks, and isolated portions of these vessels may retain a vestigial lumen for much longer. Ultimately these vessels are reduced to fibrous cords. The old course of the umbilical vein is represented in the adult by the round ligament of the liver extending from the umbilicus through the falciform ligament, and by the ligamentum venosus within the substance of the liver. The proximal portions of the umbilical arteries are retained in reduced relative size as the hypogastric or internal iliac arteries. The fibrous cords extending from these arteries on either side of the urachus toward the umbilicus represent the remains of the more distal portions of the old umbilical arteries. They are known in the adult as the obliterated branches of the hypogastric arteries, or as the lateral umbilical ligaments.

Much yet remains to be learned as to the more precise physiology of the fetal circulation and as to the interaction of various factors during the transition from intra-uterine to postnatal conditions. Nevertheless, with our present knowledge it is quite apparent that the changes in the circulation which occur following birth involve no revolutionary disturbances of the load carried by different parts of the heart. The fact that the pulmonary circulation is already so well developed before birth means that the changes which must occur following birth are far less profound than was formerly believed; and the compensatory mechanisms at the foramen ovale and the ductus arteriosus which have been functioning all during fetal life are entirely competent to effect the final postnatal rebalancing of the circulation with a minimum of functional disturbance. It is still true that as individuals we crowd into a few crucial moments the change from water living to air living that in phylogeny must have been spread over eons of transitional amphibious existence. But as we learn more about this change in manner of living, it becomes apparent that we should marvel more at the completeness and the perfection of the preparations for its smooth accomplishment, and dwell less on the old theme of the revolutionary character of the changes involved.

CHAPTER 12

The Histogenesis of Bone and the Development of the Skeletal System

I. Histogenesis of Bone

Histologically bone belongs to the group of tissues known as the connective and supporting tissues. In spite of their widely varying adult conditions these tissues are all similar in that the secreted parts, rather than the cells themselves, carry out the functional rôle characteristic of the tissues. It is the secreted, fibrous portion of the binding connective tissues which ties together various other tissues and organs; it is the secreted matrix of cartilage and of bone which affords rigid support and protection to soft parts and furnishes a lever system on which the muscles may be brought into play.

The cellular elements of these tissues must not be overlooked, however, in emphasizing the functional importance of the cell products. The cells are, so to speak, the power behind, in that they extract the appropriate raw materials from the circulation, elaborate them within their cytoplasm, and deposit the characteristic secretion as an end-product. Moreover after the fiber is formed or the matrix is laid down, it is dependent on the cells for maintenance in a healthy active condition.

Embryologically the entire connective-tissue group arises from mesenchymal cells. It is not surprising, in view of their closely related functions and their derivation from a common type of ancestral cell, that one type of connective tissue may be converted into or replaced by another. This facility for changing the type of specialization is sometimes referred to as plasticity.

The plasticity of the connective-tissue series is well exemplified in the development of bone. Bone does not form in vacant spaces. It is always laid down in an area already occupied by some less highly specialized member of the connective-tissue family. The formation of some bones begins in areas already occupied by connective tissue—such bones are said to be intramembranous in origin, or are spoken of as *membrane bones*. Other bones are laid down in areas already occupied by cartilage. In this case they are said to be endochondral in origin, or, are called *cartilage bones*. It should be clearly borne in mind that these terms apply solely to the method by which a bone develops and do not imply any differences in histological structure, once the bone is fully formed.

Likewise we should know at the outset what histologists mean when they speak of cancellous bone and compact bone. These terms refer not to the method of origin of the bone but to its density when fully formed. Developmentally all bone goes through the spongy or cancellous stage. Some bones later become compact, others remain cancellous. Most bones are compact in some areas and cancellous in others.

The subject of bone development can be presented more simply if we take up first the formation of primary cancellous bone intramembranously; then the method by which this same type of spongy bone is formed within cartilage, and finally the changes by which cancellous bone, formed in either of the above ways, may become secondarily compact.

Intramembranous Formation of Primary Cancellous Bone. In an area where intramembranous bone formation is about to begin we find an abundance of mesenchymal cells congregated and numerous small blood vessels present. The mesenchymal cells soon exhibit a tendency to cluster together in more or less elongated groups here and there throughout the area. If we study a group of this type which has been aggregated for a short time we can make out the beginning of a definite plan of organization. Near the axis of the cord delicate fibers appear, produced by the secretory activity of the cells. As this fibrous strand becomes more definite, the cells tend to become ranged against it (Fig. 151, A). In so doing they retract the cytoplasmic processes which are so characteristic of undifferentiated mesenchymal cells and become rounded. In this stage we have essentially a connective tissue in which the fibrous strands are for the most part rather widely separated from one another, and in which each strand has, lined up against it, the cells responsible for its production.

The actual deposition of bone matrix begins very soon after the establishment of these primordial strands of mesenchymal cells and fibers. In fact one usually finds the formation of bone beginning on the older part of a strand while the strand itself is still being extended at one end by the aggregation of more mesenchymal cells (Fig. 151,

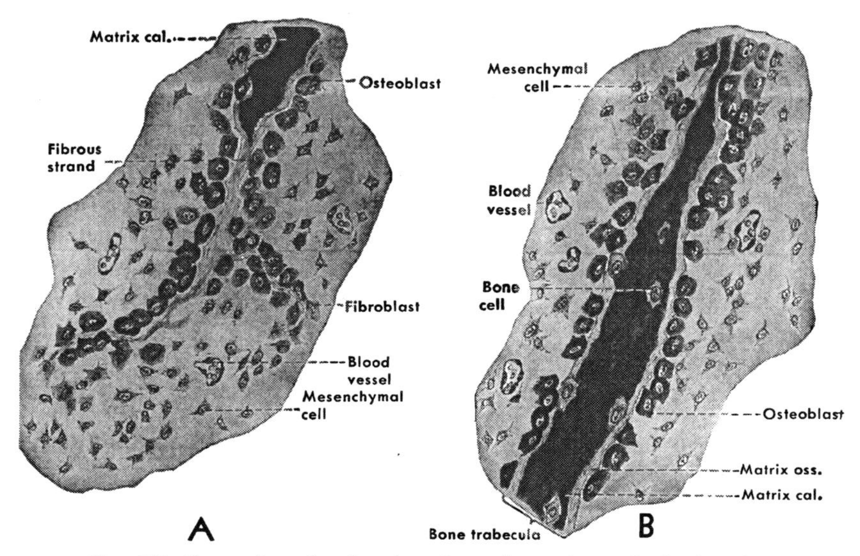

FIG. 151. Formation of trabeculae of membrane bone. Projection drawings from the mandible of a pig embryo 130 mm. in length (cf. Fig. 182).

Abbreviations: Matrix cal., ossein matrix impregnated with calcium salts; Matrix oss., ossein matrix not yet impregnated with calcium salts.

A). When the mesenchymal cells ranged against the fibrous axis of such a strand become active in the secretion of calcareous material they are spoken of as *osteoblasts*. We should not lose sight of the fact that they are the same cells which formed the fibrous axis of the original strands, given a new name in deference to their further specialization and altered internal chemistry.

In studying the deposition of *bone matrix* one must bear in mind its dual nature. The matrix consists of an organic fibrous framework which is impregnated by a subsequent deposit of inorganic calcium compounds. We may liken the matrix of bone to reinforced concrete. In the making of a road or a wall, a meshwork of steel is first placed in the forms and concrete is then poured in. The steel gives the finished structure tensile strength and a certain amount of elasticity, the concrete gives form and hardness. So in bone the organic fibers (*ossein fibers*) impart strength and resilience, while the calcium salts with which the fibers are impregnated give to the completed matrix body and rigidity.

The two steps in the deposition of bone matrix may be demonstrated readily in areas where active bone formation is going on, owing to the fact that the presence of calcium compounds in a tissue markedly increases its affinity for stains. Even after most of the calcium salts have been removed from the ossein framework by treatment of the tissue with acids (decalcification) to permit the making of sections, the staining reaction is still apparent. This indicates that the ossein fibers in which calcium has once been deposited are more or less permanently changed chemically even though all the calcium possible is subsequently removed.

If we look at a strand on which the osteoblasts have been active for a time (Fig. 151, B) we see, next to the osteoblasts, a zone of bone matrix which takes very little stain. This is the newly deposited organic portion of the matrix as yet unimpregnated with calcium salts. It consists of a feltwork of minute fibers so delicate and so closely matted together that it is very difficult in ordinary preparations to see the individual fibers at all. Slightly farther from the osteoblasts the matrix is densely stained (Fig. 151, B). This part of the matrix has been impregnated with calcium salts, chiefly phosphates and carbonates, and has thereby been converted into true bone matrix. The calcium utilized by the osteoblasts in this process is brought to them by the blood stream where it is carried in soluble form, probably in organic linkage. It is interesting to note in this connection that the presence of calcium and of phosphates in the blood is not in itself all that is necessary for this process. There must be present also sufficient vitamin D, which in some way facilitates the extraction by the osteoblasts of these raw materials from the blood and their deposition in insoluble form as part of the bone matrix. The absence of vitamin D from the system results in the formation of bone matrix deficient in calcium salts and therefore lacking in rigidity—a condition not infrequent in pigs. Stock raisers have miscalled this condition rheumatism but it is really the same condition known medically as *rickets*.

In the deposition of the matrix, the fibrous core of the original strand serves as a sort of axis on which the first matrix is laid down. When such a strand is completely invested by bone matrix it is called a *trabecula* (little beam). As the osteoblasts continue to secrete and thereby thicken the trabecula, the accumulation of their own product forces them farther and farther away from the axial strand about which the first of the matrix was formed. The new matrix added is not laid down uniformly. It is possible to make out in it markings which are suggestive of the growth rings of a tree. Apparently the osteoblasts work more or less in cycles, depositing a succession of thin layers of matrix. Each of these layers of the matrix is called a *lamella* (Fig. 152). As the row of osteoblasts is forced back with the deposit of each succeeding lamella, not all the cells free themselves from their secretion. Here and there a cell is left behind. As its former fellows

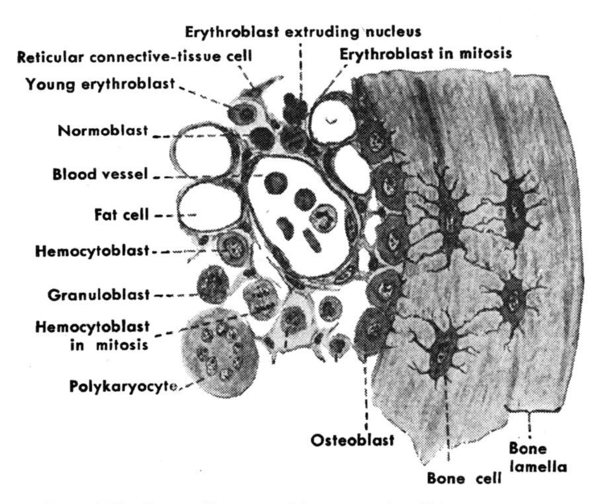

FIG. 152. A small area of bone and adjacent marrow as seen in highly magnified decalcified sections. The drawing has been schematized somewhat to emphasize the relations of the cytoplasmic processes of the osteoblasts and the bone cells so important in nutrition. In the adjacent marrow developmental stages of various types of blood cells have been suggested.

continue to pile up new matrix, it becomes completely buried (Fig. 151, B). An osteoblast so caught and buried is called a bone cell (*osteocyte*), and the space in the matrix which it occupies is called a *lacuna*. The bone cells, thus entrapped, of necessity cease to be active bone formers, but they play a vital part in the maintenance of the bone already formed. They have delicate cytoplasmic processes radiating into the surrounding matrix through minute canaliculi. The processes of one cell come into communication with the processes of its neighbors (Fig. 152). Thus the bone cells nearer to blood vessels

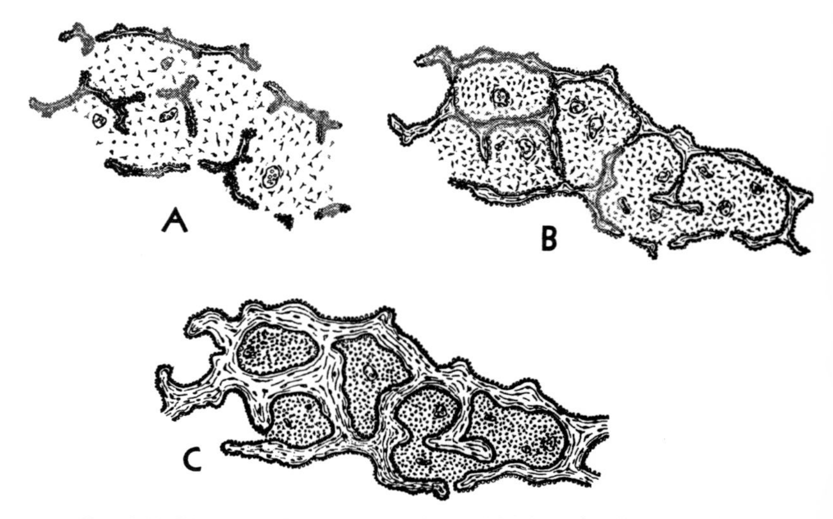

FIG. 153. Diagrams showing stages in establishing of a characteristic area of primary cancellous bone by extension and coalescence of originally separate trabeculae.

absorb and hand on materials to their more remote fellows which in turn utilize these materials in maintaining a healthy condition in the organic part of the bone matrix. It is the senescence of these cells with the consequent lowering of their efficiency and the resultant deterioration of the ossein component of the matrix which is in part responsible for the decreased resiliency of the bones in advanced age.

As the various trabeculae in an area of developing bone grow, they inevitably come in contact with each other and fuse. Thus trabeculae, at first isolated, soon come to constitute a continuous system (Fig. 153). Because of its resemblance to a latticework (Latin—cancellus), bone in this condition, where the trabeculae are slender and the spaces between them extensive, is known as *cancellous bone.* The spaces between the trabeculae are known as marrow spaces.

Endochondral Bone Formation. As the term implies, endochondral bone formation goes on within cartilage. It cannot be stated too strongly that cartilage does not, in this process, become converted into bone. Cartilage is destroyed and bone is formed where the cartilage used to be. The actual bone formation is essentially the same as in the case of membrane bone. The phenomena of special interest in connection with this type of bone development are those involved in the destruction of the cartilage preliminary to the formation of bone.

Cartilage Formation. To trace the process logically we must start back with the formation of cartilage. The first indication of impending chondrogenesis is the aggregation of an exceedingly dense mass of mesenchymal cells. This cell mass gradually takes on the shape of the cartilage to be formed. The histogenetic changes involved are not at first conspicuous. During the period of preliminary massing the cells have been migrating in from surrounding regions and also increasing the local congestion by rapid proliferation. As they are packed in together they lose their processes and become rounded (Fig. 154, A, 1). When it seems as if no more cells could possibly be crowded in, the course of events changes. The cells begin to separate from one another. This is due to the fact that they have become active in

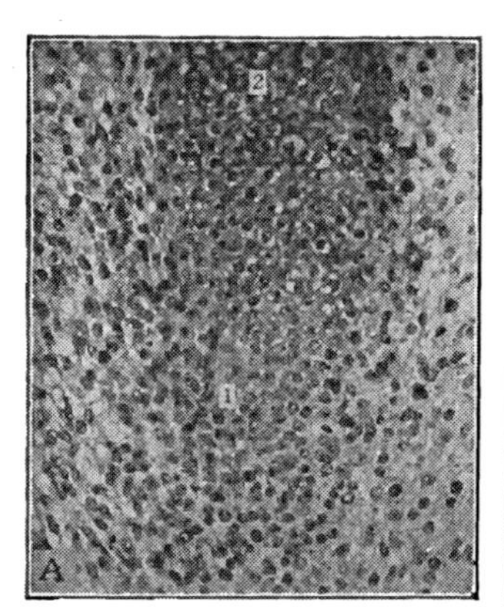

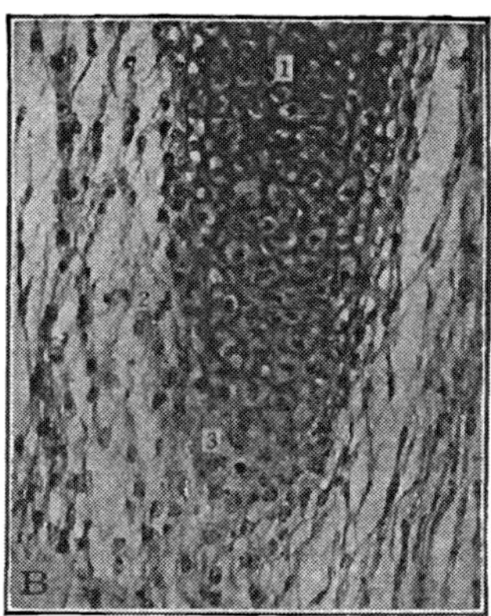

Fig. 154. Photomicrographs of developing cartilage. The areas photographed were from the margins of the paranasal cartilage of pig embryos between 25 and 30 mm. in length. For location of cartilage in head see figure 175.

A, Early stage showing: at (1) the massing of mesenchymal cells which were about to be incorporated in the growing margin of the cartilage; and at (2) an area where matrix formation is already beginning.

B, Slightly more advanced stage of the same cartilage showing: at (1) increase in the amount and density of the matrix in the center of the growing cartilage; at (2) concentration of the surrounding mesenchyme to form the perichondrium; and at (3) the addition of new cartilage matrix peripherally.

secreting. It is the accumulation of the secretion of the cells which gradually forces them farther and farther apart until they come to lie isolated from one another in the matrix they have produced (Fig. 154, A, 2). Such a method of increase in mass, where there are many scattered growth centers contributing independently to the increase in bulk of the whole, is known as *interstitial growth.* This interstitial growth of young cartilage stands in sharp contrast to the *appositional growth* of such rigid substances as bone or dentine or enamel where the matrix is laid down in successive layers one upon another. Obviously interstitial growth implies plasticity of the substance produced. Were the substance produced unyielding, the very activity of a number of growth centers within it would soon crowd those growth centers to obliteration.

As the cartilage matrix is increased in amount its affinity for basic stains becomes more marked, due probably to increase in concentration of the characteristic substance in it known chemically as chondrin. At the same time the matrix becomes more rigid with a resultant checking of interstitial growth. The cells continue to secrete to a certain extent, however, as evidenced by the fact that in mature cartilage the matrix immediately surrounding the cells becomes more dense than the rest of the matrix. This area of denser matrix around the lacuna in which the cell lies is known as the *capsule.* As the cartilage grows older the capsules become more conspicuous and many of them come to contain more than one cell. These nests of cells in a common capsule are the result of cell divisions, following which the daughter cells are held imprisoned in the original capsule of the mother cell—further evidence of the loss of plasticity in the matrix.

The formation of a matrix so rigid that interstitial growth is checked, takes place first centrally in an area of developing cartilage. When the center has become too rigid for interstitial growth to continue, appositional growth begins to take place peripherally. While the cartilage has been increasing in mass it has been acquiring a peripheral investment of compacted mesenchyme. This investing layer of mesenchyme soon becomes specialized into a connective-tissue covering called the *perichondrium.* The layer of the perichondrium next to the cartilage is less fibrous than the outer layer and the cells in it continue to proliferate rapidly and become active in the secretion of cartilage matrix. For this reason it is known as the *chondrogenetic layer* of the perichondrium. It is through the activity of the chondrogenetic layer that the cartilage continues to grow

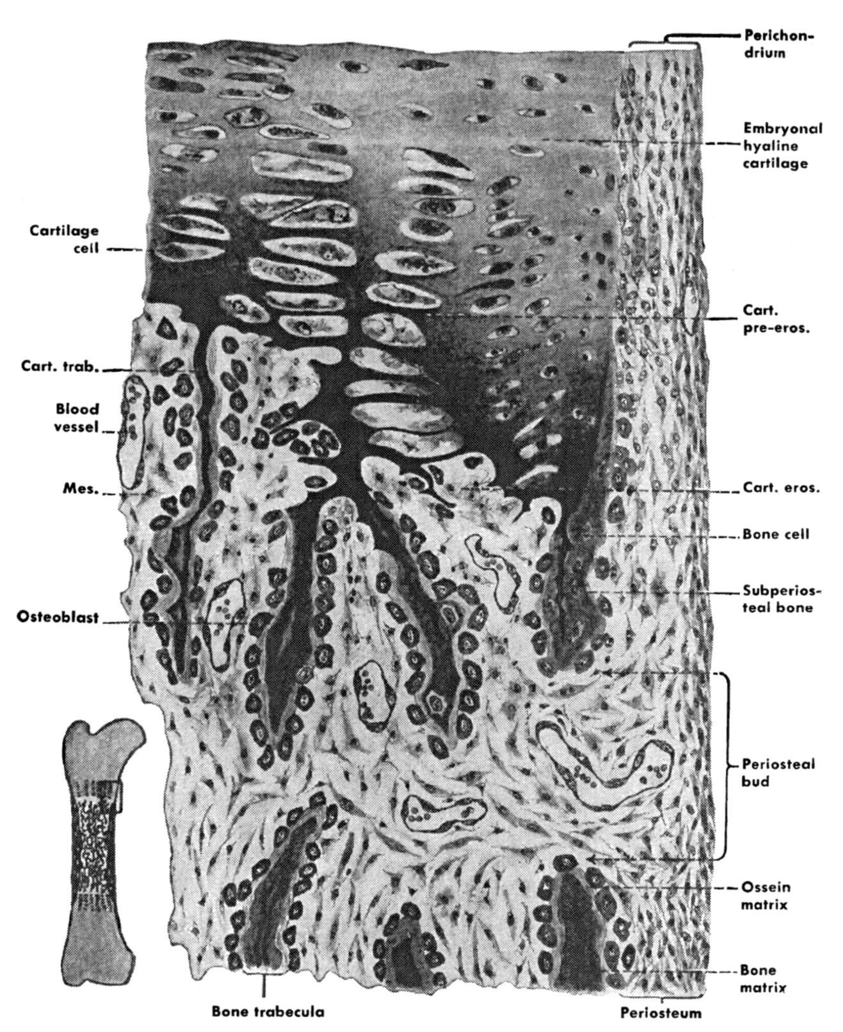

FIG. 155. Drawing showing periosteal bud and an area of endochondral bone formation from the radius of a 125 mm. sheep embryo. The small sketch indicates the location of the area drawn in detail.

Abbreviations: Cart. eros., area from which cartilage has recently been eroded; Cart. pre-eros., area with cartilage cells enlarged and arranged in rows presaging erosion; Cart. trab., remnant of cartilage matrix which has become calcified and serves as an axis or core about which bone lamellae are deposited to form a bone trabecula; Mes., mesenchymal cell.

peripherally, by apposition, long after interstitial growth has ceased in the matrix first formed.

CARTILAGE EROSION. When a mass of cartilage is about to be replaced by bone, very striking changes in its structure take place. The cells which have hitherto been secreting cartilage matrix begin to destroy the matrix. The lacunae become enlarged and a curious arrangement of the cartilage cells becomes evident. The cells erode the cartilage in such a manner that they become lined up in rows (Fig. 155). This process of destruction continues until the cartilage is extensively honeycombed. Meanwhile the tissue of the perichondrium overlying the area of cartilage erosion becomes exceedingly active. There is rapid cell proliferation and the new cells, carrying blood vessels with them, begin to invade the honeycombed cartilage (Fig. 155).

THE DEPOSITION OF BONE. It is a striking fact that during its growth cartilage is devoid of blood vessels, the nearest vessels to it being those in the perichondrium. The invasion of cartilage by blood vessels definitely determines its disintegration as cartilage, and at the same time is the initial step in the formation of bone. For this reason the enveloping layer of connective tissue, up to this time called perichondrium because of its relation to the cartilage, is now called *periosteum* because of the relations it will directly acquire to the bone about to be formed. This change will not be confusing if we stop to think that both these terms are merely ones of relation, which translated mean, respectively, that tissue which surrounds cartilage, and that tissue which surrounds bone. The important fact to bear in mind is that this enveloping layer of tissue is of mesenchymal origin and therefore contains cells of the stock that may develop into any of the connective-tissue family to which bone as well as cartilage belongs. When, therefore, a mass of periosteal tissue (*periosteal bud*, Fig. 155) grows into an area of honeycombed cartilage it carries in potentially bone-forming cells. These cells come to lie along the strand-like remnants of cartilage, just as in membrane bone formation osteoblasts ranged themselves along fibrous strands. The actual deposition of bone proceeds in the same manner endochondrally as it does intramembranously. The only difference is that in one case a strand-like remnant of cartilage serves as an axis for the trabecula, whereas in the other case deposition begins on a fibrous strand. Extensions and fusions of the growing trabeculae soon result in the establishment of typical cancellous bone similar to that formed intramembranously.

The Formation of Compact Bone from Primary Cancellous Bone. The difference between cancellous bone and compact bone is architectural rather than histological. The fundamental composition of the bone matrix, its lamellation, and the relations of the bone cells to the matrix, are the same in both cases. It is the way in which the lamellae are arranged that distinguishes these two types of bone from each other. In cancellous bone the disposition of lamellae is such that it leaves large marrow spaces between the trabeculae. In compact bone there has been a secondary deposit of concentrically arranged lamellae in the marrow spaces which greatly increases the density of the bone as a whole.

The essential differences between the two, and the way in which cancellous bone may become converted into compact bone, can be illustrated by a simple schematic diagram. Figure 156, 1, shows the arrangement of lamellae and marrow spaces in primary cancellous bone. The osteoblasts which have formed the trabeculae still lie along them on the surface toward the marrow cavity. If such an area is to become compact, these osteoblasts enter on a period of renewed activity and deposit a series of concentric lamellae in the marrow cavity. Frequently if the marrow spaces are irregular there is a preliminary rounding out of them by local resorption of the bone already formed (Fig. 156, 2). This is then followed by the deposition

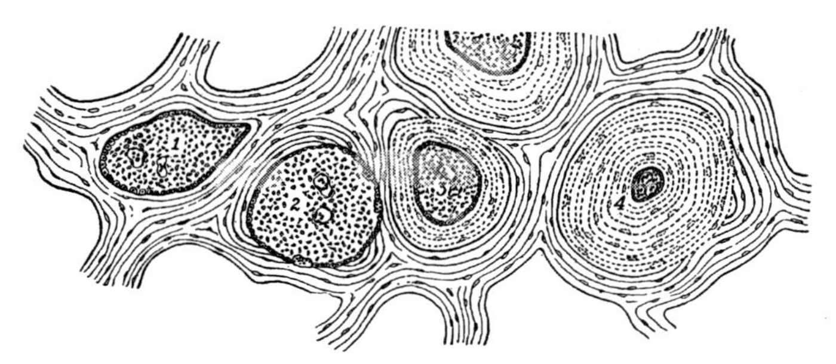

FIG. 156. Diagram showing transformation of cancellous to compact bone. The solid lines indicate the lamellae of primary cancellous bone; the dotted lines show the subsequently added concentric (Haversian) lamellae which nearly obliterate the marrow spaces of cancellous bone. The sequence of events is indicated by the numbers. Note that irregularly shaped spaces in the cancellous bone may be rounded out by absorption before the concentric lamellae are laid down.

of the concentrically arranged lamellae, sometimes called *Haversian lamellae* after the man who first described them in detail (Fig. 156, 3). In this process the original marrow spaces are reduced to small canals (*Haversian canals*) into which have been crowded the blood vessels which formerly lay in the marrow cavities (Fig. 156, 4). These canals maintain intercommunication with each other in the substance of the bone, constituting a network of pathways over which the bone receives its vascular supply. As compared with the marrow spaces of cancellous bone, however, they are very small; and the gross appearance of a bone which has undergone this secondary deposit of concentric lamellae amply justifies characterizing it as "compact."

II. The Development of the Skeletal System

In dealing with the development of the skeletal system we must recognize at the outset that the subject is far too extensive to be covered here with anything like completeness. It is not difficult, however, to become acquainted with the outstanding features in the development of two or three characteristic bones, as, for example: the sequence of events in the formation of a flat bone; the steps involved in the establishment and growth of a long bone; the way separate ossification centers appear in a common primordial cartilage mass and give rise to the various parts of a vertebra. Familiarity with such type processes gives one an understanding of the factors operative in the development of the skeleton as a whole and a background sufficient to permit ready and intelligent following up of developmental details in specific bones in which one may become interested.

Development of Flat Bones. The flat bones, such as the bones of the cranium and face, are for the most part of intramembranous origin. We are, therefore, already familiar with the early steps in their development from our study of the histogenesis of membrane bone (Figs. 151 and 153). After a mass of primary cancellous bone has been laid down in a configuration which suggests that of the adult bone being formed, there appears about this mass a peripheral concentration of mesenchyme (Fig. 157, A). This periosteal concentration of mesenchymal tissue contains potentially bone-forming cells which soon become active and lay down a dense layer of parallel lamellae about the spongy center of the growing bone (Fig. 157, B). Anatomically this dense peripheral portion is known as the *outer table* of the bone. The inner portion, which in the flat bones usually remains cancellous, is called the *diploë*. The original mesenchymal tissue which

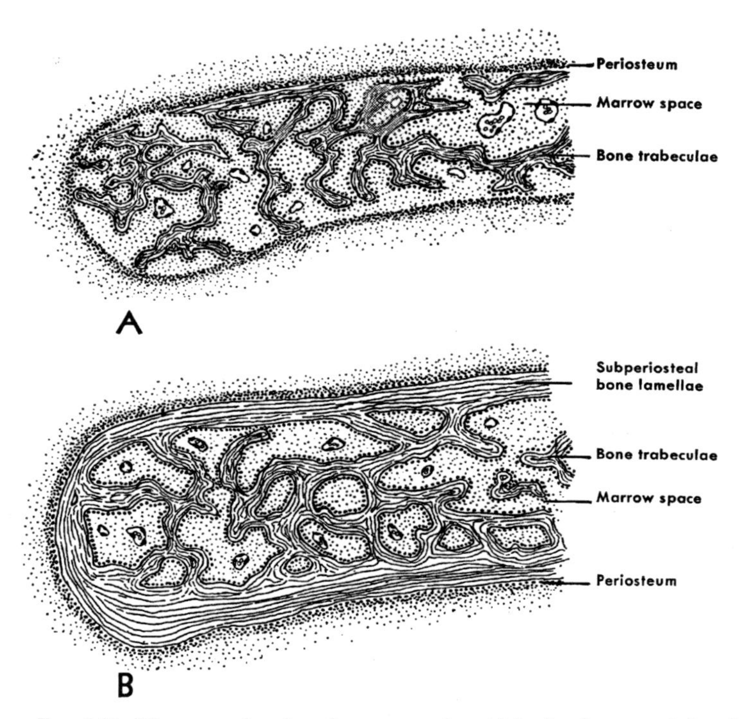

Fig. 157. Diagrams showing the manner in which the dense peripheral layer of a flat bone is formed by the deposition of subperiosteal lamellae about an area of primary cancellous bone.

remains in the marrow spaces of the diploë develops into characteristic "red bone marrow" rich in blood-forming elements (Fig. 152).

The story of the growth of tbe mandible, a membrane bone which starts after the manner of flat bones but which later takes on a very elaborate shape and finally becomes largely compact, can be gleaned by a comparative study of figures 178, 180, and 184.

Development of Long Bones. The long bones are characteristically of endochondral origin. The cartilage in which they are performed is a temporary miniature of the adult bone. Ordinarily there are several ossification centers involved in the formation of long bones. The first one to appear is that in the shaft or diaphysis. The location of this center is shown schematically in figure 158, A. Such

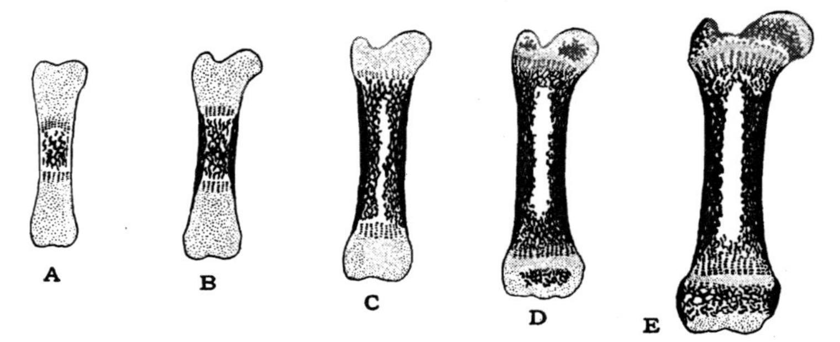

FIG. 158. Diagrams showing the progress of ossification in a long bone. The stippled areas represent cartilage; the black areas indicate bone.

A, Primary ossification center in shaft. B, Primary center plus shell of subperiosteal bone. C, Entire shaft ossified. D, Ossification centers have appeared in the epiphyses. E, Entire bone ossified except for the epiphyseal cartilage plates.

details as the cartilage erosion which preceded its appearance and the manner in which the deposit of bone was initiated have already been considered (Fig. 155). Our interest now is in the relation of such an endochondral ossification center to other centers, and to the bone as a whole.

Almost coincidently with the beginning of bone formation within the cartilage the overlying periosteum begins to add bone externally (Fig. 158, B). In view of the fact that the bone-forming tissue carried into the eroded cartilage arose from the periosteum, this activity of the periosteum itself is not surprising. Moreover we have already encountered this same phenomenon of periosteal bone formation in the outer table of flat bones.

The formation of bone which starts at about the middle of the shaft soon extends toward either end until the entire shaft is involved (Fig. 158, C), leaving the two ends (*epiphyses*) still cartilage. Toward the end of fetal life ossification centers appear in the epiphyses. The number and location of these epiphyseal centers vary in different long bones. There is always at least one center in each epiphysis and there may be two or more. Not uncommonly there are two centers in one epiphysis and one in the other, as illustrated in figure 158, D.

Between the bone formed in the diaphysis and that formed in the epiphysis there persists a mass of cartilage known as the *epiphyseal plate* which is of vital importance in the growth in length of the bone.

We should expect from the rigidity of bone matrix that interstitial growth could not account for its increase in length. This was long ago demonstrated experimentally by exposing a developing bone and driving into it three small silver pegs, two in the shaft and one in the epiphysis. The distance between the pegs being recorded, the incision was closed and development allowed to proceed until a marked increase had occurred in the length of the bone. On again exposing the pegs, the two in the shaft were found to be exactly the same distance apart as when they were driven in, but the distance between the pins in the shaft and that in the epiphysis had increased by an amount corresponding to the increase in length of the bone. This indicates clearly that the epiphyseal plates constitute a sort of temporary, plastic union between the parts of the growing bone. Continued increase in the length of the shaft is accomplished by the addition of new bone at the cartilage plate. These epiphyseal plates persist during the entire postnatal growth period. Only when the skeleton has acquired its adult size do they finally become eroded and replaced by bone which joins the epiphyses permanently to the diaphysis.

As the bone increases in length there is a corresponding increase in its diameter. The manner in which this takes place is also susceptible of experimental demonstration. If madder leaves, or some of the alizarin compounds extracted from them, be fed to a growing animal, the bone formed during the time the feeding is continued is colored red. If the madder is discontinued, bone of normal color is again formed; but the color still remains in the bone laid down while madder was being added to the diet. Thus it is possible, by keeping a record of alternate periods of feeding and withholding madder and comparing these records with the resulting zones of coloration in a bone, to obtain very accurate information on the progress of bone growth and resorption. Applied to the development of long bones this method shows their increase in diameter to be due to continued appositional growth beneath the periosteum. As the bone is added to peripherally there is a corresponding resorption centrally. This central resorption results in the formation of a cavity in the axis of the long bone which is called the *marrow canal* (Fig. 158, C). With the further increase in the diameter of a bone, its marrow canal becomes correspondingly enlarged. A significant mechanical fact might be cited in this connection. Engineers have determined that the strongest rod which can be made from a given weight of steel is obtained by molding it into tubular form. The development of an essentially tubular shaft by progressive increase in the size of the marrow cavity gives a long bone maximum strength with minimum weight.

The Formation of the Vertebrae. The development of the vertebrae is of interest to the student primarily because it exemplifies so excellently a fundamental embryological phenomenon—the origin of separate parts from an undifferentiated primordial tissue mass, and the subsequent association of these parts to form an organized structure. In studying young embryos we traced the history of the mesodermic somites through their early differentiation. It will be recalled that from the ventro-mesial face of each somite there arises a group of mesenchymal cells called collectively a sclerotome (Fig. 42). These cells migrate from either side toward the mid-line and become aggregated about the notochord. From these masses of cells the entire vertebral column is destined to arise.

The first significant change which takes place in these primordial masses is the clustering of sclerotomal cells derived in part from each of the two adjacent somites into groups which are located opposite the intervals between the myotomes. In studying series of transverse sections this arrangement is easy to overlook unless the density of the cells about the notochord is carefully noted in passing from section to section. It shows very clearly, however, in frontal sections (Fig. 159). Each of these cell clusters is the primordium of the *centrum* of a vertebra. Once formed they rapidly become more dense and more definitely circumscribed (Fig. 160). Soon after the centrum takes shape, paired mesenchymal concentrations extending dorsally and laterally from the centrum establish the primordia of the *neural arches* and of the ribs (Fig. 161).

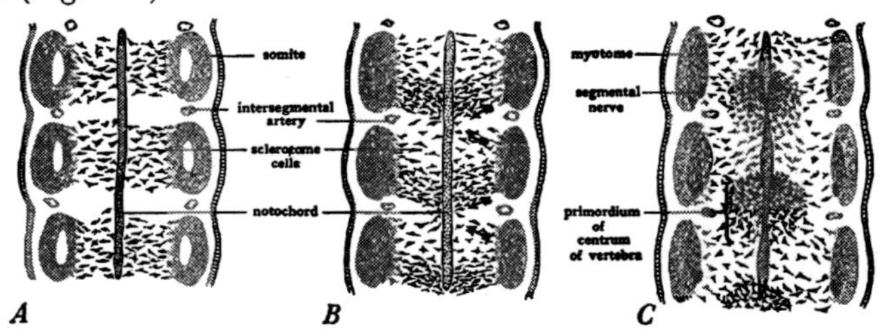

FIG. 159. Semi-schematic coronal sections through the dorsal region of young embryos to show how the vertebrae became intermyotomal in position. Note that the primordium of a centrum is formed by cells originating from the sclerotomes of both the adjacent pairs of somites.

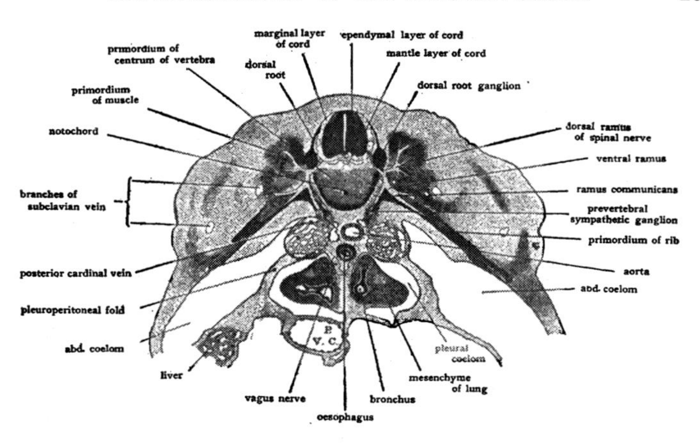

FIG. 160. Transverse section from pig embryo of 17 mm. cut at the level of the lungs to show the structures in the dorsal body-wall. (After Minot.)

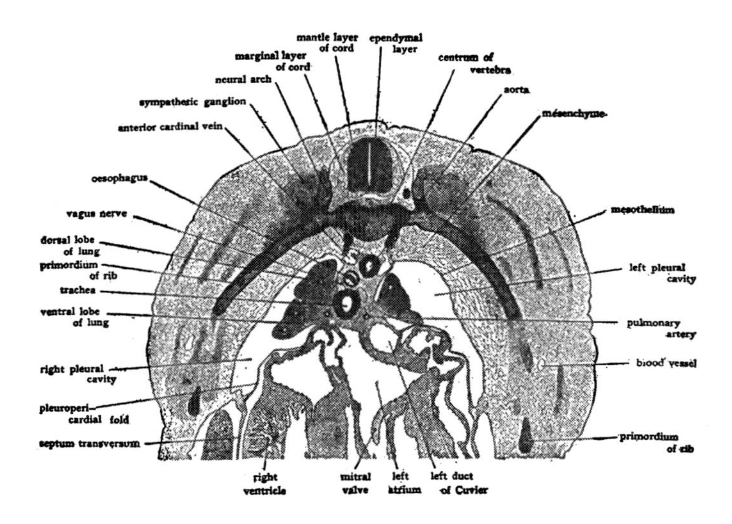

FIG. 161. Transverse section of 20 mm. pig embryo cut at the level of the lungs to show the developing vertebra and ribs. (After Minot.)

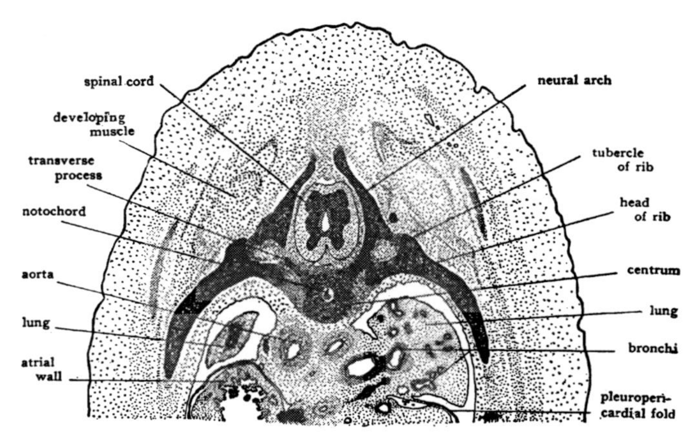

FIG. 162. Transverse section from 40 mm. pig embryo cut at the level of the lungs to show the developing vertebra and ribs.

The stage in which the various parts of the vertebrae are sketched in mesenchymal concentrations, is frequently spoken of as the blastemal stage. It is rapidly followed by the cartilage stage. Conversion to cartilage begins in the blastemal mass first in the region of the centrum and then chondrification centers appear in each neural and each costal process (Fig. 161). These spread rapidly until all the centers fuse and the entire mass is involved (Fig. 162). The cartilage miniature of the vertebra thus formed is at first a single piece showing no lines of demarcation where the original centers of cartilage formation became confluent, and no foreshadowing of the separate parts of which it will be made up after the cartilage has been replaced by bone. Shortly before ossification begins the rib cartilage becomes separated from the vertebra, but the vertebra itself remains in one piece throughout the cartilage stage (Fig. 162).

The locations of the endochondral ossification centers which appear in a vertebral cartilage are indicated schematically in figure 163. It readily can be seen how the spreading of these centers of bone formation will establish the conditions which exist in an adult vertebra. The median ossification center gives rise to the centrum. The centers in the neural processes extend dorsally to complete the neural arch. The spinous process in most of the vertebrae is formed by a prolongation of these same centers to meet dorsal to the neural canal.

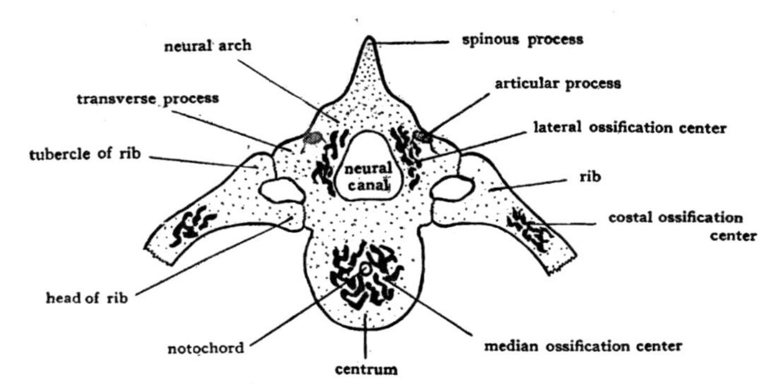

FIG. 163. Diagram indicating the location of the endochondral ossification centers which appear in the vertebrae and ribs. (The thoracic vertebrae of the pig have very long spinous processes. In these vertebrae an additional ossification center not shown in the above generalized diagram appears in the spinous process.)

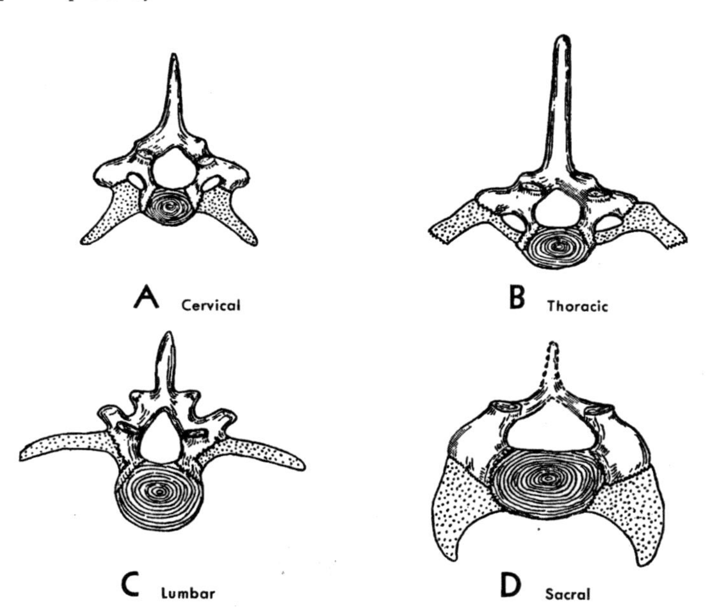

FIG. 164. Diagram of four types of vertebrae indicating the parts derived from the different ossification centers shown in figure 163. The part formed by the median center in centrum is concentrically ringed; the parts arising from the costal centers are stippled; parts derived from the lateral centers in the neural arches are indicated in line-shading.

In forms such as the pig the spinous processes of the more anterior thoracic vertebrae are very long. In these vertebrae additional ossification centers appear in the spinous process and fuse with those in the

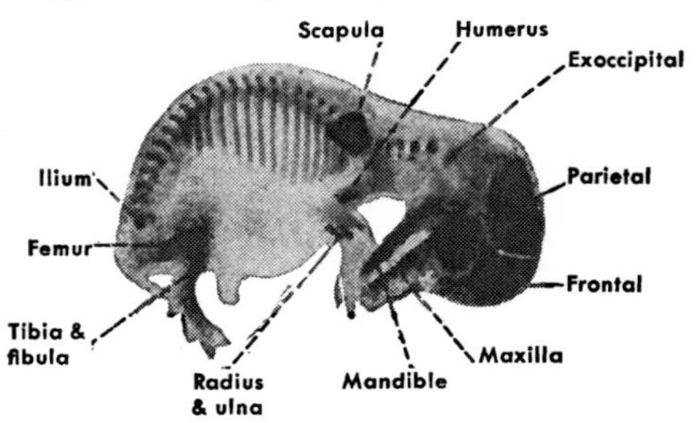

FIG. 165. Photograph (× 1½) showing the ossification centers which have appeared in pig embryos of 35 mm. This and the two following figures were made by photographing in transmitted light embryos in which all the uncalcified tissues had been rendered transparent by treatment with potassium hydroxide and glycerine.

neural processes. The transverse processes with which the tubercles of the ribs articulate are formed by the lateral extension of the primary ossification centers in the neural processes. These same centers extend ventrally also, and meet the centrum (cf. Figs. 163 and 164).

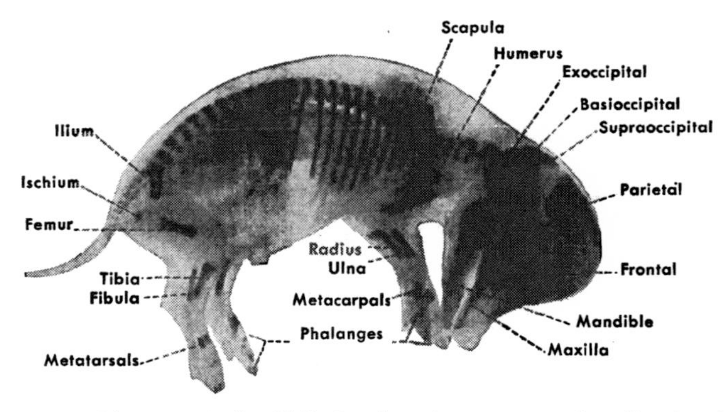

FIG. 166. Photograph (× 1½) showing the progress of ossification in the skeleton of a 65 mm. pig embryo.

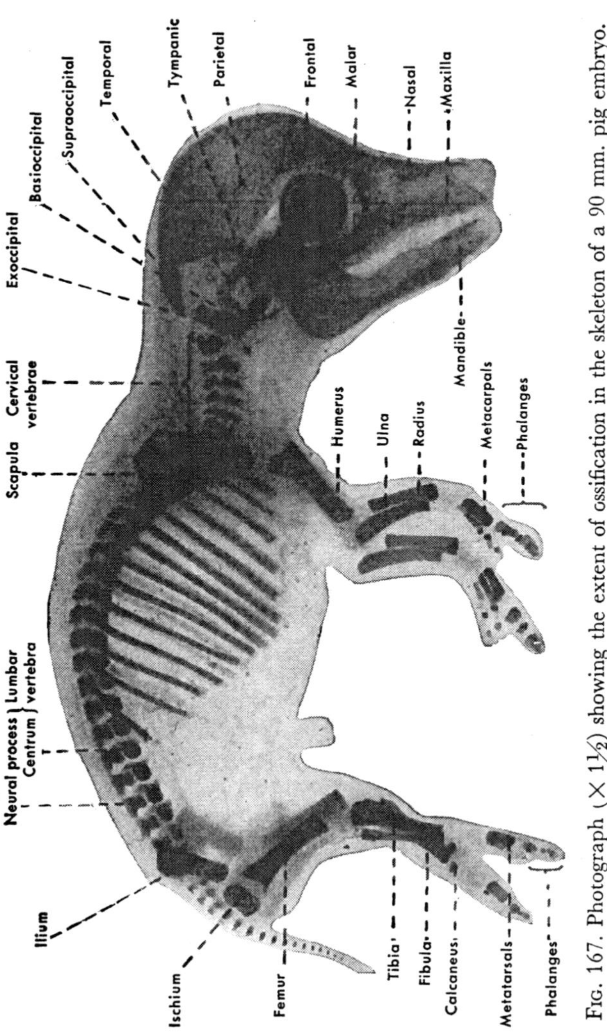

FIG. 167. Photograph ($\times 1\frac{1}{2}$) showing the extent of ossification in the skeleton of a 90 mm. pig embryo.

The shaft of the rib is formed by extension of its primary ossification center (Fig. 163). After birth, secondary epiphyseal centers appear in the tubercle and head of the rib. These centers are separated from the shaft by persistent cartilage plates in the manner described in discussing the development of long bones. Fusion of the secondary epiphyseal centers with the shaft of the rib does not take place until the skeleton has acquired its adult dimensions.

The foregoing discussion has been based on a thoracic vertebra in which the relations of the rib to the vertebra show most clearly. All the vertebrae have the costal element represented, although it is greatly reduced and modified in other regions than the thoracic. A study of figure 164, in which the components of vertebrae from the cervical, thoracic, lumbar, and sacral regions are schematically indicated, will make the homologies apparent. With these homologies in mind it is sufficiently evident, without going into further detail, how all these vertebrae arise by a process similar to that described for the thoracic vertebrae.

The Progress of Ossification in the Skeleton as a Whole. It would carry us beyond the scope of this book to take up the development of specific bones. Each has its own story involving the formation of the connective tissue or the cartilage mass which precedes it; local erosion centers if it be preformed in cartilage; number, location, and time of appearance of ossification centers; growth in length and diameter; development of epiphyses; time of fusion of epiphyses and diaphysis; and finally the development of muscle ridges and articular facets. Without entering into a discussion of details of this sort, it is possible nevertheless to follow the general progress of ossification in the skeletal system as a whole. Embryos which have been treated with potassium hydroxide and then cleared in glycerine clearly show the various ossification centers. In such preparations the areas where calcium salts have been deposited stand out white in reflected light and opaque in transmitted light. Figures 165–167, which are photographs of preparations of this type, can be used to trace the history of the more important bones. It should perhaps be stated explicitly that these figures are included primarily to give a general view of the progress of ossification and secondarily to afford a readily available source of reference for following up points of interest that may arise. It is not a profitable use of the student's time to attempt to memorize the ossification centers which have appeared in embryos of any given age.

CHAPTER 13

The Development of the Face and Jaws and the Teeth

I. The Face and Jaws

The Stomodaeum. In studying the early development of the digestive tract we saw that the primitive gut first appeared as a cavity which was blind at both its anterior and posterior ends (Fig. 37). Its opening in the future oral region is established by the meeting of an ectodermal depression, the stomodaeum, with the cephalically growing anterior end of the gut. The stomodaeal depression, even as late as the time the oral plate ruptures and establishes communication between the anterior end of the gut and the outside world, is very shallow (Fig. 40). The deep oral cavity characteristic of the adult is formed by the forward growth of structures about the margins of the stomodaeum. Some idea of the extent of this forward growth can be gained from the fact that the tonsillar region of the adult is at about the level occupied by the stomodaeal plate when it ruptures. The growth of the structures bordering the stomodaeum, then, not only gives rise to the superficial parts of the face and jaws, but actually builds out the walls of the oral cavity itself.

The Jaws. Because the face of a young embryo is pressed against the thorax it is difficult to study unless the entire head is cut off and mounted separately. Preparations of this kind observed under a dissecting microscope by strong reflected light show the surface configuration of the facial region very clearly. The most conspicuous landmarks are the stomodaeal depression, which in view of its fate we may now call the oral cavity, and the olfactory pits. In embryos as small as 7 mm. most of the structures which take part in the formation of the face and jaws are already clearly distinguishable (Fig. 168). In the mid-line cephalic to the oral cavity is a rounded overhanging prominence known as the *frontal process*. On either side of the frontal process are horseshoe-shaped elevations surrounding the olfactory pits. The median limbs of these elevations are known as the *naso-medial processes* and the lateral limbs are called the *naso-lateral processes.*

Growing toward the mid-line from the cephalo-lateral angles of the oral cavity are the *maxillary processes.* In lateral views of the head (Figs. 31 and 32) it will be seen that the maxillary processes and the mandibular arch merge with each other at the angles of the mouth. Thus the structures which border the oral cavity cephalically are: the unpaired frontal process in the mid-line, the paired nasal processes on either side of the frontal, and the paired maxillary

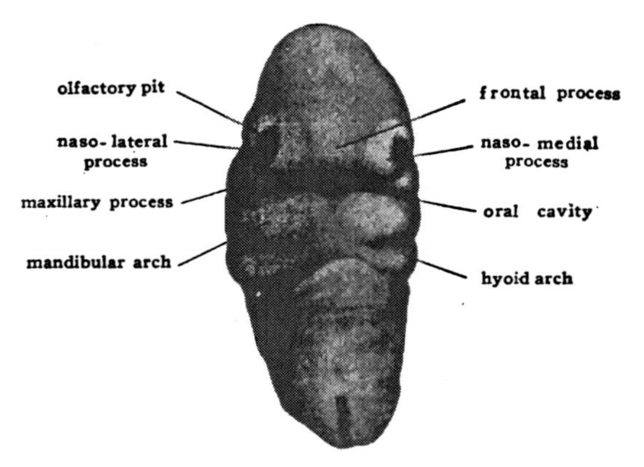

Fig. 168. Face of 7 mm. pig embryo photographed × 15. Note especially the unmistakably paired character of the thickenings which later fuse in the mid-line to complete the mandibular arch.

processes at the extreme lateral angles. From these primitive tissue masses the upper jaw and the nose are derived.

The caudal boundary of the oral cavity is less complex, being constituted by the mandibular arch alone. In very young embryos (Fig. 168) the origin of the mandibular arch from paired primordia is still clearly evident. Appearing first on either side of the mid-line are marked local thickenings due to the rapid proliferation of mesenchymal tissue. Until these thickenings have extended from either side to meet in the mid-line there remains a conspicuous mesial notch. With their fusion, the arch of the lower jaw is completed (Figs. 169–172).

In 10–12 mm. embryos (Fig. 169) very marked progress can be seen in the development of the facial region. The maxillary processes are much more prominent and have grown toward the mid-line, crowding the nasal processes closer to each other. The nasal processes have grown so extensively that the frontal process between them is completely overshadowed (cf. Figs. 168 and 169). The growth of the

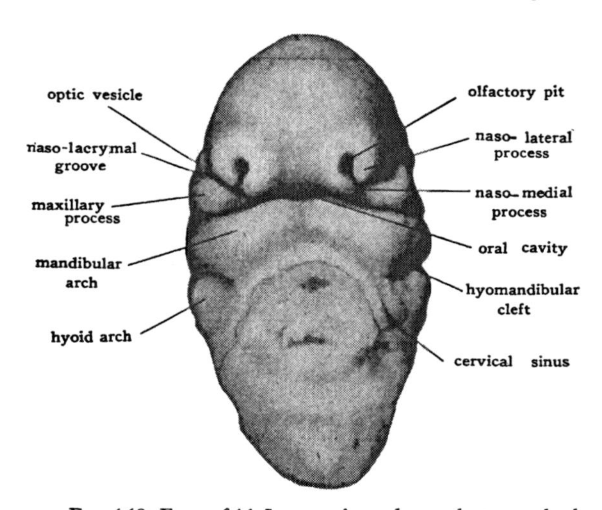

FIG. 169. Face of 11.5 mm. pig embryo photographed × 12. Fusion of the right and left components of the mandibular arch is practically complete. Both the medial and lateral limbs of the horseshoe-shaped nasal processes have undergone conspicuous enlargement. Note especially the approximation of each naso-medial process to the maxillary process of the same side.

medial limbs of the nasal processes has been especially marked and they appear almost in contact with the maxillary processes on either side.

The groundwork for the formation of the upper jaw is now well laid down. Its arch is completed by the fusion of the two naso-medial processes with each other in the mid-line, and with the maxillary processes laterally (Fig. 170). The premaxillary bones carrying the incisor teeth are formed, later, in the part of the upper jaw which is of naso-medial origin. The maxillary bones, carrying all the upper teeth posterior to the incisors, are developed in the part of the arch arising from the maxillary processes.

Nasal Chambers. The olfactory pits have by this time become much deepened, not only by the growth of the nasal processes about them, but also by extension of the original pits themselves which soon break through into the oral cavity (Figs. 93 and 97, C). We may now

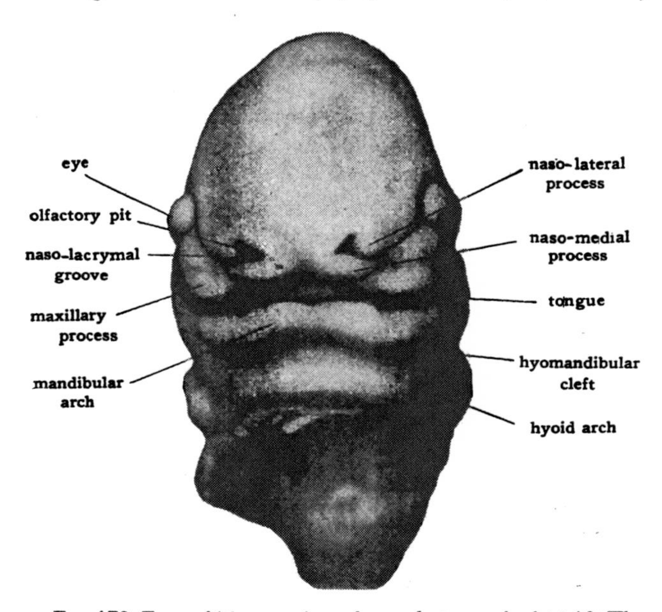

FIG. 170. Face of 16 mm. pig embryo photographed × 10. The naso-medial processes have fused with the maxillary processes on either side, and with each other in the mid-line, thus completing the arch of the upper jaw.

speak of the external openings of the nasal pits as the nostrils (*external nares*) and their new openings into the oral cavity as posterior nares or *nasal choanae*. The septum of the nose is formed by fusion in the mid-line of the original naso-medial processes; the upper part of the bridge of the nose is derived from the frontal process; and the alae of the nose arise from the naso-lateral processes (Fig. 172).

Naso-lacrimal Duct. Where the naso-lateral process and the maxillary process meet each other there is formed for a time a well-

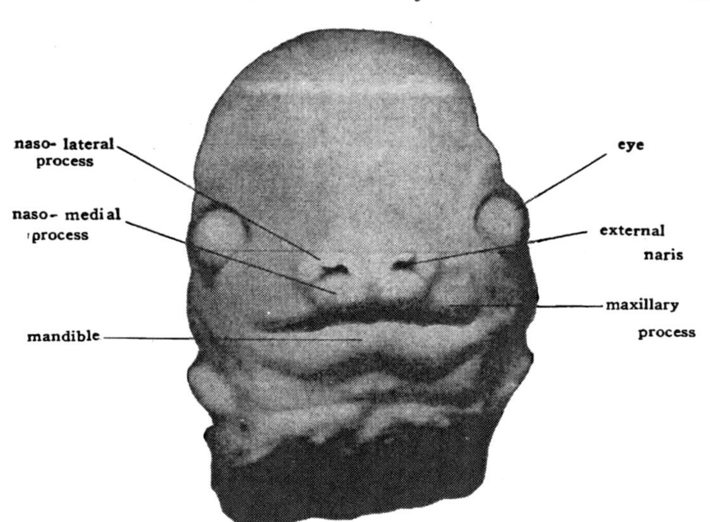

Fig. 171. Face of 17.5 mm. pig embryo photographed × 10. The originally separate processes have now largely lost their identity in the series of fusions which have taken place in the formation of the face.

marked groove, which extends to the mesial angle of the eye (Fig. 169). This is known as the *naso-lacrimal groove.* It soon closes over superficially (Fig. 171), and it is usually stated that the deep portion of the original groove is converted into a tube, the *naso-lacrimal duct*, or tear duct, which drains the fluid from the conjunctival sac of the eye into the nose. Recently Politzer has maintained that the naso-lacrimal duct arises as an independent epithelial downgrowth from the conjunctival sac which follows closely along the line of closure of the old naso-optic furrow.

Tongue. While these changes are going on externally, the tongue is being formed in the floor of the mouth. Anatomically the tongue is usually described as consisting of a freely movable part called its body, and a less freely movable portion, called its root, by which it is attached in the oro-pharyngeal floor. The body of the tongue arises from a small median elevation, the *tuberculum impar*, and paired *lateral lingual primordia.* These elevations appear very early in development on the inner face of the first branchial (mandibular) arch (Fig. 173, B). The tuberculum impar grows slowly and is soon crowded in on by the more rapidly growing lateral lingual primordia which form the great bulk of the body of the tongue (Fig. 173, C).

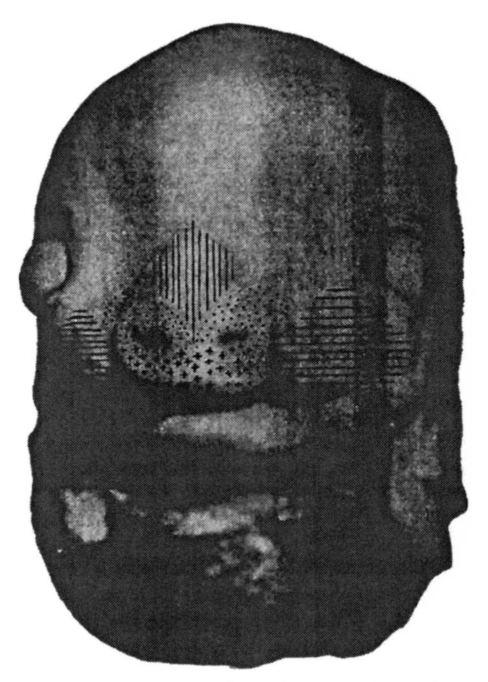

FIG. 172. Face of 21.5 mm. pig embryo photographed × 10. The characteristic features of the adult face are even at this early stage clearly recognizable. The regions of the upper jaw and nose which have arisen from originally distinct primordia are differentiated by shading. Vertical hatching indicates origin from frontal process; stippling, from naso-lateral processes; small crosses, from naso-medial processes; horizontal hatching, from maxillary processes. The entire lower jaw is derived from the mandibular arch.

Arising in the pharyngeal floor at the bases of the second and third branchial arches is an elevation known as the *copula* (i.e., yoke) because of the way it joins these arches together (Fig. 173, B). The copula, supplemented by some tissue from the adjacent basal portions of branchial arches 2, 3, and 4, gives rise to the root of the tongue.

All the various elevations which thus take part in the formation

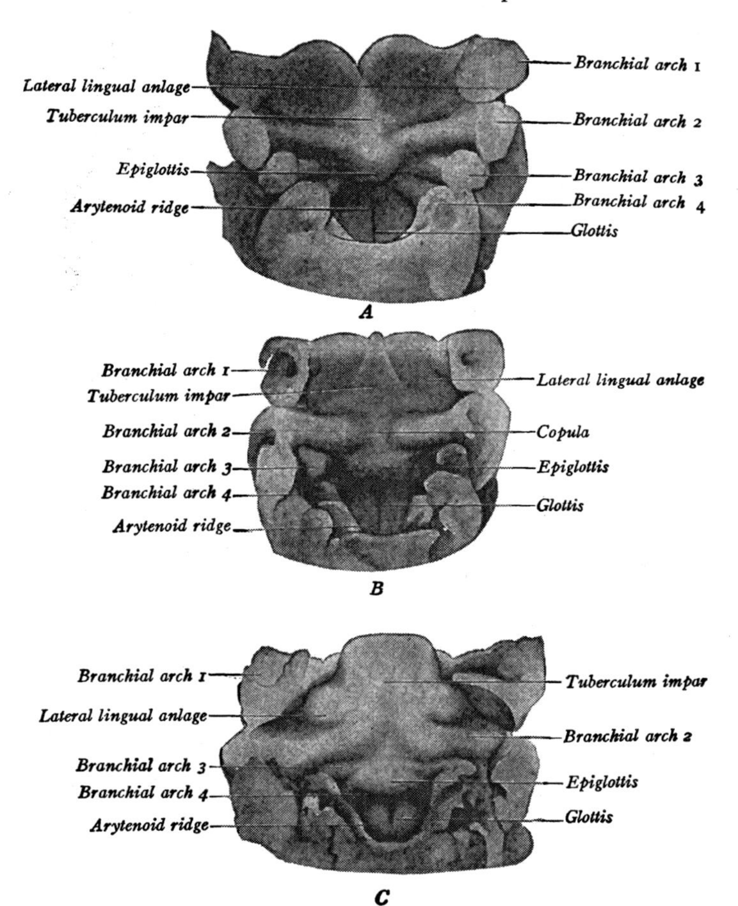

FIG. 173. Dissections of pig embryos made to expose the floor of the mouth and show the development of the tongue. (After Prentiss.) A, 7 mm.: B, 9 mm.; C, 13 mm. (All figures $\times$ 12.)

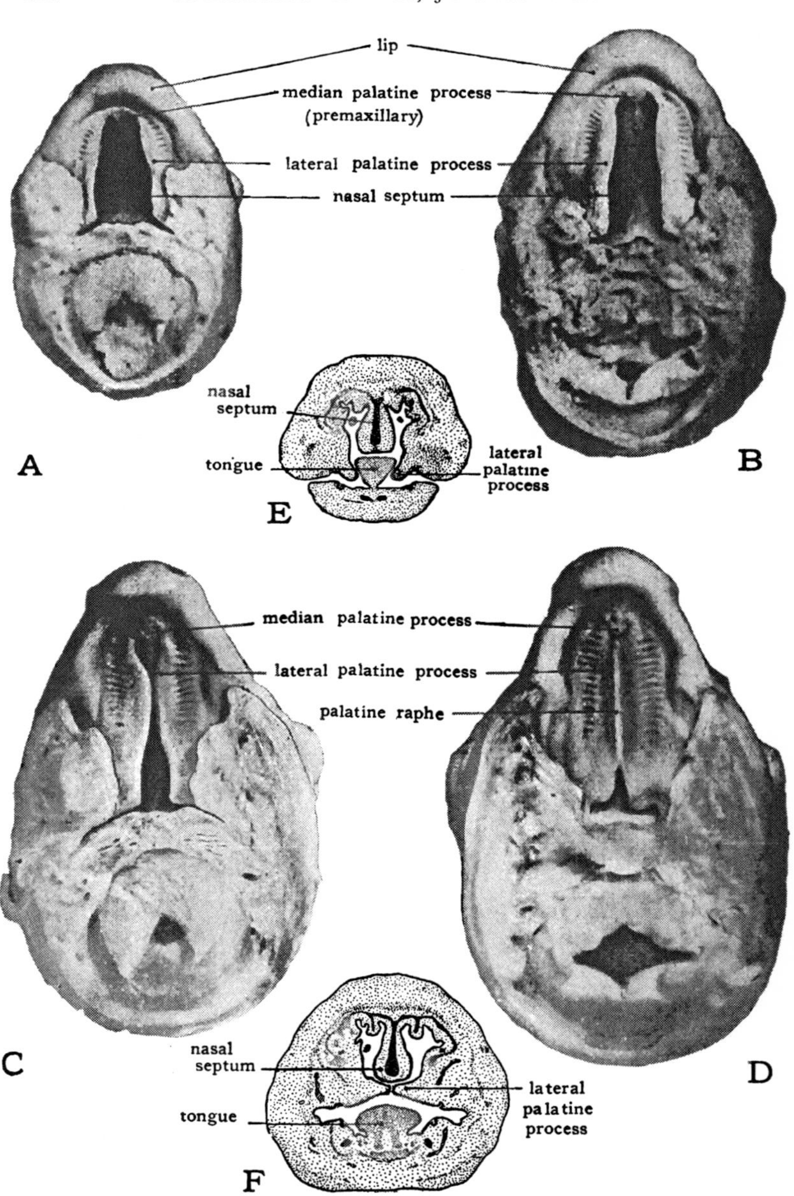

FIG. 174. For legend see opposite page.

of the tongue must be thought of as composed of an outer covering and the underlying mesodermal tissue which causes the covering to bulge into the lumen. The covering tissue arises in situ from the lining of the branchial arches involved. The sensory innervation of the surface of the tongue is, therefore, just what one would expect from the basic relations of the cranial nerves to the branchial arches. The epithelium of the body of the tongue gets its sensory supply from the lingual branch of the mandibular division of the trigeminal (V) nerve (Fig. 97, A, B), and from the chorda tympani branch of the seventh nerve. The root of the tongue receives its sensory fibers from the glossopharyngeal (Fig. 94) and vagus nerves.

The skeletal muscle that makes up the main mass of the tongue beneath the mucosal covering is derived from mesodermal cell masses that are believed to migrate into the pharyngeal floor from the myotomes of the occipital somites. Ontogenetically, in mammalian embryos this migration is exceedingly difficult to trace, for the cells of myotomal origin early mingle indistinguishably with the local mesenchymal cells. Nevertheless the way the hypoglossal nerve (XII), which is the cranial nerve arising at the level of these occipital myotomes, grows in with the developing lingual muscles (Figs. 93, 94, and 97, A, B) and innervates them, furnishes strong circumstantial evidence for this interpretation of tongue muscle origin and migration.

Palate. Coincidently also the palatal shelf is being formed in the upper jaw and separating off the more cephalic portion of the original stomodaeal chamber. Since it is into this cephalic portion of the cavity that the nasal pits break through (Figs. 93 and 97, C), the formation of the palatal shelf in effect prolongs the nasal chambers backwards so they open eventually into the region where the oral cavity becomes continuous with the pharynx.

The palate as well as the arch of the upper jaw is contributed to by both the naso-medial processes and the maxillary processes. From the premaxillary region a small triangular median portion of the palate is formed (Fig. 174). The main portion of the palate is derived from the maxillary processes. From them shelf-like outgrowths arise

FIG. 174. Photographs (× 6) of dissections of pig embryos made to expose the roof of the mouth and show the development of the palate. A, 20.5 mm.; B, 25 mm.; C, 26.5 mm.; D, 29.5 mm.

The diagrams of transverse sections are set in to show the relations before (E) and after (F) the retraction of the tongue from between the palatine processes.

on either side and extend toward the mid-line (Fig. 174, A–C). When these palatal shelves first start to develop, the tongue lies between them (Fig. 174, E). As development progresses the tongue drops down (Fig. 174, F); the palatal shelves are extended toward the mid-line and finally fuse with each other medially and with the premaxillary process anteriorly to complete the palate (Fig. 174, D). At the same time the nasal septum grows toward the palate and becomes fused to its cephalic face (Figs. 174, F, and 178). Thus the separation of

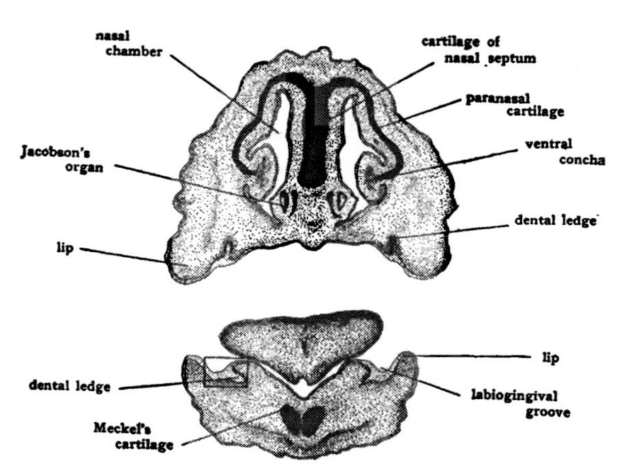

FIG. 175. Transverse section of snout of 28 mm. pig embryo (× 12). The area included in the rectangle is shown in detail in the following figure.

right and left nasal chambers from each other is accomplished at the same time that the nasal region as a whole is separated from the oral.

II. The Development of the Teeth

The Dental Ledge. Local changes leading toward tooth formation can be made out in the jaws of embryos as small as 15 mm. or even less. By the time a size of 28–30 mm. has been attained, a definite thickening of the oral epithelium can readily be seen on both the upper and the lower jaw. This band of epithelial cells which pushes into the underlying mesenchyme around the entire arc of each jaw is known as the *labio-dental ledge* (*labio-dental lamina*) (Figs. 175 and 176). Shortly after its first appearance, cross-sections show this ledge of epithelial cells to be differentiating into two parts, a more distal part which by its ingrowth marks off the elevation which is to become

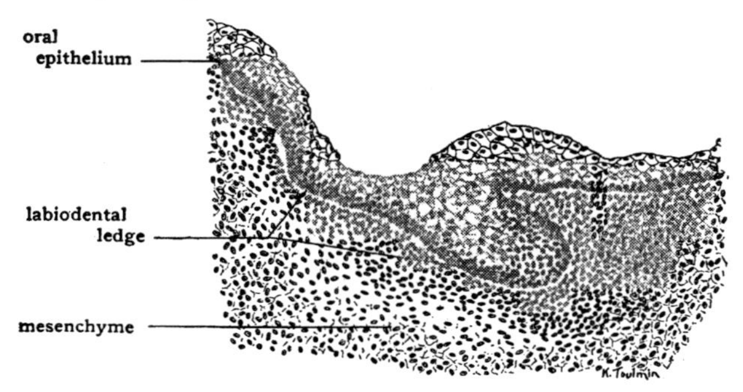

Fig. 176. Drawing (× 130) showing labio-dental ledge of 28 mm. pig embryo. For location of area represented see preceding figure.

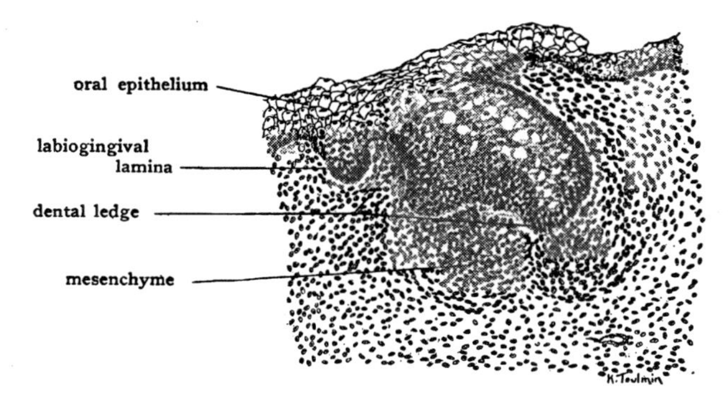

Fig. 177. Drawing (× 130) showing differentiation of the labio-dental ledge into labio-gingival lamina and dental ledge. The ingrowth of the labio-gingival lamina initiates the separation of the lip from the gum (gingiva). From the dental ledge a series of local bud-like outgrowths are formed, each of which gives rise to the enamel cap of a tooth.

The region shown is the same as that in the preceding figure but from a slightly older (37 mm.) embryo.

the lip from that which is to become the gum, and a more proximal part which is destined to grow into the gum and give rise to the enamel-forming organs of the teeth. The part of the original labio-dental ledge which separates the lip from the gum (gingiva) is known as the *labio-gingival lamina*, and the part of the original ledge which is to take part in tooth formation is known as the *dental ledge* or *dental lamina* (Fig. 177).

Enamel Organs. Soon after the dental ledge is established, local buds arise from it at each point where a tooth is destined to be formed. Since these cell masses give rise to the enamel crown of the tooth they are termed *enamel organs*. As would be expected, the enamel organs

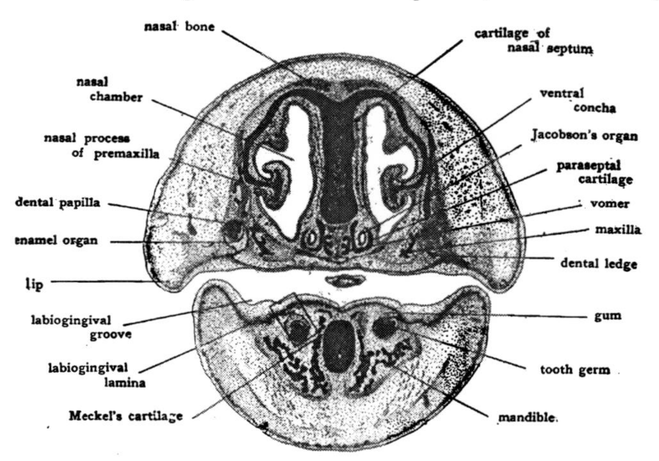

FIG. 178. Drawing (× 10) of a transverse section of the snout of a 71 mm. pig embryo. The area included in the rectangle is shown in detail in the following figure.

for the milk teeth are budded off from the dental ledge first, but the cell clusters which later give rise to the enamel of the permanent teeth are formed at a surprisingly early time (Fig. 180). They remain dormant, however, during the growth period of the milk teeth and begin to develop actively only after the jaws have enlarged sufficiently to accommodate the permanent dentition.

The histogenetic processes involved in the formation of milk teeth and permanent teeth are essentially the same. It is, therefore, sufficient to trace them in the case of the milk teeth only, keeping in mind that the same process is repeated later in life in the formation of the permanent teeth.

In a section of the developing mandible which cuts the dental ledge at a point where an enamel organ is being formed, the shape of the enamel organ suggests that of an irregularly shaped, inverted goblet, the section of the dental ledge appearing somewhat like a distorted stem (Fig. 178). The epithelial cells lining the inside of the goblet early take on a columnar shape. Because they constitute the layer which secretes the enamel cap of the tooth, they are called *ameloblasts* (enamel

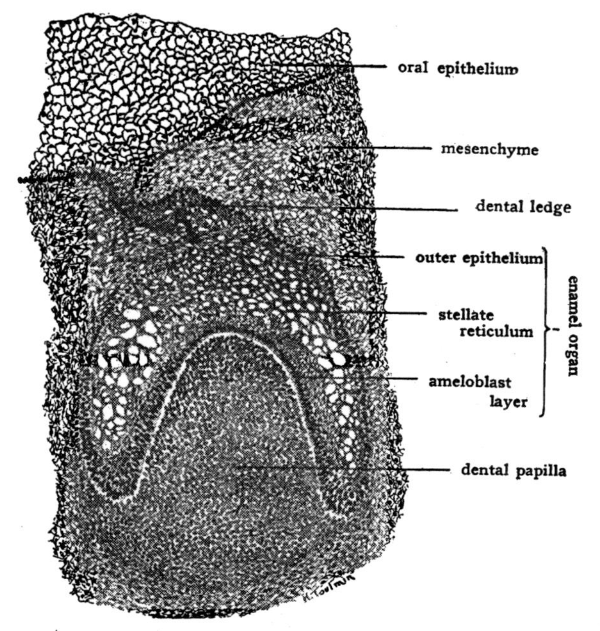

FIG. 179. Tooth germ from the lower jaw of a 71 mm. pig embryo. (Projection drawing × 100.) For location of area represented see preceding figure.

formers) (Fig. 179). The outer layer of the enamel organ is made up of closely packed cells which are at first polyhedral in shape but which soon, with the rapid growth of the enamel organ, become flattened. They constitute the so-called *outer epithelium of the enamel organ* (Fig. 179). Between the outer epithelium and the ameloblast layer is a loosely aggregated mass of cells called collectively, because of their characteristic appearance, the *enamel pulp* or the *stellate reticulum* (Fig. 179).

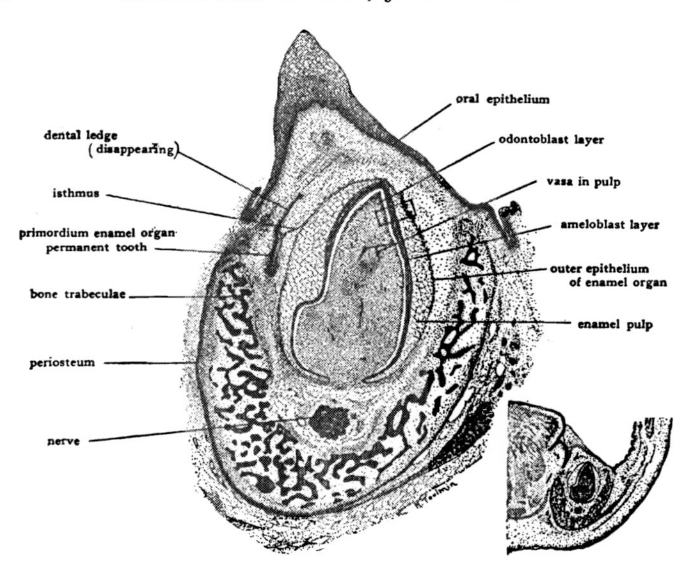

FIG. 180. Developing tooth from lower jaw of a 120 mm. pig embryo (× 14). The small sketch including half of the tongue (left) and part of the lip (right) gives the relations of the region drawn. The area in the rectangle is shown in detail in the following figure.

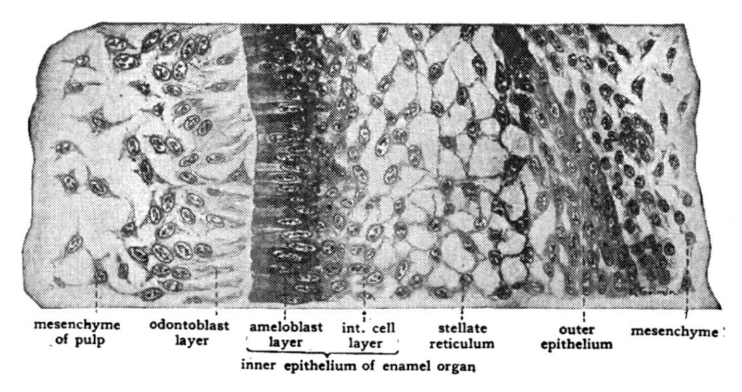

FIG. 181. Projection drawing (× 350) of segment of enamel organ and adjacent pulp from a 120 mm. pig embryo to show ameloblast and odontoblast layers. For location of area represented see preceding figure.

The Dental Papilla. Inside the goblet-shaped enamel organ there is caught a mass of mesenchymal cells which are said to constitute the *dental papilla* (Fig. 179). The cells of the dental papilla proliferate rapidly and soon form a very dense aggregation. The outer cells of this mass are destined to secrete the dentine of the tooth and the inner cells to give rise to the pulp of the tooth.

A little later in development the enamel organ begins to assume the shape characteristic of the crown of the tooth it is to lay down (Fig. 180). At the same time the outer cells of the dental papilla take on a columnar form similar to that of the ameloblasts (Fig. 181). They are now called *odontoblasts* (dentine formers) because they are about to become active in secreting the dentine of the tooth.

In the central portion of the dental papilla vessels and nerves are beginning to make their appearance so that the picture is already suggestive of the condition seen in the pulp of an adult tooth. Meanwhile the growth of the dental papilla toward the gum has crowded the stellate reticulum of the enamel organ in the crown region so it is nearly obliterated (Fig. 180). This brings the ameloblasts of this region much closer to the many small blood vessels which lie in the surrounding mesenchyme. The approach of the ameloblasts to the neighboring vascular supply would appear to be significant, since it is precisely here at the tip of the crown where the ameloblasts first begin to secrete enamel (Fig. 182).

By the time the enamel organ has been well established the dental ledge has lost its connection with the oral epithelium, although traces of it can still be identified in the mesenchyme at the lingual side of the tooth germ (Fig. 180). The cluster of cells which is destined to give rise to the enamel organ of the permanent tooth of this level can be seen budding off from the ledge close to the point from which the enamel organ of the milk tooth arose (Fig. 180).

Formation of Dentine. With these preparatory developments complete, the tooth-forming structures are, so to speak, ready to go about the fabrication of dentine and enamel. As is the case with bone, enamel and dentine are both composed of an organic basis in which inorganic compounds are deposited. We may use the same comparison that was used in describing bone: that of the familiar use in construction operations of a steel meshwork into which concrete is poured, the steel giving the finished structure some degree of elasticity and increasing the tensile strength while the concrete gives body and solidity. In the case of such hard structures in the body as bone, dentine, and

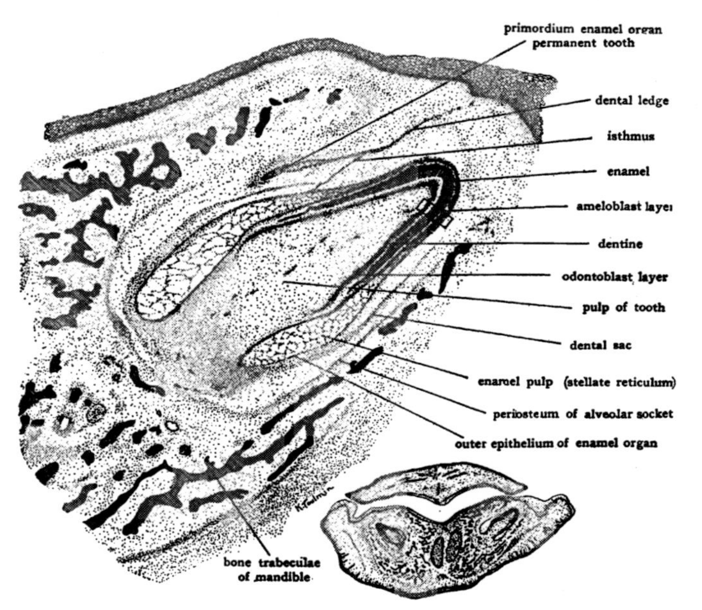

FIG. 182. Developing tooth from lower jaw of a 130 mm. pig embryo (× 30). The small sketch gives the relations of the regions drawn. The area in the rectangle is shown in detail in the following figure.

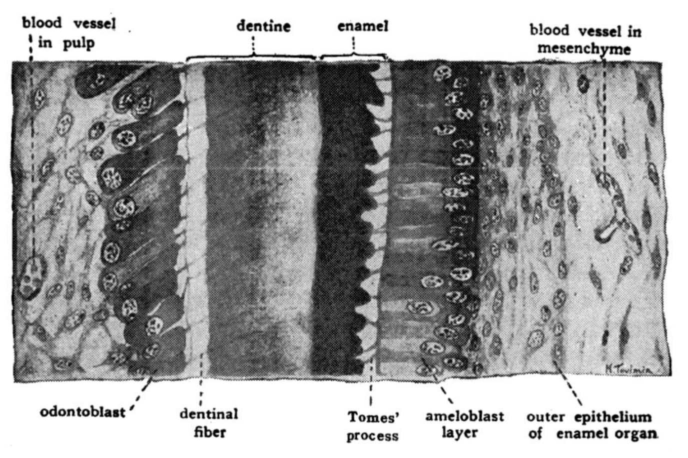

FIG. 183. Projection drawing (× 350) of small segment of developing incisor from 130 mm. pig embryo to show formation of enamel and dentine. For location of arca represented see preceding figure.

enamel, these interlacing organic strands in the matrix give the tissue its resilience and tensile strength, and the calcareous compounds deposited in the organic framework give form and hardness.

Although bone, dentine, and enamel are similar in having both organic and inorganic constituents in their matrix they are quite different in detail, both as to composition and microscopical structure. Bone has approximately 45 per cent of organic material while dentine has but 30 per cent and adult enamel 5 per cent or less. There is also considerable difference in the kind and proportion of inorganic compounds present in each. Structurally they are totally unlike. Bone matrix is formed in lamellae and has cells scattered through it. Dentine is formed without lamellation and has its cellular elements lying against one face and sending long processes into tubules in the matrix. Enamel is prismatic in structure and the cells which form it lie against its outer surface while it is being deposited, but are destroyed in the eruption of the tooth.

The first dentine is deposited against the inner face of the enamel organ, the odontoblasts drawing their raw materials from the small vessels in the pulp and secreting their finished product toward the enamel organ. It is significant in this connection that in an active odontoblast the nucleus, which is the metabolic center of the cell, has gravitated toward the source of supplies and come to lie in the extreme pulpal end of the cell (Fig. 183). Also, the end of the odontoblast toward the enamel organ, where the elaborated product of the cell is being accumulated preparatory to its extrusion, can be seen to take the stain especially intensely. Although our knowledge of intracellular chemistry is as yet exceedingly fragmentary and we do not know the exact chemical nature of the product in this stage, the staining reaction is clearly indicative of the presence of calcium compounds of some sort.

If attention is turned now to the recently formed dentine, two zones distinctly different in staining reaction can be seen. The zone nearer the cells is pale, taking but little stain (Fig. 183). This zone consists of the recently deposited organic part of the matrix not as yet impregnated with calcareous material. The zone nearer the enamel organ will be found, by contrast, very intensely stained. This is the older part of the dentine matrix which has had the organic framework impregnated with calcareous material.

As the odontoblasts continue to secrete additional dentine matrix the accumulation of their own product inevitably forces the cell layer back, away from the material previously deposited. Apparently strands of their cytoplasm become embedded in the material first laid down and are then pulled out to form the characteristic processes of the odontoblasts known as the *dentinal fibers* (Fig. 183). As the layer of secreted material becomes thicker and the cells are forced farther from the material first deposited, these dentinal fibers become progressively longer. Even in adult teeth where the dentine may be as much as 2 mm. in thickness they extend from the odontoblasts which line the pulp chamber to the very outer part of the dentine. These dentinal fibers are believed to be concerned with maintaining the organic portion of the dentine matrix in a healthy condition. When the pulp is removed from a tooth, taking with it the odontoblasts, we know that the dentine undergoes degenerative changes which involve, among other things, increase in brittleness. This would seem to be attributable to the degeneration of the organic framework of a matrix no longer nourished by the odontoblasts.

Formation of Enamel. While the dentine is being laid down by the cells of the odontoblast layer, the enamel cap of the tooth is being formed by the ameloblast layer of the enamel organ. As was the case with the odontoblasts, the active cells of the ameloblast layer are columnar in shape and their nuclei, too, lie in the ends of the cells toward the source of supplies, in this case the small vessels in the adjacent mesenchyme (Fig. 183). The amount of organic material laid down as the framework of enamel is much less than is the case with either bone or dentine, and it is therefore more difficult to make out its precise character and arrangement. It is, nevertheless, possible to see in decalcified sections, delicate fibrous strands projecting from the tips of the ameloblasts into the areas of newly formed enamel (Fig. 183). It seems probable that these strands (*Tomes' processes*) are in some way involved in the formation of the organic matrix of enamel. The problem of tracing the relations of Tomes' processes to the organic framework of enamel is greatly complicated by the fact that where the ameloblasts have deposited calcium compounds the calcium has rendered the organic part of the matrix so avid in its affinity for stains that it is not possible to discern fine structural details because of the very density of the resulting coloration (Fig. 183). This reaction of the tissue to stains persists even after the inorganic calcium compounds have been removed by decalcification, indicating that the organic framework itself has been chemically altered by the calcium deposited in it.

In spite of these difficulties in getting at the exact nature and arrangement of the organic matrix of enamel, it is quite possible to see the genesis of its fundamental prismatic structure. Each ameloblast builds up beneath itself a minute rod or prism of calcareous material. These prisms are placed with their long axes approximately at right angles to the dento-enamel junction. Collectively these *enamel prisms* form an exceedingly hard cap over the crown of the tooth which in its structural arrangement suggests a paving of polygonal bricks laid on end. There is sufficient difference in the rate at which the different

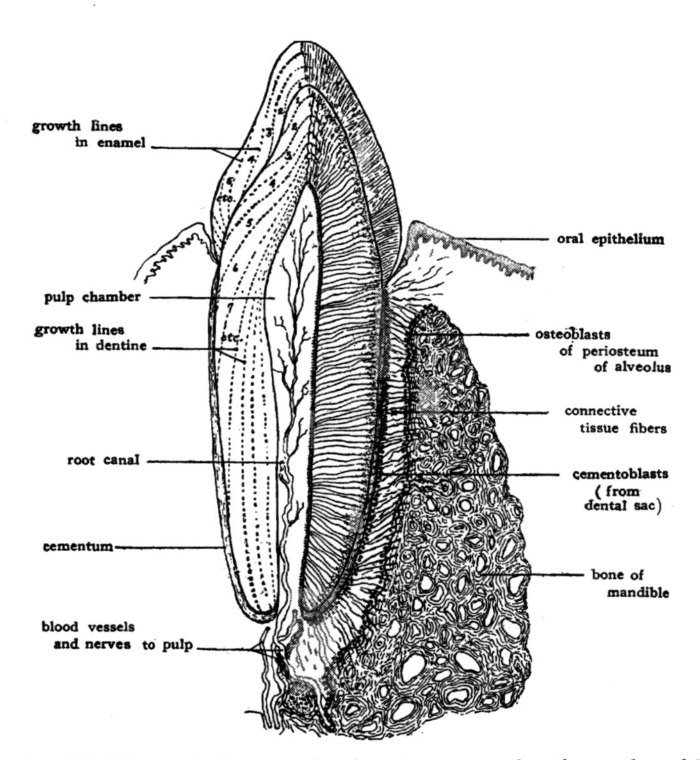

FIG. 184. Schematic diagram showing the topography of a tooth and its relations to the bone of the jaw. The numbered zones indicate empirically the sequence of deposition of the dentine and enamel. The so-called growth lines in the dentine and enamel follow the general contours indicated by the dotted lines in the figure but are much more numerous.

ameloblasts work so that in actively growing enamel the surface is jagged and irregular due to the varying extent to which the different prismatic elements have been calcified (Fig. 183).

Both enamel formation and dentine formation begin at the tip of the crown and progress toward the root of the tooth (Figs. 182 and 184). But the entire crown is well formed before the root is much more than begun. The progressive increase in the length of the root is an important factor in the eruption of the tooth, for as the root increases in length the previously formed crown must move closer to the surface of the gum. Even when the crown of the tooth begins to erupt the root is still incomplete, and it does not acquire its full length until the crown has entirely emerged.

The Formation of Cementum. The so-called cementum of the tooth is virtually a bone encrustation of its root. No cementum is formed until the tooth has acquired nearly its full growth and its definitive position in the jaw. But the first indications of specialization in the tissue destined to give rise to it can be seen long before the cementum itself appears.

Outside the entire tooth germ, between it and the developing bone of the jaw, there occurs a definite concentration of mesenchyme. The concentration becomes evident first at the base of the dental papilla and extends thence crownwards about the developing tooth, which it eventually completely surrounds.

This mesenchymal investment is known as the *dental sac* (Fig. 182). In the eruption of the tooth the portion of the dental sac over the crown is destroyed, but the deeper portion of the sac persists and becomes closely applied to the growing root. At about the time the tooth has acquired its final position in the jaw, the cells of the dental sac begin to form the cementum. Histologically and chemically cementum is practically identical with subperiosteal bone. When we consider the manner of origin of the dental sac and of the periosteum of the bone socket (alveolar socket) in which the root of the tooth lies, and see how they arise side by side from the same sort of tissue, this seems but natural. The dental sac is essentially a layer of periosteal tissue facing the root of the tooth and back to back with the periosteal tissue of the alveolar socket (Fig. 182).

The Attachment of Tooth in the Jaw. The attachment of the tooth in its socket is brought about by the development, between the dental sac and the periosteum of the tooth socket, of an exceedingly tough fibrous connective tissue. As the periosteum of the alveolus adds new lamellae of bone to the jaw on the one side, and the dental sac adds lamellae of cementum to the root of the tooth on the other, the fibers of this connective tissue are caught in the new lamellae. Thus the tooth comes to be supported by fibers which are literally calcified into the cementum of the tooth at one end and into the bone of the jaw at the other (Fig. 184). The mechanism involved is precisely the same as that which occurs in the burying of tendon fibers in a growing bone, where the buried ends of the fibers are known as the penetrating fibers of Sharpey.

Replacement of Deciduous Teeth by Permanent Teeth. The replacement of the temporary or "milk" (deciduous) dentition by the permanent teeth is a process which varies in detail for each tooth. The general course of events is, however, essentially similar in all cases. The enamel organ of the permanent tooth arises from the dental ledge near the point of origin of the corresponding deciduous tooth (Fig. 180). With the disappearance of the dental ledge, the permanent tooth germ comes to lie in a depression of the alveolar socket on the lingual side of the developing deciduous tooth (Fig. 185).

When the jaws approach their adult size the hitherto latent primordia of the permanent teeth begin to go through the same histo-

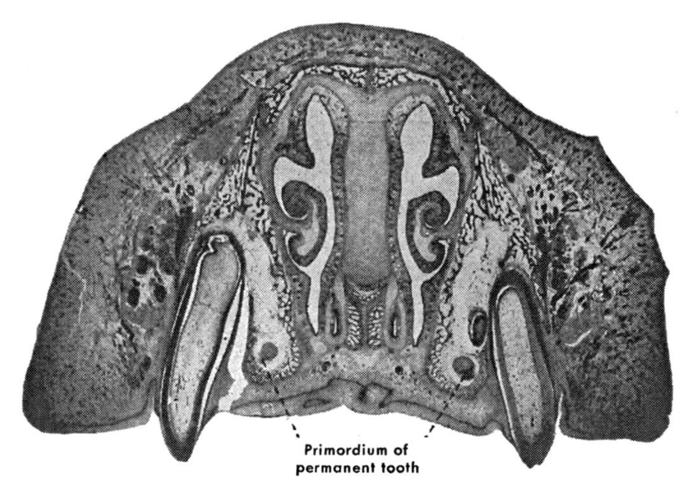

Fig. 185. Photomicrograph (× 5) of upper jaw of 160 mm. pig embryo showing the milk cuspids just breaking through the gum.

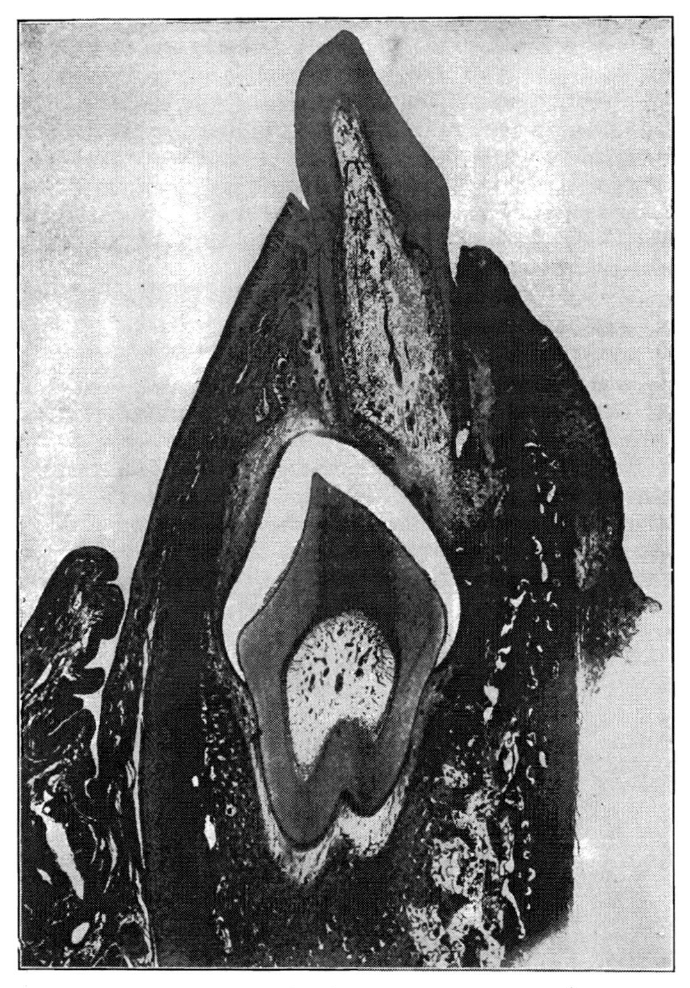

FIG. 186. Photomicrograph (× 6) of section through the jaw of a puppy showing a deciduous tooth nearly ready to drop out and the developing permanent tooth deeply embedded in the jaw below it. The space about the crown of the permanent tooth was occupied in the living condition by enamel. Fully formed enamel, being approximately 97 per cent inorganic in composition, is almost completely destroyed by the decalcification with acids which must be carried out before such material can be sectioned. (From a preparation loaned by Dr. S. W. Chase.)

genetic changes we have already traced in the case of the temporary teeth. As the permanent tooth increases in size, the root of the deciduous tooth is resorbed and the permanent tooth comes to lie underneath its remaining portion (Fig. 186). Eventually nearly the entire root of the deciduous tooth is destroyed and its loosened crown drops out, making way for the eruption of the corresponding permanent tooth.

Bibliography

Although this does not purport to be a complete bibliography on the development of the pig, I have tried to make it comprehensive. At least one reference has been included on every phase of the subject concerning which I could find published information. Under each of the main subject headings some of the articles referred to have extensive bibliographies of the literature in their special field. But, with the exception of a few outstanding contributions, no papers have been included which are merely of historical interest. For such articles reference should be made to the exhaustive bibliographies compiled by Minot (1893) and by Keibel (1897). Furthermore, the list is largely restricted to contributions based directly on pig embryos. Exception to this rule has been made in favor of a few general articles which are of especial assistance in acquiring a perspective on some phase of the subject. Also, a number of articles based on other forms have been included when no work appeared to have been done on corresponding phases of development in the pig. It is hoped that such a list of selected references will furnish a starting point for following up any desired line of inquiry without involving one in a discouraging multiplicity of titles.

Texts and Manuals

Arey, L. B., 1946. Developmental Anatomy. Saunders, Philadelphia, 5th Ed., ix & 616 pp.

Baumgartner, W. J., 1924. Laboratory Manual of the Foetal Pig. Macmillan, New York, vii & 57 pp.

Boyden, E. A., 1936. A Laboratory Atlas of the 13-mm. Pig Embryo. (Prefaced by younger stages of the chick embryo.) The Wistar Institute Press, Philadelphia, iv & 104 pp.

Hamilton, W. J., Boyd, J. D., and Mossman, H. W., 1945. Human Embryology. Williams and Wilkins, Baltimore, viii & 366 pp.

Hertwig, O., 1901–07. Handbuch der Vergleichenden und experimentellen Entwicklungslehre der Wirbeltiere. (Edited by Dr. Oskar Hertwig and written by numerous collaborators.) Fischer, Jena.

Huettner, A. F., 1941. Fundamentals of Comparative Embryology of the Vertebrates. Macmillan, New York, xiv & 416 pp.

Jordan, H. E., and Kindred, J. E., 1948. A Textbook of Embryology. Appleton, New York, 5th Ed., xiv & 613 pp.

Keibel, F., 1897. Normentafeln zur Entwicklungsgeschichte der Wirbelthiere. I. Des Schweines. Fischer, Jena, 114 S.

Martin, P., 1912. Lehrbuch der Anatomie der Haustiere. Schickhard & Ebner, Stuttgart. (Bd. 1, Allgemeine und vergleichende Anatomie mit Entwicklungsgeschichte, xii & 811 S.)

Minot, C. S., 1893. A Bibliography of Vertebrate Embryology. Memoirs, Boston Soc. Nat. History, Vol. 4, pp. 487–614.

Minot, C. S., 1911. A Laboratory Textbook of Embryology. The Blakiston Company, Philadelphia, 2nd Ed., xii & 402 pp.

Needham, J., 1931. Chemical Embryology. Macmillan, New York, xxii & 2021 pp.

Patten, B. M., 1929. Early Embryology of the Chick. The Blakiston Company, Philadelphia, 3rd Ed., xiii & 228 pp.

Patten, B. M., 1946. Human Embryology. The Blakiston Company, Philadelphia, xv & 776 pp.

Sisson, S., 1921. The Anatomy of the Domestic Animals. Saunders, Philadelphia, 2nd Ed., 930 pp.

Weiss, P., 1939. Principles of Development. A Text in Experimental Embryology. Holt, New York, xix & 601 pp.

Wieman, H. L., 1930. An Introduction to Vertebrate Embryology. McGraw-Hill, New York, xi & 411 pp.

Windle, W. F., 1940. Physiology of the Fetus. Saunders, Philadelphia, xiii & 249 pp.

Zeitzschmann, O., 1923–24. Lehrbuch der Entwicklungsgeschichte der Haustiere. R. Schoetz, Berlin, 542 S.

General Articles and Articles of Significance for the Interpretation of Developmental Processes

Alexander, J., 1944. The Gene—A Structure of Colloidal Dimensions. Chapt. 37, pp. 808–819, in "Colloid Chemistry," Vol. V, edited by Jerome Alexander, Reinhold Publishing Corp., New York.

Allen, W. F., 1918. Advantages of sagittal sections of pig embryos for a medical embryology course. Anat. Rec., Vol. 14, pp. 183–191.

Barth, L. G., 1944. Colloid Chemistry in Embryonic Development. Chapt. 39, pp. 851–859, in "Colloid Chemistry," Vol. V, edited by Jerome Alexander, Reinhold Publishing Corp., New York.

Chambers, R., 1944. Some Physical Properties of Protoplasm. Chapt. 41, pp. 864–875, in "Colloid Chemistry," Vol. V, edited by Jerome Alexander, Reinhold Publishing Corp., New York.

Conklin, E. G., 1914. The cellular basis of heredity and development. Pop. Sci. Monthly, Vol. 85, pp. 105–133.

Hartman, C. G., 1931. Development of the egg as seen by the physiologist. Sci. Monthly, Vol. 33, pp. 17–28.

Hedley, O. F., 1926. Quantitative study of growth of certain organs in pig fetus. Bull. Med. Coll. Va., Vol. 23, pp. 19–36.

Holtfreter, J., 1947. Changes of structure and the kinetics of differentiating embryonic cells. Jour. Morph., Vol. 80, pp. 57–92.

Holtfreter, J., 1947. Observations on the migration, aggregation and phagocytosis of embryonic cells. Jour. Morph., Vol. 80, pp. 93–111.

Kingsbury, B. F., 1926. On the so-called law of anteroposterior development. Anat. Rec., Vol. 33, pp. 73–87.

Lewis, W. H., 1947. Mechanics of invagination. Anat. Rec., Vol. 97, pp. 139–156.

Spemann, H., 1927. Organizers in animal development. Proc. Roy. Soc. London, Ser. B, Vol. 102, pp. 177–187.

Spemann, H., 1938. Embryonic Development and Induction. Yale Univ. Press, New Haven, xii & 401 pp.

The Sex Organs and Gametogenesis

Allen, E., 1923. Ovogenesis during sexual maturity. Am. Jour. Anat., Vol. 31, pp. 439–481.

Allen, E., Kountz, W. B., and Francis, B. F., 1925. Selective elimination of ova in the adult ovary. Am. Jour. Anat., Vol. 34, pp. 445–468.

Bascom, K. F., 1925. Quantitative studies of the testis. I. Some observations on the cryptorchid testes of sheep and swine. Anat. Rec., Vol. 30, pp. 225–241.

Bascom, K. F., and Osterud, H. L., 1925. Quantitative studies of the testicle. II. Pattern and total tubule length in the testicles of certain common mammals. Anat. Rec., Vol. 31, pp. 159–169.

Blandau, R. J., 1945. The first maturation division of the rat ovum. Anat. Rec., Vol. 92, pp. 449–457.

Corner, G. W., 1917. Maturation of the ovum in swine. Anat. Rec., Vol. 13, pp. 109–112.

Corner, G. W., 1919. On the origin of the corpus luteum of the sow from both granulosa and theca interna. Am. Jour. Anat., Vol. 26, pp. 117–183.

Evans, H. M., and Swezy, O., 1929. Ovogenesis and the normal follicular cycle in adult mammalia. Mem. Univ. Cal., Vol. 9, pp. 119–224.

Everett, N. B., 1945. The present status of the germ-cell problem in vertebrates. Biol. Rev., Vol. 20, pp. 45–55.

Gould, H. N., 1923. Observations on the genital organs of a sex intergrade hog. Anat. Rec., Vol. 26, pp. 241–261.

Hargitt, G. T., 1925–30. The formation of the sex glands and germ cells of mammals. Jour. Morph. & Physiol., Vols. 40, 41, 42, 49.

Hartman, C. G., 1926. Polynuclear ova and polyovular follicles in the opossum and other mammals, with special reference to the problem of fecundity. Am. Jour. Anat., Vol. 37, pp. 1–52.

Hill, R. T., Allen, E., and Kramer, T. C., 1935. Cinemicrographic studies of rabbit ovulation. Anat. Rec., Vol. 63, pp. 239–245.

Kellicott, W. E., 1913. A Textbook of General Embryology. Holt, New York, v & 376 pp.

Küpfer, M., 1920. Beiträge zur Morphologie der weiblichen Geschlechtsorgane bei den Säugetieren. Ueber das Auftreten gelber Körper am Ovarium des domestizierten Rindes und Schweines. Vierteljahrsschrift d. Naturf. Gesellsch., Zurich, Bd. 65, S. 377–433.

Latta, J. S., and Pederson, E. S., 1944. The origin of ova and follicle cells from the germinal epithelium of the ovary of the albino rat as demonstrated by selective intravital staining with India ink. Anat. Rec., Vol. 90, pp. 23–35.

Morgan, T. H., 1926. The Theory of the Gene. Yale Univ. Press, New Haven, xvi & 343 pp.

Painter, T. S., 1922. Studies in mammalian spermatogenesis. Jour. Exp. Zoöl., Vols. 35, 37, 39. Jour. Morph. & Physiol., Vol. 43.

Parker, G. H., and Bullard, C., 1913. On the size of litters and the number of nipples in swine. Proc. Am. Acad. Arts & Sciences, Vol. 49, pp. 399–426.

Patten, W., 1925. Life, evolution and heredity. Sci. Monthly, Vol. 21, pp. 122–134.

Pincus, G., and Enzmann, E. V., 1937. The growth, maturation and atresia of ovarian eggs in the rabbit. Jour. Morph., Vol. 61, pp. 351–383.

Pliske, E. C., 1940. Studies on the influence of the zona pellucida in atresia. Jour. Morph., Vol. 67, pp. 321–349.

Schmaltz, R., 1911. Die Strucktur der Geschlechtsorgane der Haussäugetiere. P. Parey, Berlin, xii & 388 S.

Smith, J. T., and Ketteringham, R. C., 1937–38. Rupture of the graafian follicles. Part I. Am. Jour. Obs. & Gyn., Vol. 33, pp. 820–827. Part II. Am. Jour. Obs. & Gyn., Vol. 36, pp. 453–460.

Stein, K. F., and Allen, E., 1942. Attempts to stimulate proliferation of the germinal epithelium of the ovary. Anat. Rec., Vol. 82, pp. 1–9.

Thanhoffer, L. de, 1934. The structure of the graafian follicle as revealed by microdissection. Zeitschr. f. Anat. u. Entwg., Bd. 102, S. 402–408.

Warwick, B. L., 1925. The effect of vasectomy on swine. Anat. Rec., Vol. 31, pp. 19–21.

Wilson, E. B., 1925. The Cell in Development and Heredity. Macmillan, New York, 3rd Ed., ix & 1232 pp.

Young, W. C., 1929. A study of the function of the epididymis. I. Is the attainment of full spermatozoon maturity attributable to some specific action of the epididymal secretion? Jour. Morph. & Physiol., Vol. 47, pp. 479–495.

Young, W. C., 1929. A study of the function of the epididymis. II. The importance of an aging process in sperm for the length of the period during which fertilizing capacity is retained by sperm isolated in the epididymis of the guinea-pig. Jour. Morph. & Physiol., Vol. 48, pp. 475–491.

Young, W. C., 1931. A study of the function of the epididymis. III. Functional changes undergone by spermatozoa during their passage through the epididymis and vas deferens in the guinea-pig. Jour. Exp. Biol., Vol. 8, pp. 151–162.

The Sexual Cycle, Fertilization, Sex Determination

Allen, E., 1926. The ovarian follicular hormone: a study of variation in pig, cow, and human ovaries. Proc. Soc. Exp. Biol. & Med., Vol. 23, pp. 383–387.

Allen, E., Danforth, C. H., and Doisy, E. A., 1939. Sex and Internal Secretions. Williams and Wilkins, Baltimore, 2nd Ed., xxxvi & 1346 pp.

Allen, E., and Doisy, E. A., 1927. Ovarian and placental hormones. Physiol. Reviews, Vol. 7, pp. 600–650.

Anopolsky, D., 1928. Cyclic changes in size of muscle fibers of the fallopian tube of the sow. Am. Jour. Anat., Vol. 40, pp. 459–469.

Blandau, R., and Money, W. L., 1944. Observations on the rate of transport of spermatozoa in the female genital tract of the rat. Anat. Rec., Vol. 90, pp. 255–260.

Burns, R. K., Jr., 1938. Hormonal control of sex differentiation. Am. Nat., Vol. 72, pp. 207–227.

Comstock, R. E., 1939. A study of the mammalian sperm cell. I. Variations in the glycolytic power of spermatozoa and their relation to motility and its duration. Jour. Exp. Zoöl., Vol. 81, pp. 147–164.

Corner, G. W., 1915. The corpus luteum of pregnancy as it is in swine. Carnegie Inst., Contrib. to Embryol., Vol. 2, pp. 69–94.

Corner, G. W., 1917. Variations in the amount of phosphatids in the corpus luteum of the sow during pregnancy. Jour. Biol. Chem., Vol. 29, pp. 141–143.

Corner, G. W., 1919. On the origin of the corpus luteum of the sow from both granulosa and theca interna. Am. Jour. Anat., Vol. 26, pp. 117–183.

Corner, G. W., 1921. Cyclic changes in the ovaries and uterus of swine, and their relations to the mechanism of implantation. Carnegie Inst., Contrib. to Embryol., Vol. 13, pp. 117–146.

Corner, G. W., 1923. Cyclic variation in uterine and tubal contraction waves. Am. Jour. Anat., Vol. 32, pp. 345–351.

Corner, G. W., 1928. Physiology of the corpus luteum. I. The effect of very early ablation of the corpus luteum upon embryos and uterus. Am. Jour. Physiol., Vol. 86, pp. 74–81.

Corner, G. W., 1942. The Hormones in Human Reproduction. Princeton Univ. Press, Princeton, xix & 265 pp.

Corner, G. W., and Allen, W. M., 1929. Physiology of the corpus luteum. II. Production of a special uterine reaction (progestational proliferation) by extracts of the corpus luteum. Am. Jour. Physiol., Vol. 88, pp. 326–339.

Corner, G. W., and Allen, W. M., 1929. Physiology of the corpus luteum. III. Normal growth and implantation of embryos after very early ablation of the ovaries, under the influence of extracts of the corpus luteum. Am. Jour. Physiol., Vol. 88, pp. 340–346.

Corner, G. W., and Amsbaugh, A. E., 1917. Oestrus and ovulation in swine. Anat. Rec., Vol. 12, pp. 287–291.

Corner, G. W., and Snyder, F. F., 1922. Observations on the structure and function of the uterine ciliated epithelium in the pig, with reference to certain clinical hypotheses. Am. Jour. Obs. & Gyn., Vol. 3, pp. 358–366.

Crew, F. A. E., 1925. Prenatal death in the pig and its effect upon the sex-ratio. Proc. Roy. Soc. Edinburgh, Vol. 46, pp. 9–14.

Hammond, J., 1934. The fertilisation of rabbit ova in relation to time: A method of controlling the litter size, the duration of pregnancy and the weight of the young at birth. Jour. Exp. Biol., Vol. 11, pp. 140–161.

Hartman, C. G., 1929. The homology of menstruation. J. A. M. A., Vol. 92, pp. 1992–1995.

Hartman, C. G., and Squier, R. R., 1931. The follicle-stimulating effect of pig anterior lobe on the monkey ovary. Anat. Rec., Vol. 50, pp. 267–273.

Keye, J. D., 1923. Periodic variations in spontaneous contractions of uterine muscle in relation to the oestrous cycle and early pregnancy. Bull. Johns Hopkins Hosp., Vol. 34, pp. 60–63.

Lewis, L. L., 1911. The vitality of reproductive cells. Agric. Exp. Sta. Oklahoma, Bull. No. 96.

Lillie, F. R., 1919. Problems of Fertilization. Univ. Chicago Press, xii & 278 pp.

Loeb, L., 1923. The mechanism of the sexual cycle, with special reference to the corpus luteum. Am. Jour. Anat., Vol. 32, pp. 305–343.

Markee, J. E., Pasqualetti, R. A., and Hinsey, J. C., 1936. Growth of intraocular endometrial transplants in spinal rabbits. Anat. Rec., Vol. 64, pp. 247–253.

Marshall, F. H. A., 1922. The Physiology of Reproduction. Longmans, Green & Co., London, 2nd Ed., xvi & 770 pp.

McKenzie, F. F., 1926. The normal oestrous cycle in the sow. Univ. Missouri Coll. Agric. Exp. Sta. Res. Bull., Vol. 86, pp. 5–41.

Papanicolaou, G. N., 1923. Oestrus in mammals from a comparative point of view. Am. Jour. Anat., Vol. 32, pp. 285–292.

Parker, G. H., 1931. Passage of sperms and of eggs through oviducts in terrestrial vertebrates. Phil. Trans. Roy. Soc. London, Ser. B, Vol. 219, pp. 381–419.

Phillips, R. W., and Andrews, F. N., 1937. The speed of travel of ram spermatozoa. Anat. Rec., Vol. 68, pp. 127–132.

Pincus, G., 1936. The Eggs of Mammals. Macmillan, New York, ix & 160 pp.

Schott, R. G., and Phillips, R. W., 1941. Rate of sperm travel and time of ovulation in sheep. Anat. Rec., Vol. 79, pp. 531–540.

Seckinger, D. L., 1923. Spontaneous contractions of the Fallopian tube of the domestic pig with reference to the oestrous cycle. Bull. Johns Hopkins Hosp., Vol. 34, pp. 236–239.

Snyder, F. F., 1923. Changes in the Fallopian tube during the ovulation cycle and early pregnancy. Bull. Johns Hopkins Hosp., Vol. 34, pp. 121–125.

Soderwall, A. L., and Blandau, R. J., 1941. The duration of the fertilizing capacity of spermatozoa in the female genital tract of the rat. Jour. Exp. Zoöl., Vol. 88, pp. 55–64.

Stockard, C. R., 1923. The general morphological and physiological importance of the oestrous problem. Am. Jour. Anat., Vol. 32, pp. 277–283.

Surface, F. M., 1908. Fecundity of swine. Biometrika, Vol. 6, pp. 433–436.

Swingle, W. W., 1926. The determination of sex in animals. Physiol. Rev., Vol. 6, pp. 28–61.

Toothill, M. C., and Young, W. C., 1931. The time consumed by spermatozoa in passing through the ductus epididymidis of the guinea-pig as determined by means of India-ink injections. Anat. Rec., Vol. 50, pp. 95–107.

Wilson, K. M., 1926. Histological changes in the vaginal mucosa of the sow in relation to the oestrous cycle. Am. Jour. Anat., Vol. 37, pp. 417–432.

Witschi, E., 1932. Physiology of embryonic sex differentiation. Am. Nat., Vol. 66, pp. 108–117.

Cleavage, Germ Layer Formation, and the Structure of Young Embryos

Assheton, R., 1899. The development of the pig during the first ten days. Quart. Jour. Micr. Sci., Vol. 41, pp. 329–359.

Clark, R. T., 1934. Studies on the physiology of reproduction in the sheep. II. The cleavage stages of the ovum. Anat. Rec., Vol. 60, pp. 135–159.

Conklin, E. G., 1909. The application of experiment to the study of the organization and early differentiation of the egg. Anat. Rec., Vol. 3, pp. 149–154.

Defrise, A., 1933. Some observations on living eggs and blastulae of the albino rat. Anat. Rec., Vol. 57, pp. 239–250.

Green, W. W., and Winters, L. M., 1946. Cleavage and attachment stages of the pig. Jour. Morph., Vol. 78, pp. 305–316.

Heuser, C. H., and Streeter, G. L., 1929. Early stages in the development of pig embryos, from the period of initial cleavage to the time of the appearance of limb-buds. Carnegie Inst., Contrib. to Embryol., Vol. 20, pp. 1–29.

Keibel, F., 1891. Ueber die Entwicklungsgeschichte des Schweines. Anat. Anz., Bd. 6, S. 193–198.

Keibel, F., 1894. Studien zur Entwicklungsgeschichte des Schweines (Sus scrofa domesticus). I. Morph. Arbeiten, Bd. 3, S. 1–139.

Keibel, F., 1895. Ueber einige Plattenmodelle junger Schwein-embryonen. Verhandlungen d. Anat. Ges., Ergänzungsheft Bd. 10, Anat. Anz., S. 199–201.

Keibel, F., 1896. Studien zur Entwicklungsgeschichte des Schweines (Sus scrofa domesticus). II. Morph. Arbeiten, Bd. 5, S. 17–168.

Kingsbury, B. F., 1920. The developmental origin of the notochord. Science, N. S., Vol. 51, pp. 190–193.

Kingsbury, B. F., 1924a. The developmental significance of the notochord (Chorda dorsalis). Zeitschr. f. Morphologie u. Anthropologie, Vol. 24, pp. 59–74.

Kingsbury, B. F., 1924b. The significance of the so-called law of cephalocaudal differential growth. Anat. Rec., Vol. 27, pp. 305–321.

Kingsbury, B. F., 1926. On the so-called law of antero-posterior development. Anat. Rec., Vol. 33, pp. 73–87.

Lewis, F. T., 1902. The gross anatomy of a 12-mm. pig. Am. Jour. Anat., Vol. 2, pp. 211–226.

Lewis, W. H., and Gregory, P. W., 1929. Cinematographs of living developing rabbit-eggs. Science, Vol. 69, pp. 226–229.

Lewis, W. H., and Hartman, C. G., 1933. Early cleavage stages of the egg of the monkey (Macacus rhesus). Carnegie Inst., Contrib. to Embryol., Vol. 24, pp. 187–201.

Lewis, W. H., and Wright, E. S., 1935. On the early development of the mouse egg. Carnegie Inst., Contrib. to Embryol., Vol. 25, pp. 113–144.

Macdonald, E., and Long, J. A., 1934. Some features of cleavage in the living egg of the rat. Am. Jour. Anat., Vol. 55, pp. 343–361.

Nicholas, J. S., and Hall, B. V., 1942. Experiments on developing rats. II. The development of isolated blastomeres and fused eggs. Jour. Exp. Zoöl., Vol. 90, pp. 441–459.

Patten, B. M., and Philpott, R., 1921. The shrinkage of embryos in the processes preparatory to sectioning. Anat. Rec., Vol. 20, pp. 393–413.

Pincus, G., 1939. The comparative behavior of mammalian eggs in vivo and in vitro. IV. The development of fertilized and artificially activated rabbit eggs. Jour. Exp. Zoöl., Vol. 82, pp. 85–131.

Robinson, A., 1892. Observations upon the development of the segmentation cavity, the archenteron, the germinal layers, and the amnion in mammals. Quart. Jour. Micr. Sci., Vol. 33, pp. 369–455.

Streeter, G. L., 1926. Development of the mesoblast and notochord in pig embryos. Carnegie Inst., Contrib. to Embryol., Vol. 19, pp. 73–92.

Thyng, F. W., 1911. The anatomy of a 7.8-mm. pig embryo. Anat. Rec., Vol. 5, pp. 17–45.

Waldo, C. M., and Wimsatt, W. A., 1945. The effect of colchicine on early cleavage of mouse ova. Anat. Rec., Vol. 93, pp. 363–375.

Wallin, E., 1917. A teaching model of a 10-mm. pig embryo. Anat. Rec., Vol. 13, pp. 295–297.

Weysse, A. W., 1894. On the blastodermic vesicle of Sus scrofa domesticus. Proc. Am. Acad. Arts & Sciences, Vol. 30, pp. 283–321.

Williams, L. W., 1908. The later development of the notochord in mammals. Am. Jour. Anat., Vol. 8, pp. 251–284.

Fetal Membranes and the Relations of the Embryos to the Uterus

Abromavich, C. E., 1926. The morphology and distribution of the rosettes on the foetal placenta of the pig. Anat. Rec., Vol. 33, pp. 69–72.

Assheton, R., 1906. The morphology of the ungulate placenta. Phil. Trans. Roy. Soc. London, Ser. B, Vol. 198, pp. 143–220.

Brambel, C. E., Jr., 1933. Allantochorionic differentiations of the pig studied morphologically and histochemically. Am. Jour. Anat., Vol. 52, pp. 397–459.

Corner, G. W., 1921. Internal migration of the ovum. Bull. Johns Hopkins Hosp., Vol. 32, pp. 78–83.

Fawcett, D. W., Wislocki, G. B., and Waldo, C. M., 1947. The development of mouse ova in the anterior chamber of the eye and in the abdominal cavity. Am. Jour. Anat., Vol. 81, pp. 413–443.

Gellhorn, A., Flexner, L. B., and Pohl, H. A., 1941. The transfer of radioactive sodium across the placenta of the sow. Jour. Cell. & Comp. Physiol., Vol. 18, pp. 393–400.

Goldstein, S. R., 1926. A note on the vascular relations and areolae in the placenta of the pig. Anat. Rec., Vol. 34, pp. 25–35.

Grosser, O., 1909. Vergleichende Anatomie und Entwicklungsgeschichte der Eihäute und der Placenta. Braumüller, Leipzig, xi & 314 S.

Grosser, O., 1927. Frühentwicklung, Eihautbildung und Placentation des Menschen und der Säugetiere. Verlag von J. F. Bergmann, München, viii & 454 pp.

Heuser, C. H., 1927. A study of the implantation of the ovum of the pig from the stage of the bilaminar blastocyst to the completion of the fetal membranes. Carnegie Inst., Contrib. to Embryol., Vol. 19, pp. 229–243.

Hubrecht, A. A. W., 1895. Die Phylogenese des Amnions und die Bedeutung des Trophoblastes. Verh. Koniklijke Akad. van Wetenschappen te Amsterdam, für 1895, S. 3–66.

Lell, W. A., Liber, K. E., and Snyder, F. F., 1931. Quantitative study of placental transmission and permeability of fetal membranes at various stages of pregnancy. Am. Jour. Physiol., Vol. 100, pp. 21–31.

Mossman, H. W., and Noer, H. R., 1947. A study of the amnion with the electron microscope. Anat. Rec., Vol. 97, pp. 253–257.

Noer, H. R., 1946. A study of the effect of flow direction on placental transmission, using artificial placentas. Anat. Rec., Vol. 96, pp. 383–389.

Robinson, A., 1904. Lectures on the early stages in the development of mammalian ova and on the formation of the placenta in different groups of mammals. Jour. Anat. & Physiol., Vol. 38, pp. 186–204.

Runner, M. N., 1947. Development of mouse eggs in the anterior chamber of the eye. Anat. Rec., Vol. 98, pp. 1–17.
Warwick, B. L., 1926. Intra-uterine migration of ova in the sow. Anat. Rec., Vol. 33, pp. 29–33.
Wislocki, G. B., 1929. On the placentation of primates, with a consideration of the phylogeny of the placenta. Carnegie Inst., Contrib. to Embryol., Vol. 20, pp. 51–80.
Wislocki, G. B., 1935. On the volume of the fetal fluids in sow and cat. Anat. Rec., Vol. 63, pp. 183–192.
Wislocki, G. B., and Dempsey, E. W., 1946. Histochemical reactions of the placenta of the pig. Am. Jour. Anat., Vol. 78, pp. 181–225.

The Nervous System and Sense Organs

Assheton, R., 1892. On the development of the optic nerve of vertebrates and the choroidal fissure of embryonic life. Quart. Jour. Micr. Sci., Vol. 34, pp. 85–103.
Bardeen, C. R., 1903. The growth and histogenesis of the cerebrospinal nerves in mammals. Am. Jour. Anat., Vol. 2, pp. 231–257.
Barnes, W., 1883–84. On the development of the posterior fissure of the spinal cord and the reduction of the central canal in the pig. Proc. Am. Acad. Arts. & Sciences, Vol. 19, pp. 97–110.
Bedford, E. A., 1904. The early history of the olfactory nerve in swine. Jour. Comp. Neur., Vol. 14, pp. 390–410.
Bradley, O. C., 1904. Neuromeres of the rhombencephalon of the pig. Review of Neurology & Psychiatry, Vol. 2, pp. 625–635.
Coghill, G. E., 1926. The growth of functional neurones and its relation to the development of behavior. Proc. Am. Philosoph. Soc., Vol. 65, pp. 51–55.
Detwiler, S. R., 1936. Neuroembryology. An Experimental Study. Macmillan, New York, x & 218 pp.
Dowd, L. W., 1929. The development of the dentate nucleus in the pig. Jour. Comp. Neur., Vol. 48, pp. 471–498.
Gradenigo, G., 1887. Die embryonale Anlage des Mittelohrs; die morphologische Bedeutung der Gehörknöchelchen. Mitth. aus dem Embryolog. Inst. d. Universität. Wien, 1887, S. 85–232.
Hardesty, I., 1904. On the development and nature of the neuroglia. Am. Jour. Anat., Vol. 3, pp. 229–268.
Hardesty, I., 1905. On the occurrence of the sheath cells and the nature of the axone sheaths in the central nervous system. Am. Jour. Anat., Vol. 4, pp. 329–354.
Held, H., 1909. Die Entwicklung des Nervengewebes bei den Wirbeltieren. J. A. Barth, Leipzig, ix & 378 S.
Herrick, C. J., 1893. The development of the medullated nerve fibres. Jour. Comp. Neur., Vol. 3, pp. 11–16.
Herrick, C. J., 1909. The criteria of homology in the peripheral nervous system. Jour. Comp. Neur., Vol. 19, pp. 203–209.
Herrick, C. J., 1925. Morphogenetic factors in the differentiation of the nervous system. Physiol. Reviews, Vol. 5, pp. 112–130.

Heuser, C. H., 1913. The development of the cerebral ventricles in the pig. Am. Jour. Anat., Vol. 15, pp. 215–252.

Hoskins, E. R., 1914. On the vascularization of the spinal cord of the pig. Anat. Rec., Vol. 8, pp. 371–391.

Johnston, J. B., 1909. The morphology of the fore-brain vesicle in vertebrates. Jour. Comp. Neur., Vol. 19, pp. 457–539.

Kallius, E., 1894. Untersuchungen über die Netzhaut die Säugethiere. Anat. Hefte, Bd. 3, S. 527–582.

Kappers, C. U. A., und Fortuyn, A. B. D., 1921. Vergleichende Anatomie des Nervensystems. (Theil II. Des Kleinhirns, Des Mittle- und Zwischenhirns und des Vorderhirns. S. 626–1329.) Bohn, Haarlem.

Kastschenko, N., 1887. Das Schicksal der embryonalen Schlundspalten bei Säugetieren. Arch. f. mikr. Anat., Bd. 30, S. 1–26.

Kessler, L., 1877. Zur Entwicklung des Auges der Wirbelthiere. Vogel, Leipzig 112 S.

Kingsbury, B. F., 1922. The fundamental plan of the vertebrate brain. Jour. Comp. Neur., Vol. 34, pp. 461–491.

Kingsbury, B. F., and Adelmann, H. B., 1924. The morphological plan of the head. Quart. Jour. Micr. Sci., Vol. 68, pp. 239–285.

Köllicker, A., 1904. Die Entwicklung und Bedeutung des Glaskörpers. Zeitschr. f. wiss. Zoöl., Bd. 76, S. 1–25.

Krausse, R., 1890. Entwicklungsgeschichte der häutigen Bogengänge. Arch. f. mikr. Anat., Bd. 35, S. 287–304.

Kuntz, A., 1909. A contribution to the histogenesis of the sympathetic nervous system. Anat. Rec., Vol. 3, pp. 458–465.

Kuntz, A., 1922. Experimental studies on the histogenesis of the sympathetic nervous system. Jour. Comp. Neur., Vol. 34, pp. 1–36.

Kupffer, K. v., 1905. Die Morphogenie des Centralnervensystems. Hertwig's Handbuch, Bd. 2, Teil 3, Kap. VIII, S. 1–394.

Locy, W. A., 1895. Contribution to the structure and development of the vertebrate head. Jour. Morph., Vol. 11, pp. 497–594.

Paterson, A. M., 1891. Development of the sympathetic nervous system in mammals. Phil. Trans. Roy. Soc. London, Ser. B., Vol. 181, pp. 159–186.

Prentiss, C. W., 1910. The development of the hypoglossal ganglia of pig embryos. Jour. Comp. Neur., Vol. 20, pp. 265–282.

Prentiss, C. W., 1913. On the development of the membrana tectoria with reference to its structure and attachments. Am. Jour. Anat., Vol. 14, pp. 425–460.

Rabl, C., 1899. Über den Bau und die Entwicklung der Linse. Theil III. Die Linse der Säugethiere. Rückblick und Schluss. Zeitschr. f. wiss. Zoöl., Bd. 67, S. 1–138.

Rabl, C., 1900. Über den Bau und die Entwicklung der Linse. Wilhelm Englemann, Leipzig, vi & 324 S.

Retzius, G., 1881–1884. Das Gehörorgan der Wirbelthiere. Theil II. Das Gehörorgan der Reptilien, der Vögel und der Säugethiere. Stockholm, viii & 368 S.

Sauer, F. C., 1935a. Mitosis in the neural tube. Jour. Comp. Neur., Vol. 62, pp. 377–405.

Sauer, F. C., 1935b. The cellular structure of the neural tube. Jour. Comp. Neur., Vol. 63, pp. 13–23.

Sauer, F. C., 1939. Development of beta crystallin in the pig and prenatal weight of the lens. Growth, Vol. 3, pp. 381–386.

Seefelder, R., 1910. Beiträge zur Histogenese und Histologie der Netzhaut, des Pigmentepithels und des Sehnerven. Arch. f. Ophthal., Bd. 73, S. 419–537.

Shambaugh, G. E., 1907. A restudy of the minute anatomy of structures in the cochlea, with conclusions bearing on the solution of the problem of tone perception. Am. Jour. Anat., Vol. 7, pp. 245–257.

Shambaugh, G. E., 1926. The development of the membranous labyrinth. Arch. Otolaryng., Vol. 3, pp. 233–236.

Shaner, R. F., 1932. The development of the nuclei and tracts of the midbrain. Jour. Comp. Neur., Vol. 55, pp. 493–511.

Shaner, R. F., 1934a. The development of a medial motor nucleus and an accessory abducens nucleus in the pig. Jour. Anat., Vol. 68, pp. 314–317.

Shaner, R. F., 1934b. The development of the nuclei and tracts related to the acoustic nerve in the pig. Jour. Comp. Neur., Vol. 60, pp. 5–19.

Takahashi, K., 1931. Pri la genezo de la papilo de nerve optica ce Sus scrofa domesticus. Folia Anatomica Japonica, Bd. 9, S. 149–167.

Tuckerman, F., 1888. Note on the papilla foliata and other taste areas of the pig. Anat. Anz., Bd. 3, S. 69–73.

Tuttle, A. H., 1884. The relation of the external meatus, tympanum and Eustachian tube to the first visceral cleft. Proc. Am. Acad. Arts & Sciences, Vol. 19, pp. 111–132.

Van Campenhout, E., 1935. Origine du ganglion acoustique chez le porc. Arch. Biol., T. 46, pp. 273–286.

Waterman, A. J., 1938. The development of the inner ear rudiment of the rabbit embryo in a foreign environment. Am. Jour. Anat., Vol. 63, pp. 161–219.

Weed, L. W., 1917. The development of the cerebrospinal spaces in pig and in man. Carnegie Inst., Contrib. to Embryol., Vol. 5, No. 14, 116 pp.

Windle, W. F., Fish, M. W., and O'Donnell, J. E., 1934. Myelogeny of the cat as related to development of fiber tracts and prenatal behavior patterns. Jour. Comp. Neur., Vol. 59, pp. 139–165.

The Circulatory System

Barclay, A. E., Franklin, K. J., and Prichard, M. M. L., 1944. The Foetal Circulation and Cardiovascular System, and the Changes That They Undergo at Birth. Blackwell Scientific Publications, Ltd., Oxford, xvi & 275 pp.

Barcroft, J., 1946. Researches on Pre-natal Life. Blackwell Scientific Publications, Ltd., Oxford, Vol. I, xiii & 292 pp.

Barry, A., 1942. The intrinsic pulsation rates of fragment of the embryonic chick heart. Jour. Exp. Zoöl., Vol. 91, pp. 119–130.

Begg, A. S., 1920. Absence of the vena cava inferior in a 12-mm. pig embryo, associated with the drainage of the portal system into the cardinal system. Am. Jour. Anat., Vol. 27, pp. 395–403.

Boas, J. E. V., 1887. Ueber die Arterienbogen der Wirbelthiere. Morph. Jahrb., Bd. 13, S. 115–118.

Born, G., 1888. Über die Bildung der Klappan, Ostien und Scheidewände im Säugetierherzen. Anat. Anz., Bd. 3, S. 606–612.

Born, G., 1889. Beiträge zur Entwicklungsgeschichte des Säugethierherzens. Arch. f. mikr. Anat., Bd. 33, S. 284–377.

Bremer, J. L., 1902. I. The origin of the pulmonary arteries in mammals. Am. Jour. Anat., Vol. I, pp. 137–144.

Bremer, J. L., 1909. II. On the origin of the pulmonary arteries in mammals. Anat. Rec., Vol. 3, pp. 334–340.

Butler, E. G., 1927. The relative rôle played by the embryonic veins in the development of the mammalian vena cava posterior. Am. Jour. Anat., Vol. 39, pp. 267–353.

Congdon, E. D., and Wang, H. W., 1926. The mechanical processes concerned in the formation of the differing types of aortic arches of the chick and the pig and in the divergent early development of their pulmonary arches. Am. Jour. Anat., Vol. 37, pp. 499–520.

Davis, D. M., 1910. Studies on the chief veins in early pig embryos and the origin of the vena cava inferior. Am. Jour. Anat., Vol. 10. pp. 461–472.

Doan, C. A., Cunningham, R. S., and Sabin, F. R., 1925. Experimental studies on the origin and maturation of avian and mammalian red blood-cells. Carnegie Inst., Contrib. to Embryol., Vol. 16, pp. 163–226.

Emmel, V. E., 1914. Concerning certain cytological characteristics of the erythroblasts in the pig embryo and the origin of non-nucleated erythrocytes by a process of cytoplasmic constriction. Am. Jour. Anat., Vol. 16, pp. 127–206.

Evans, H. M., 1909. On the development of the aortae, cardinal and umbilical veins and other blood-vessels of vertebrate embryos from capillaries. Anat. Rec., Vol. 3, pp. 498–518.

Flint, J. M., 1903. The angiology, angiogenesis, and organogenesis of the submaxillary gland. Am. Jour. Anat., Vol. 2, pp. 417–444.

Frazer, J. E., 1917. Formation of Pars Membranacea Septi. Jour. Anat. & Physiol., Vol. 51, pp. 19–29.

Goss, C. M., 1935. Double hearts produced experimentally in rat embryos. Jour. Exp. Zool., Vol. 72, pp. 33–49.

Goss, C. M., 1938. The first contractions of the heart in rat embryos. Anat. Rec., Vol. 70, pp. 505–524.

Goss, C. M., 1942. The physiology of the embryonic mammalian heart before circulation. Am. Jour. Physiol., Vol. 137, pp. 146–152.

Gregg, R. E., 1946. An arterial anomaly in the fetal pig. Anat. Rec., Vol. 95, pp. 53–65.

Henser, C. H., 1923. The branchial vessels and their derivatives in the pig. Carnegie Inst., Contrib. to Embryol., Vol. 15, pp. 121–139.

Hill, E. C., 1907. On the gross development and vascularization of the testis. Am. Jour. Anat., Vol. 6, pp. 439–459.

His, W., 1900. Lecithoblast und Angioblast der Wirbeltiere. Abhandl. der Math. phys. Klasse der Königl. Säch. Gesellsch. d. Wissenschaften, Bd. 26, S. 173–328.

Hochstetter, F., 1906. Die Entwicklung des Blutgefässsystems. Hertwig's Handbuch, Bd. 3, Teil 2, S. 21–166.

Hofmann, L. v., 1914. Die Entwicklung der Kopfarterien bei Sus scrofa domesticus. Morph. Jahrb., Bd. 48, S. 645–671.

Hogue, M. J., 1937. Studies of heart muscle in tissue cultures. Anat. Rec., Vol. 67, pp. 521–535.

Jordan, H. E., 1916. The microscopic structure of the yolk-sac of the pig embryo with especial reference to the origin of erythrocytes. Am. Jour. Anat., Vol. 19, pp. 277–303.

Jordan, H. E., 1919a. The histogenesis of blood-platelets in the yolk-sac of the pig embryo. Anat. Rec., Vol. 15, pp. 391–406.

Jordan, H. E., 1919b. The histology of the umbilical cord of the pig, with special reference to the vasculogenic and hemopoietic activity of its extensively vascularized connective tissue. Am. Jour. Anat., Vol. 26, pp. 1–27.

Kellogg, H. B., 1928. The course of the blood flow through the fetal mammalian heart. Am. Jour. Anat., Vol. 42, pp. 443–465.

Kellogg, H. B., 1929. Studies on the fetal circulation of mammals. Am. Jour. Physiol., Vol. 91, pp. 637–648.

Kimball, P., 1928. A comparative study of the vas subintestinale in the vertebrates. Am. Jour. Anat., Vol. 42, pp. 371–398.

Kramer, T. C., 1942. The partitioning of the truncus and conus and the formation of the membranous portion of the interventricular septum in the human heart. Am. Jour. Anat., Vol. 71, pp. 343–370.

Lehmann, H., 1905. On the embryonic history of the aortic arches in mammals. Anat. Anz., Bd. 26, S. 406–424.

Lewis, F. T., 1902. The development of the vena cava inferior. Am. Jour. Anat., Vol. 1, pp. 229–244.

Lewis, F. T., 1906. The fifth and sixth aortic arches and the related pharyngeal pouches in the rabbit and pig. Anat. Anz., Bd. 28, S. 506–513.

Lockwood, C. B., 1888. The early development of the pericardium, diaphragm and great veins. Phil. Trans. Roy. Soc. London, Ser. B, Vol. 179, pp. 365–384.

McClendon, J. F., 1913. Preparation of material for histology and embryology, with an appendix on the arteries and veins in a thirty-millimeter pig embryo. Anat. Rec., Vol. 7, pp. 51–61.

McClure, C. F. W., 1921. The endothelial problem. Anat. Rec., Vol. 22, pp. 219–237.

McClure, C. F. W., and Huntington, G. S., 1929. The mammalian vena cava posterior. Am. Anat. Mem., No. 15, 56 pp. and 46 plates.

Minot, C. S., 1898. On the veins of the Wolffian bodies in the pig. Proc. Boston Soc. Nat. Hist., Vol. 28, pp. 265–274.

Morrill, C. V., 1916. On the development of the atrial septum and the valvular apparatus in the right atrium of the pig embryo with a note on the fenestration on the anterior cardinal veins. Am. Jour. Anat., Vol. 20, pp. 351–374.

Odgers, P. N. B., 1938. The development of the pars membranacea septi in the human heart. Jour. Anat., Vol. 72, pp. 247–259.

Paff, G. H., 1936. Transplantation of sino-atrium to conus in the embryonic heart in vitro. Am. Jour. Physiol., Vol. 117, pp. 313–317.

Parker, G. H., and Tozier, C. H., 1898. The thoracic derivatives of the posterior cardinal veins of swine. Bull. Museum of Comp. Zoöl., Harvard, Vol. 31, pp. 133–144.

Patten, B. M., 1930. The changes in circulation following birth. Am. Heart Journal, Vol. 6, pp. 192–205.

Patten, B. M., 1931. The closure of the foramen ovale. Am. Jour. Anat., Vol. 48, pp. 19–44.

Patten, B. M., 1939. Microcinematographic and electrocardiographic studies of the first heart beats and the beginning of the circulation in living embryos. Proc. Inst. of Med. of Chicago, Vol. 12, pp. 366–380.

Patten, B. M., and Kramer, T. C., 1933. The initiation of contraction in the embryonic chick heart. Am. Jour. Anat., Vol. 53, pp. 349–375.

Patten, B. M., Sommerfield, W. A., and Paff, G. H., 1929. Functional limitations of the foramen ovale in the human foetal heart. Anat. Rec., Vol. 44, pp. 165–178.

Pohlman, A. G., 1909. The course of the blood through the heart of the fetal mammal, with a note on the reptilian and amphibian circulations. Anat. Rec., Vol. 3. pp. 75–109.

Rathke, H., 1843. Ueber die Entwicklung der Arterien welche bei den Säugethieren von dem Bogen der Aorta ausgehen. Arch. f. Anat. u. Physiol., Jg. 1843, S. 276–302.

Reagan, F., 1912. The fifth aortic arch of mammalian embryos and the nature of the last pharyngeal evagination. Am. Jour. Anat., Vol. 12, pp. 493–514.

Reagan, F. P., 1917. Experimental studies on the origin of vascular endothelium and of erythrocytes. Am. Jour. Anat., Vol. 21, pp. 39–175.

Reagan, F. P., 1919. On the later development of the azygos veins of swine. Anat. Rec., Vol. 17, pp. 111–126.

Reagan, F. P., 1927. The supposed homology of vena azygos and vena cava inferior considered in the light of new facts concerning their development. Anat. Rec., Vol. 35, pp. 129–148.

Reagan, F. P., 1929. A century of study upon the development of the eutherian vena cava inferior. The Quarterly Review of Biology, Vol. 4, pp. 179–212.

Reinke, E. E., 1910. Note on the presence of the fifth aortic arch in a 6-mm. pig embryo. Anat. Rec., Vol. 4, pp. 453–459.

Robinson, A., 1903. The early stages of the development of the pericardium. Jour. Anat. & Physiol., Vol. 37, pp. 1–17.

Sabin, F. R., 1915. On the fate of the posterior cardinal veins and their relation to the development of the vena cava and azygos in pig embryos. Carnegie Inst., Contrib. to Embryol., Vol. 3, pp. 5–32.

Sabin, F. R., 1917. Origin and development of the primitive vessels of the chick and the pig. Carnegie Inst., Contrib. to Embryol., Vol. 6, pp. 61–124.

Schaeffer, J. P., 1914. The behavior of elastic tissue in the post-fetal occlusion and obliteration of the ductus arteriosus (Botalli) in Sus scrofa. Jour. Exp. Med., Vol. 19, pp. 129–143.

Shaner, R. F., 1928. The development of the muscular architecture of the ventricles of the pig's heart, with a review of the adult heart and a note on two abnormal mammalian hearts. Anat. Rec., Vol. 39, pp. 1–36.

Shaner, R. F., 1929. The development of the atrioventricular node, bundle of His, and sino-atrial node in the calf, with a description of a third embryonic node-like structure. Anat. Rec., Vol. 44, pp. 85–99.

Smith, H. W., 1909. On the development of the superficial veins of the body wall in the pig. Am. Jour. Anat., Vol. 9, pp. 439–462.

Stienon, L., 1926. Recherches sur l'origine du noeud sinusal dans le coeur des mammifères. Archives de Biologie, T. 36, pp. 523–539.

Stockard, C. R., 1915. An experimental analysis of the origin of blood and vascular endothelium. Amer. Anat. Mem., No. 7, 174 pp.

Thienes, C. H., 1925. Venous system associated with the liver of a 6-mm. pig embryo. Anat. Rec., Vol. 31, pp. 149–158.

Whitehead, W. H., 1942. A working model of the crossing caval blood streams in the fetal right atrium. Anat. Rec., Vol. 82, pp. 277–280.

Windle, W. F., 1940. Circulation of blood through the fetal heart and lungs and changes occurring with respiration at birth. Quart. Bull., Northwestern Univ. Med. School, Vol. 14, pp. 31–36.

Witte, L., 1919. Histogenesis of the heart muscle of the pig in relation to the appearance and development of the intercalated discs. Am. Jour. Anat., Vol. 25, pp. 333–347.

Woollard, H. L., 1922. The development of the principal arterial stems in the forelimb of the pig. Carnegie Inst., Contrib. to Embryol., Vol. 14, pp. 139–154.

Yoshida, T., 1932. On the development of the heart primordia. II. Observations upon the development of the interatrial septum of the pig. Okayama-Igakkai-Zasshi, Vol. 44, pp. 438–460.

Young, A. H., and Robinson, A., 1898. The development and morphology of the vascular system in mammals. The posterior end of the aortae and the iliac arteries. Jour. Anat. & Physiol., Vol. 32, pp. 605–607.

Lymphatic Vessels and Organs

Badertscher, J. A., 1915. The development of the thymus in the pig. I. Morphogenesis. Am. Jour. Anat., Vol. 17, pp. 317–338. II. Histogenesis. Am. Jour. Anat., Vol. 17, pp. 437–494.

Baetjer, W. A., 1908. The origin of the mesenteric sac and thoracic duct in the embryo pig. Am. Jour. Anat., Vol. 8, pp. 303–310.

Bell, E. T., 1905. The development of the thymus. Am. Jour. Anat., Vol. 5, pp. 29–61.

Cash, J. R., 1917. On the development of the lymphatics in the heart of the embryo pig. Anat. Rec., Vol. 13, pp. 451–464.

Cash, J. R., 1921. On the development of the lymphatics in the stomach of the embryo pig. Carnegie Inst., Contrib. to Embryol., Vol. 13, pp. 1–15.

Clark, A. H., 1912. On the fate of the jugular lymph sacs and the development of the lymph channels in the neck of the pig. Am. Jour. Anat., Vol. 14, pp. 47–62.

Cunningham, R. S., 1916. On the development of the lymphatics of the lung in the embryo pig. Carnegie Inst., Contrib. to Embryol., Vol. 4, pp. 45–68.

Fischelis, P., 1885. Beiträge zur Kenntnis der Entwicklungsgeschichte der Gl. thyreoidea und Gl. thymus. Arch. f. mikr. Anat., Bd. 25, S. 405–440.

Fox, H., 1908. The pharyngeal pouches and their derivatives in the mammalia. Am. Jour. Anat., Vol. 8, pp. 187–250.

Heuer, G. J., 1909. The development of the lymphatics in the small intestine of the pig. Am. Jour. Anat., Vol. 9, pp. 93–118.

Holyoke, E. A., 1936. The role of the primitive mesothelium in the development of the mammalian spleen. Anat. Rec., Vol. 65, pp. 333–349.

Huntington, G. S., 1910a. The phylogenetic relations of the lymphatic and blood vascular systems in the vertebrates. Anat. Rec., Vol. 4, pp. 1–14.

Huntingon, G. S., 1910b. The genetic principles of the development of the systemic lymphatic vessels in the mammalian embryo. Anat. Rec., Vol. 4, pp. 399–424.

Kampmeier, O. F., 1912. The development of the thoracic duct in the pig. Am. Jour. Anat., Vol. 13, pp. 401–476.

Kastschenko, N., 1887. Das Schicksal der embryonalen Schlundspalten bei Säugetieren. Arch. f. mikr. Anat., Bd. 30, S. 1–26.

Levin, P. M., 1930. The development of the tonsil of the domestic pig. Anat. Rec., Vol. 45, pp. 189–201.

McClure, C. F. W., 1915a. On the provisional arrangement of the embryonic lymphatic system. An arrangement by means of which a centripetal flow of lymph toward the venous circulation is controlled and regulated in an orderly and regular manner, from the time lymph begins to collect in the intercellular spaces until it is forwarded to the venous circulation. Anat. Rec., Vol. 9, pp. 281–296.

McClure, C. F. W., 1915b. The development of the lymphatic system in the light of the more recent investigations in the field of vasculogenesis. Anat. Rec., Vol. 9, pp. 563–579.

Rand, R., 1917. On the relation of the head chorda to the pharyngeal epithelium in the pig embryo: a contribution to the development of the bursa pharyngea and the tonsilla pharyngea. Anat. Rec., Vol. 13, pp. 465–491.

Reichert, F. L., 1921. On the fate of the primary lymph-sacs in the abdominal region of the pig, and the development of the lymph-channels in the abdominal and pelvic regions. Carnegie Inst., Contrib. to Embryol., Vol. 13, pp. 17–39.

Sabin, F. R., 1902. On the origin of the lymphatic system from the veins and the development of the lymph hearts and thoracic duct in the pig. Am. Jour. Anat., Vol. 1, pp. 367–389.

Sabin, F. R., 1904. On the development of the superficial lymphatics in the skin of the pig. Am. Jour. Anat., Vol. 3, pp. 183–195.

Sabin, F. R., 1905. The development of the lymphatic nodes in the pig and their relation to the lymph hearts. Am. Jour. Anat., Vol. 4, pp. 355–389.

Sabin, F. R., 1912. On the origin of the abdominal lymphatics in mammals from the vena cava and the renal veins. Anat. Rec., Vol. 6, pp. 335–342.

Sabin, F. R., 1916. The origin and development of the lymphatic system. Johns Hopkins Hosp. Rep., Vol. 17, pp. 347–440.

Theil, G. A., and Downey, H., 1921. The development of the mammalian spleen, with special reference to its hematopoietic activity. Am. Jour. Anat., Vol. 28, pp. 279–339.

Zottermann, A., 1911. Die Schweinthymus als eine Thymus ectoentodermalis. Anat. Anz., Bd. 38, S. 514–530.

The Digestive and Respiratory Systems, and the Body Cavities and Mesenteries

Boyden, E. A., 1926. The accessory gall bladder. An embryological and comparative study of aberrant biliary vesicles occurring in man and the domestic animals. Am. Jour. Anat., Vol. 38, pp. 177–231.

Bremer, J. L., 1932. Accessory bronchi in embryos; their occurrence and probable fate. Anat. Rec., Vol. 54, pp. 361–374.

Chamberlain, R. V., 1909. On the mode of disappearance of the villi from the colon of mammals. Anat. Rec., Vol. 3, pp. 282–283.

Clements, L. P., 1938. Embryonic development of the respiratory portion of the pig's lung. Anat. Rec., Vol. 70, pp. 575–595.

Corner, G. W., 1914. The structural unit and growth of the pancreas of the pig. Am. Jour. Anat., Vol. 16, pp. 207–236.

Felix, W., 1892. Zur Leber- und Pankreasentwicklung. Arch. f. Anat. u. Phys., Anat. Abt., f. 1892, S. 281–323.

Flint, J. M., 1902. The development of the reticulated basement membranes in the submaxillary gland. Am. Jour. Anat., Vol. 2, pp. 1–11.

Flint, J. M., 1903. The angiology, angiogenesis, and organogenesis of the submaxillary gland. Am. Jour. Anat., Vol. 2, pp. 417–444.

Flint, J. M., 1906. The development of the lungs. Am. Jour. Anat., Vol. 6, pp. 1–138.

Ham, A. W., and Baldwin, K. W., 1941. A histological study of the development of the lung with particular reference to the nature of alveoli. Anat. Rec., Vol. 81, pp. 363–379.

Hammar, G. A., 1893. Einige Plattenmodelle zur Beleuchtung der früheren embryonalen Leberentwickelung. Arch. f. Anat. u. Phys., Anat. Abt., f. 1893, S. 123–156.

Hammar, G. A., 1897a. Ueber einige Hauptzüge der ersten embryonalen Leberentwickelung. Anat. Anz., Bd. 13, S. 233–247.

Hammar, G. A., 1897b. Einiges üeber die Duplicität der vertralen Pancreasanlage. Anat. Anz., Bd. 13, S. 247–249.

Herzfeld, P., 1889. Ueber das Jacobson'sche Organ des Menschen und der Säugethiere. Zoölog. Jahrb. (Abt. f. Anat. u. Ontogenie), Bd. 3, S. 551–574.

Huber, G. C., 1912. On the relation of the chorda dorsalis to the anlage of the pharyngeal bursa or median pharyngeal recess. Anat. Rec., Vol. 6, pp. 373–404.

Johnson, F. P., 1919. The development of the lobule of the pig's liver. Am. Jour. Anat., Vol. 25, pp. 299–331.

Lewis, F. T., 1911. The bi-lobed form of the ventral pancreas in mammals. Am. Jour. Anat., Vol. 12, pp. 389–400.

Lewis, F. T., and Thyng, F. W., 1908. The regular occurrence of intestinal diverticula in embryos of the pig, rabbit and man. Am. Jour. Anat., Vol. 7, pp. 505–519.

Lineback, P. E., 1916. The development of the spiral coil in the large intestine of the pig. Am. Jour. Anat., Vol. 20, pp. 483–503.

Mall, F. P., 1891. Development of the lesser peritoneal cavity in birds and mammals. Jour. Morph., Vol. 5, pp. 165–179.

Mall, F. P., 1906. A study of the structural unit of the liver. Am. Jour. Anat., Vol. 5, pp. 227–308.

Max, C., 1931. Das Wachstum der Leberzellen und die Entwicklung der Leberläppchen beim Schweine. Anat. Anz., Bd. 72, S. 219–227.

Ravn, E., 1889. Ueber die Bildung der Scheidewand zwishen Brust- und Bauchhöhle in Säugetier-embryonen. Arch. f. Anat. u. Phys., Anat. Abt., S. 123–154.

Ravn, E., 1899. Ueber die Entwicklung des Septum transversum. Anat. Anz., Bd. 15, S. 528–534.

Reuter, C., 1896. Ueber die Entwicklung der Kaumusculatur beim Schwein. Anat. Hefte f. 1896–97, S. 239–262.

Robinson, A., 1903. The early stages of the development of the pericardium. Jour. Anat. & Physiol., Vol. 37, pp. 1–17.

Swaen, A., 1896 and 1897. Recherches sur le développement du foie, du tube digestive, de l'arrière cavité, du péritoine et du mésentère. Jour. de l'Anat. et de la Physiol., Part I, T. 32, pp. 1–84; Part II, 1, T. 33, pp. 32–99; Part II, 2, T. 33, pp. 222–258; Part II, 3, T. 33, pp. 525–585.

Thyng, F. W., 1908. Models of the pancreas in embryos of the pig, rabbit, cat and man. Am. Jour. Anat., Vol. 7, pp. 489–503.

Uskow, N., 1883. Ueber die Entwickelung des Zwerchfells, des Pericardiums und des Coeloms. Arch. f. mikr. Anat., Bd. 22, S. 143–218.

Whitehead, W. H., Windle, W. F., and Becker, R. F., 1942. Changes in lung structure during aspiration of amniotic fluid and during air-breathing at birth. Anat. Rec., Vol. 83, pp. 255–265.

Windle, W. F., Becker, R. F., Barth, E. E., and Schulz, M. D., 1939. Proof of fetal swallowing, gastrointestinal peristalsis and defecation in amnio. Proc. Am. Physiol. Soc.—Am. Jour. Physiol., Vol. 126, pp. P429–P430.

Wlassow, 1895. Zur Entwicklung des Pankreas beim Schwein. Morph. Arbeiten, Bd. 4, S. 67–76.

The Urogenital System

Allen, B. M., 1904. The embryonic development of the ovary and testis of the mammals. Am. Jour. Anat., Vol. 3, pp. 89–154.

Angle, E. J., 1918. Development of the Wolffian body in Sus scrofa domesticus. Trans. Am. Micr. Soc., Vol. 37, pp. 215–238.

Balfour, F. M., 1876. On the origin and history of the urogenital organs of vertebrates. Jour. Anat. & Physiol., Vol. 10, pp. 17–48.

Bascom, K. F., and Osterud, H. L., 1927. Quantitative studies of the testis. III. A numerical treatment of the development of the pig testis. Anat. Rec., Vol. 37, pp. 63–82.

Bremer, J. L., 1916. The interrelations of the mesonephros, kidney, and placenta in different classes of animals. Am. Jour. Anat., Vol. 19, pp. 179–209.

Brody, H., and Bailey, P. L., Jr., 1939. Unilateral renal agenesia in a fetal pig. Anat. Rec., Vol. 74, pp. 159–163.

Corner, G. W., 1920. A case of true lateral hermaphroditism in a pig with functional ovary. Carnegie Inst., Contrib. to Embryol., Vol. 11, pp. 137–142.

Felix, W., und Bühler, A., 1906. Die Entwicklung der Harn- und Geschlechts organe. Hertwig's Handbuch, Bd. 3, Teil I, K. II.

Flexner, L. B., and Gersh, I., 1937. The correlation of oxygen consumption, function and structure in the developing metanephros of the pig. Carnegie Inst., Contrib. to Embryol., Vol. 26, pp. 121–127.

Gersh, I., 1937. The correlation of structure and function in the developing mesonephros and metanephros. Carnegie Inst., Contrib. to Embryol., Vol. 26, pp. 33–58.

Gruenwald, P., 1942. The development of the sex cords in the gonads of man and mammals. Am. Jour. Anat., Vol. 70, pp. 359–397.

Hamburger, O., 1890. Ueber die Entwicklung der Säugethierniere. Arch. f. Anat. u. Phys., Anat. Abt., f. 1890, Suppl., S. 15–51.

Henneberg, B., 1922. Anatomie und Entwicklung der Äusseren Genitalorgane des Schweines und vergleichend-anatomische Bemerkungen. I. Weibliches Schwein. Zeitschr. f. Anat. u. Entwg., Bd. 63, S. 431–494.

Henneberg, B., 1925. Anatomie und Entwicklung der Äusseren Genital organe des Schweines und vergleichend-anatomische Bemerkungen. II. Männliches Schwein. Zeitschr. f. Anat. u. Entwg., Bd. 75, S. 265–318.
Henneberg, B., 1926. Beitrag zur ontogenetischen Entwicklung des Scrotums und der Labia maiora. Zeitschr. f. Anat. u. Entwg., Bd. 81, S. 198–219.
Hill, E. C., 1905. On the first appearance of the renal artery and the relative development of kidneys and Wolffian bodies in pig embryos. Bull. Johns Hopkins Hosp., Vol. 16, pp. 60–64.
Hill, E. C., 1906. On the gross development and vascularization of the testis. Am. Jour. Anat., Vol. 6, pp. 439–459.
Huber, G. C., 1905. On the development and shape of uriniferous tubules of certain of the higher mammals. Am. Jour. Anat., Supplement to Vol. 4, pp. 1–98.
Kambmeier, O. F., 1926. The metanephros or so-called permanent kidney in part provisional and vestigial. Anat. Rec., Vol. 33, pp. 115–120.
Kingsbury, B. F., 1913. The morphogenesis of the mammalian ovary: Felis domestica. Am. Jour. Anat., Vol. 15, pp. 345–387.
Kitahara, Y., 1923. Über die Entstehung der Zwischenzellen der Keimdrüsen des Menschen und der Säugetiere und über deren physiologische Bedeutung. Arch. f. Entwcklngsmechn. d. Organ., Bd. 52, S. 550–615.
Klaatsch, H., 1890. Ueber den Descensus testiculorum. Morph. Jahrb., Bd. 16, S. 587–646.
Lewis, F. T., 1920. The course of the Wolffian tubules in mammalian embryos. Am. Jour. Anat., Vol. 26, pp. 423–435.
Lockwood, C. B., 1888. Development and transition of the testis, normal and abnormal. Jour. Anat. & Physiol., Part I, Vol. 21, pp. 635–664. Part II, Vol. 22, pp. 38–77. Part III, Vol. 22, pp. 461–478. Part IV, Vol. 22, pp. 505–541.
MacCallum, J. B., 1902. Notes on the Wolffian body of higher mammals. Am. Jour. Anat., Vol. 1, pp. 245–260.
Price, D., 1947. An analysis of the factors influencing growth and development of the mammalian reproductive tract. Physiol. Zool., Vol. 20, pp. 213–247.
Schreiner, K. E., 1902. Ueber die Entwicklung der Amniotenniere. Zeitschr. f. wiss. Zoöl., Bd. 71, S. 1–188.
Selye, H., 1943. Factors influencing development of scrotum. Anat. Rec., Vol. 85, pp. 377–385.
Weinberg, E., 1929. A note on the origin and histogenesis of the mesonephric duct in mammals. Anat. Rec., Vol. 41, pp. 373–386.
Whitehead, R. H., 1904. The embryonic development of the interstitial cells of Leydig. Am. Jour. Anat., Vol. 3, pp. 167–182.

The Ductless Glands

Badertscher, J. A., 1918. The fate of the ultimobranchial bodies in the pig (Sus scrofa). Am. Jour. Anat., Vol. 23, pp. 89–131.
Born, G., 1883. Ueber die Derivate der embryonalen Schlundbogen und Schlundspalten bei Säugethieren. Arch. f. mikr. Anat., Bd. 22, S. 271–318.
Emmart, E. W., 1936. A study of the histogenesis of the thymus in vitro. Anat. Rec. Vol. 66, pp. 59–73.

Fischelis, P., 1885. Beiträge zur Kenntnis der Entwicklungsgeschichte der Gl. thyreoidea und Gl. thymus. Arch. f. mikr. Anat., Bd. 25, S. 405–440.

Flint, J. M., 1900. The bloodvessels, angiogenesis, reticulum and histology of the adrenal. Johns Hopkins Hosp. Rep., Vol. 9, pp. 153–229.

Fox, H., 1908. The pharyngeal pouches and their derivatives in the mammalia. Am. Jour. Anat., Vol. 8, pp. 187–250.

Gilbert, M. S., 1935. Some factors influencing the early development of the mammalian hypophysis. Anat. Rec., Vol. 62, pp. 337–359.

Godwin, M. C., 1939. The mammalian thymus. IV. The development in the dog. Am. Jour. Anat., Vol. 64, pp. 165–201.

Holt, E., 1921. Absence of the pars buccalis of the hypophysis in a 40-mm. pig. Anat. Rec., Vol. 22, pp. 207–215.

Jacoby, M., 1895. Studien zur Entwicklungsgeschichte der Halsorgane der Säugethiere und des Menschen. Histor. Krit. Beobachtungen über die Entwicklung der Kiemendarm-derivate. Gustav Schade, Berlin, 70 S.

Kastschenko, N., 1887. Das Schicksal der embryonalen Schlundspalten bei Säugethieren (zur Entwicklungsgeschichte des mittleren und äusseren Ohres, der Thyreoidea und des Thymus. Carotidenanlage). Arch. f. mikr. Anat., Bd. 30, S. 1–26.

Minervini, R., 1904. Des Capsules surrénales: Développement, structure, fonctions. Jour. de l'Anat. et de la Phys., T. 40, pp. 449–492 and pp. 634–667.

Mitsukuri, K., 1882. On the development of the suprarenal in mammalia. Quart. Jour. Micr. Sci., Vol. 22, pp. 17–29.

Moody, R. O., 1910. Some features of the histogenesis of the thyreoid gland in the pig. Anat. Rec., Vol. 4, pp. 429–452.

Nelson, W. O., 1933. Studies on the anterior hypophysis. I. The development of the hypophysis in the pig (Sus scrofa). II. The cytological differentiation in the anterior hypophysis of the foetal pig. Am. Jour. Anat., Vol. 52, pp. 307–332.

Poll, H., 1906. Die vergleichende Entwicklungsgeschichte der Nebennierensysteme der Wirbeltiere. Hertwig's Handbuch, Bd. 3, Teil 1, K. II, 2.

Rumph, P., and Smith, P. E., 1926. The first occurrence of secretory products and of a specific structural differentiation in the thyroid and anterior pituitary during the development of the pig foetus. Anat. Rec., Vol. 33, pp. 289–298.

Shanklin, W. M., 1944. Histogenesis of the pig neurohypophysis. Am. Jour. Anat., Vol. 74, pp. 327–353.

Smith, P. E., and Dortzbach, C., 1929. The first appearance in the anterior pituitary of the developing pig foetus of detectable amounts of the hormones stimulating ovarian maturity and general body growth. Anat. Rec., Vol. 43, pp. 277–297.

Snyder, F. F., 1928. The presence of melanophore-expanding and uterus-stimulating substance in the pituitary body of early pig embryos. Am. Jour. Anat., Vol. 41, pp. 399–409.

Soulié, A. H., 1903. Recherches sur le développement des capsules surrénales chez les vertébrés supérieurs. Jour. de l'Anat. et de la Physiol., T. 39, pp. 197–293.

Stieda, L., 1881. Untersuchungen über die Entwicklung der Glandula thymus, Glandula thyreoidea und Glandula carotica. Engelmann, Leipzig, 38 S.

Warren, J., 1917. The development of the pineal region in mammalia. Jour. Comp. Neur., Vol. 28, pp. 75–135.

Weymann, M. F., 1922. The beginning and development of function in the suprarenal medulla of pig embryos. Anat. Rec., Vol. 24, pp. 299–313.

Whitehead, R. H., 1903. The histogenesis of the adrenal in the pig. Jour. Am. Anat., Vol. 2, pp. 349–360.

Connective Tissues, Skeletal, and Muscular Systems

Asai, T., 1914. Beiträge zur Histologie und Histogenese der quergestreiften Muskulatur der Säugetiere. Arch. f. mikr. Anat., Bd. 86, S. 8–68.

Augier, M. A., 1923. Notocorde et épithélium pharyngien chez sus scrofa domesticus. Comptes Rendus de L'Assoc. des Anat., T. 18, pp. 57–65.

Bardeen, C. R., 1900. The development of the musculature of the body wall in the pig, including its histogenesis and its relation to the myotomes and to the skeletal and nervous apparatus. Johns Hopkins Hosp. Rep., Vol. 9, pp. 367–399.

de Beer, G. R., 1937. The Development of the Vertebrate Skull. Clarendon Press, Oxford, xxiii & 552 pp.

Butcher, E. O., 1933. The development of striated muscle and tendon from the caudal myotomes in the albino rat, and the significance of myotomic-cell arrangement. Am. Jour. Anat., Vol. 53, pp. 177–189.

Carey, E. J., 1921. Studies in the dynamics of histogenesis. VI. Resistances to skeletal growth as stimuli to chondrogenesis and osteogenesis. Am. Jour. Anat., Vol. 29, pp. 93–115.

Carey, E. J., 1922a. Studies in the dynamics of histogenesis. Intermittent traction and contraction of differential growth, as a stimulus to myogenesis. XI. The dynamics of the pectoralis major muscle tendon. Anat. Rec., Vol. 24, pp. 89–96.

Carey, E. J., 1922b. Direct observations on the transformation of the mesenchyme in the thigh of the pig embryo (Sus scrofa), with especial reference to the genesis of the thigh muscles, of the knee- and hip-joints, and of the primary bone of the femur. Jour. Morph., Vol. 37, pp. 1–77.

Glücksmann, A., 1939. Studies on bone mechanics in vitro. II. The rôle of tension and pressure in chondrogenesis. Anat. Rec., Vol. 73, pp. 39–55.

Godlewski, E., 1902. Die Entwicklung des Skelet- und Herzmuskelgewebes der Säugethiere. Arch. f. mikr. Anat. u. Entwg., Bd. 60, S. 111–156.

Hanson, F. B., 1919a. The ontogeny and phylogeny of the sternum. Am. Jour. Anat., Vol. 26, pp. 41–115.

Hanson, F. B., 1919b. The development of the sternum in Sus scrofa. Anat. Rec., Vol. 17, pp. 1–23.

Hanson, F. B., 1920. The development of the shoulder-girdle of Sus scrofa. Anat. Rec., Vol. 18. pp. 1–21.

Huber, E., 1931. Evolution of Facial Musculature and Facial Expression. The Johns Hopkins Press, Baltimore, xii & 184 pp.

Ingalls, T. H., 1941. Epiphyseal growth: Normal sequence of events at the epiphyseal plate. Endocrinology, Vol. 29, pp. 710–719.

Isaacs, R., 1919. The structure and mechanics of developing connective tissue. Anat. Rec., Vol. 17, pp. 243–270.

Kibrick, E. A., Becks, H., Marx, W., and Evans, H. M., 1941. The effect of different dose levels of growth hormone on the tibia of young hypophysectomized female rats. Growth, Vol. 4, pp. 437–447.

Kingsbury, B. F., 1920. The developmental origin of the notochord. Science, Vol. 51, pp. 190–193.

Lacroix, P., 1945. On the origin of the diaphysis. Anat. Rec., Vol. 92, pp. 433–439.

McGill, C., 1907–08. The histogenesis of smooth muscle in the alimentary canal and respiratory tract of the pig. Internat. Monatschr. f. Anat. u. Physiol., Bd. 24, S. 209–245.

McGill, C., 1910. The early histogenesis of striated muscle in the oesophagus of the pig and the dogfish. Anat. Rec., Vol. 4, pp. 23–47.

Mead, C. S., 1909. The chondrocranium of an embryo pig, Sus scrofa. A contribution to the morphology of the mammalian skull. Am. Jour. Anat., Vol. 9, pp. 167–210.

Murray, P. D. F., 1936. Bones. A Study of the Development and Structure of the Vertebrate Skeleton. Cambridge Univ. Press, London, x & 203 pp.

Nauck, E. T., 1926. Entwicklung des Schultergelenkes beim Schwein; Wachsplattenmodelle (als Ergäzung zum Vortrag über das Coracoideum der Säuger). Verhandl. Anat. Ges., Bd. 35, S. 260–261.

Parker, W. K., 1874. On the structure and development of the skull of the pig. Phil. Trans. Roy. Soc. London, Ser. B, Vol. 164, pp. 289–336.

Ruth, E. B., 1932. A study of the development of the mammalian pelvis. Anat. Rec., Vol. 53, pp. 207–225.

Sawin, P. B., 1945. Morphogenetic studies of the rabbit. I. Regional specificity of hereditary factors affecting homoeotic variations in the axial skeleton. Jour. Exp. Zoöl., Vol. 100, pp. 301–329.

Sawin, P. B., 1946. Morphogenetic studies of the rabbit. III. Skeletal variations resulting from the interaction of gene determined growth forces. Anat. Rec., Vol. 96, pp. 183–200.

Shields, R. T., 1923. On the development of tendon sheaths. Carnegie Inst., Contrib. to Embryol., Vol. 15, pp. 53–61.

Silberberg, M., and Silberberg, R., 1946. Further investigations on the effect of the male sex hormone on endochondral ossification. Anat. Rec., Vol. 95, pp. 97–117.

Stearns, M. L., 1940. Studies on the development of connective tissue in transparent chambers in the rabbit's ear. Part I, Am. Jour. Anat., Vol. 66, pp. 133–176; Part II, Am. Jour. Anat., Vol. 67, pp. 55–97.

Warkany, J., and Nelson, R. C., 1942. Skeletal abnormalities induced in rats by maternal nutritional deficiency. Arch. Path., Vol. 34, pp. 375–384.

Weed, I. G., 1936. Cytological studies of developing muscle with special reference to myofibrils, mitochondria, Golgi material and nuclei. Zeitschr. f. Zellforsch. u. mikr. Anat., Bd. 25, S. 516–540.

Williams, L. W., 1908. The later development of the notochord in mammals. Am. Jour. Anat., Vol. 8, pp. 251–284.

Teeth, Hair, and Hoofs

Adloff, P., 1901. Zur Entwickelungsgeschichte des Zahnsystems von Sus scrofa domest. Anat. Anz., Bd. 19, S. 481–490.

Beams, H. W., and King, R. L., 1933. The Golgi apparatus in the developing tooth, with special reference to polarity. Anat. Rec., Vol. 57, pp. 29–39.

Bevelander, G., 1941. The development and structure of the fiber system of dentin. Anat. Rec., Vol. 81, pp. 79–97.

Bevelander, G., and Johnson, P. L., 1945. The histochemical localization of alkaline phosphatase in the developing tooth. Jour. Cell. & Comp. Physiol., Vol. 26, pp. 25–33.

Bevelander, G., and Johnson, P. L., 1946. The histochemical localization of glycogen in the developing tooth. Jour. Cell. & Comp. Physiol., Vol. 28, pp. 129–137.

Bild, A., 1902. Die Entwickelungsgeschichte des Zahnsystems bei Sus domesticus und das Verhältnis der Lippenfurchenanlage zur Zahnleiste. Anat. Anz., Bd. 20, S. 401–410.

Chase, S. W., 1932. Histogenesis of the enamel. Jour. Am. Dental Assn., Vol. 19, pp. 1275–1289.

Glasstone, S., 1935. The development of tooth germs in vitro. Jour. Anat., Vol. 70, pp. 260–266.

Hampp, E. G., 1940. Mineral distribution in the developing tooth. Anat. Rec., Vol. 77, pp. 273–291.

Held, H., 1926. Über die Bildung des Schmelzgewebes. Zeitschr. f. mikr. Anat. Forsch., Bd. 5, S. 668–687.

Hirsch, M., 1921. Der Lückzahn von Sus domesticus, ein Beitrag zur Entwicklungsgeschichte des Gebisses von Sus domesticus und zur Kenntnis des Wesens der Dentitionen. Anat. Anz., Bd. 54, S. 321–330.

Jasswoin, G., 1924. Über die Histogenese der Dentingrundsubstanz der Säugetiere. Arch. f. mikr. Anat., Bd. 102, S. 291–310.

Morse, A., and Greep, R. O., 1947. Alkaline glycerophosphatase in the developing teeth of the rat: Its localization and activity characteristics as influenced by pH of the substrate and length of incubation time. Anat. Rec., Vol. 99, pp. 379–395.

Rein, G., 1882. Untersuchungen über embryonale Entwicklungsgeschichte der Milchdrüse. Arch. f. mikr. Anat., Bd. 20, S. 431–501 und Bd. 21, S. 678–694.

Saunders, J. B. de C. M., Nuckolls, J., and Frisbie, H. F., 1942. Amelogenesis. A histologic study of the development, formation and calcification of the enamel in the molar tooth of the rat. Jour. Am. Coll. Dentists, Vol. 9, pp. 107–136.

Schmidt, V., 1925. Studien über die Histogenese der Haut und ihrer Anhangsgebilde bei Säugetieren und beim Menschen. I. Die Histogenese des Hufes bei Schweineembryonen. Zeitschr. f. mikr. Anat. Forsch., Bd. 3, S. 500–557.

Schour, I., and Steadman, S. R., 1935. The growth pattern and daily rhythm of the incisor of the rat. Anat. Rec., Vol. 63, pp. 325–333.

Schultze, O., 1892. Ueber die erste Anlage des Milchdrüsenapparates. Anat. Anz., Bd. 7, S. 265–270.

Thoms, H., 1896. Untersuchungen über Bau, Wachsthum und Entwicklung des Hufes der Artiodactylen, insbesondere des Sus scrofa. Deutsche Thieraerzliche Wochenschr., Jahrgang. 4, S. 379–383.

Zeitzschmann, O., 1924. Die Entwicklung des Systems der äusseren Haut. (b) Die Haare. (Schwein.) Lehrbuch der Entwicklungsgeschichte der Haustiere, S. 186–194.

Twins, Double Monsters, Anomalies

Baumgartner, W. J., 1928. A double monster pig—Cephalothoracopagus monosymmetros. Anat. Rec., Vol. 37, pp. 303–316.

Berge, S., 1941. The inheritance of paralysed hind legs, scrotal hernia and atresia ani in pigs. Jour. Heredity, Vol. 32, pp. 271–274.

Bishop, M., 1921. The nervous system of a two-headed pig embryo. Jour. Comp. Neur., Vol. 32, pp. 379–428.

Bishop, M., 1923. The arterial system of a two-headed pig embryo. Anat. Rec., Vol. 26, pp. 205–222.

Carey, E., 1917. The anatomy of a double pig, Syncephalus thoracopagus, with especial consideration of the genetic significance of the circulatory apparatus. Anat. Rec., Vol. 12, pp. 177–192.

Chidester, F. R., 1914. Cyclopia in mammals. Anat. Rec., Vol. 8, pp. 355–366.

Corner, G. W., 1921. Abnormalities of the mammalian embryo occurring before implantation. Carnegie Inst., Contrib. to Embryol., Vol. 13, pp. 61–66.

Corner, G. W., 1922. The morphological theory of monochorionic twins as illustrated by a series of supposed early twin embryos of the pig. Bull. Johns Hopkins Hosp., Vol. 33, pp. 389–392.

Corner, G. W., 1923. The problem of embryonic pathology of mammals with observations upon intra-uterine mortality in the pig. Am. Jour. Anat., Vol. 31, pp. 523–545.

Denison, H., 1908. Notes on pathological changes found in the embryo pig and its membranes. Anat. Rec., Vol. 2, pp. 253–256.

Dutta, S. K., 1930. Notes on the cyclopian eye and other deformities of the head in a pig. (Sus cristatus Wagn.) Allahabad Univ. Studies, Sci. Sec., Vol. 7, pp. 53–103.

Fitzpatrick, F. L., 1928. The dissection of an abnormally developed foetal pig, with notes on the possible origins of such "freaks." Proc. Iowa Acad. Sci., Vol. 35, pp. 319–325.

Hughes, W., 1927. Sex-intergrades in foetal pigs. Biol. Bull., Vol. 52, pp. 121–136.

Jordan, H. E., Davis, J. S., and Blackford, S. D., 1923. The operation of a factor of spatial relationship in mammalian development, as illustrated by a case of quadruplex larynx and triplicate mandible in a duplicate pig monster. Anat. Rec., Vol. 26, pp. 311–318.

Kingsbury, B. F., 1909. Report of a case of hermaphroditism (H. Verus lateralis) in Sus scrofa. Anat. Rec., pp. 278–281.

Nordby, J. E., 1929. Congenital skin, ear, and skull defects in a pig. Anat. Rec., Vol. 42, pp. 267–280.

Pohlman, A. G., 1919. Double ureters in human and pig embryos. Anat. Rec., Vol. 15, pp. 369–373.

Schwalbe, E., 1906–1913. Die Morphologie der Missbildungen des Menschen und der Tiere. Fischer, Jena, Vol. 1, xvi & 230 pp., Vol. 2, xx & 410 pp., Vol. 3, Abt. 1, 270 pp., Abt. 2, 858 pp., Anhang, 266 pp.

Streeter, G. L., 1924. Single-ovum twins in the pig. Am. Jour. Anat., Vol. 34, pp. 183–194.

Thuringer, J. M., 1919. The anatomy of a dicephalic pig (Monosomus diprosopus). Anat. Rec., Vol. 15, pp. 359–367.

Warkany, J., and Roth, C. B., 1948. Congenital malformations induced in rats by maternal vitamin A deficiency. II. Effect of varying the preparatory diet upon the yield of abnormal young. Jour. Nutrition, Vol. 35, pp. 1–11.

Williams, S. R., and Rauch, R. W., 1917. The anatomy of a double pig (Syncephalus thoracopagus). Anat. Rec., Vol. 13, pp. 273–280.

Index

To facilitate the use of this book in connection with others in which the terminology may differ somewhat, many synonyms which were not used in the text have been put into the index and cross-referenced to the alternative terms used in this book; for example, Wolffian body, a term not used in this text, is frequently applied to the mesonephros. It appears in the index thus: Wolffian body (= mesonephros, q.v.).

Both figure and page references are given in the index. The figure references are preceded by the letter *f*.

CPSIA information can be obtained at www.ICGtesting.com
Printed in the USA

268822BV00001BA/15/P

9 781446 540